Mathematical Foundations of System Safety Engineering

This is a square high density pin array. The length of these pins is much greater than the distance that separates them. If one of these pins should be bent, there is a high probability that it will contact a neighbouring pin and create a short circuit. In spite of this risk, the use of such pin arrays is common in the electronics industry because they save space.

Richard R. Zito

Mathematical Foundations of System Safety Engineering

A Road Map for the Future

 Springer

Richard R. Zito
Richard R. Zito Research LLC
Tucson, AZ, USA

ISBN 978-3-030-26243-3   ISBN 978-3-030-26241-9   (eBook)
https://doi.org/10.1007/978-3-030-26241-9

This Springer imprint is published by the registered company Springer Nature Switzerland AG
The registered company address is: Gewerbestrasse 11, 6330 Cham, Switzerland

To my former supervisor and mentor Rene Fitzpatrick (1956–2015). He challenged everyone, because he believed that everyone rises to a challenge.
—Richard R. Zito

Preface

The eminent biologist David Attenborough once said, "There are some four million (species of) animals and plants - four million solutions to the problems of staying alive." In a sense, that is what this book is about - solutions to the problems of staying alive. Since this book deals primarily with complex nonliving hardware/firmware/software systems, the expression "staying alive" should be interpreted as *continuing to perform a desired function*. Of course, collateral loss of human life as well as environmental damage can also occur during the failure of complex nonliving systems. Furthermore, even the casual observer cannot help noticing that catastrophic failures of nuclear reactors, submarines, aircraft, and other potentially dangerous complex systems occur far more frequently than anyone would want. Ideally, such disasters should never occur! In spite of engineers' best efforts, there seems to be no way to stop such draconian accidents. This record of failures naturally begs the question, "Is there any way to build a perfect failure-proof complex system?" Or, at least, is there any systematic, scientific, way to design a complex system that minimizes risk? And, how does one define "perfect" and "risk" for real systems? This book will attempt to answer these difficult questions. Traditionally, the task of "mishap" (accident) prediction, as well as detection, and correction of system "hazards" (flaws that have not yet become mishaps) has been a heuristic activity of personnel who have gained their experience via "lessons learned" over (hopefully) many years of analyzing accidents. This hit-and-miss approach to system safety is not what this book is about. Here, mathematical tools will be described that provide certain, or at least probabilistic, solutions to the problems of prediction, detection, and correction. As such, this book provides a definite course of study, instead of an apprenticeship of indefinite length and effectiveness. Furthermore, it is a course of study that opens a new branch of engineering, and a new branch of scientific inquiry.

Tucson, USA   Richard R. Zito

Physical Constants and Mathematical Formulae

Physical Constants

Gravitational Constant, G	6.67×10^{-11} N-m^2/kg^2
Ideal Gas Constant, R	0.001987 kcal/mole K
Faraday Constant, $\mathcal{F}$	23.06 kcal/V mole
Boltzmann's Constant, k	1.38×10^{-23} J/molecule K
Proton Rest Mass, m_p	1.67×10^{-27} kg
Equatorial Radius of the Earth, R_E	6.378×10^6 m
Mass of the Earth, M_E	5.983×10^{24} kg
Radius of the Moon, R_M	1738 km

Transcendental Numbers

Pi, π	3.14159
Natural base, e	2.71828
Euler–Mascheroni Constant, γ	0.57721
Logarithmic Conversion Factor (common to natural)	2.30258

Infinite Series

$$\sum_{N}^{\infty} a\, r^N = a/(1-r) \text{ where} |r| < 1, a \neq 0,\ r \neq 0 \text{ (Geometric Series)}$$

Derivatives (Fundamental Forms)

$da/dx = 0$, where "a" is a constant
$dx/dx = 1$
$d(a\, f(x))/dx = a\, df(x)/dx$, where "a" is a constant
$d(f(x) + g(x) + \ldots)/dx = df(x)/dx + dg(x)/dx + \ldots$
$d(uv)/dx = u\,(dv/dx) + v\,(du/dx)$, where u and v may be functions of independent variable x
$dx^N/dx = Nx^{N-1}$, where N is a real number
$de^{ax}/dx = ae^x$, where "a" is a constant

Indefinite Integrals (Fundamental Forms)

$\int dx = x + C$, where C is a constant of integration
$\int a\, dx = a \int dx$, where "a" is a constant
$\int (f(x) + g(x) + \ldots)\, dx = \int f(x)\, dx + \int g(x)\, dx + \ldots$
$\int u\, dv = uv - \int v\, du$, where u and v may be functions of independent variable x
$\int x^N\, dx = \{x^{N+1}/(N+1)\} + C$, where C is a constant of integration and N is real but $\neq -1$
$\int x^{-1}\, dx = \{\ln x\} + C$, where C is a constant of integration

Separation of Variables

$\iint f(x)\, g(y)\, dxdy = \left[\int f(x)\, dx\right]\left[\int g(y)\, dy\right]$ if x and y are independent variables

Differential Equations

Linear Homogeneous Equation of 2nd Order: $u'' + p(z)u' + q(z)u = 0$
(a solution exists if $p(z)$ and $q(z)$ are continuous over an open interval containing z and $u(z)$ is twice differentiable with respect to z)

Contents

Chapter 1
Introduction

1.1 Heuristic Versus Analytic Methods

In order to put this book in its proper perspective with respect to system safety science as a whole, the author can think of no better focusing mechanism than to contrast the assumptions used here with those of Leveson's work [1]. In this well-researched book, Leveson outlines seven "Assumptions" that are claimed to be false and suggests seven "New Assumptions" as replacements. Each of these will be examined in turn.

Assumption 1 Safety is increased by increasing system or component reliability. If components or systems do not fail, then accidents will not occur.

New Assumption 1 High reliability is neither necessary or sufficient for safety.

Leveson provides two examples that illustrate this point. The first example involves reliability without safety. The loss of the Mars Polar Lander was attributed to system noise generated during deployment of the landing legs. This noise was interpreted by the onboard computer as an indication that landing had occurred. Consequently, the decent engines were turned off prematurely, resulting in a crash landing. Everything worked correctly, but the system was poorly designed. According to Leveson, the system was "reliable but unsafe". In fact, a system that is susceptible to temporary "soft failures", like noise interference or radiation-induced bit flips, is still considered to have *failed* in the sense that it was *unable to perform its intended function*. All space flight hardware is expected to be tested in both white and pink noise environments. So, this system is both *unreliable and unsafe*. The quantification of the notion of *intent* is not as hopeless as one might think, *provided the analyst really knows what he or she wants*! Chapters 10, 13, and 14 of this book will deal with design *intent*.

The next example concerns safety without reliability. It is argued that operators at Three Mile Island were "reliable", in the sense that they followed procedure, but their unwillingness to deviate from the procedure in a way that could have brought

© Springer Nature Switzerland AG 2020
R. R. Zito, *Mathematical Foundations of System Safety Engineering*,
https://doi.org/10.1007/978-3-030-26241-9_1

the reactor under control was "unsafe". In this book (Chap. 3), *all* human behavior of *man–machine chimeras*[*] will be considered both *unreliable and unsafe*, in the sense that people under pressure will make the wrong decision(s) with high probability. This is not to say that human oversight is not important, because people provide the kind of insight, flexible improvisation, and imaginative solutions that no machine or software can provide in an emergency. However, assuming *unit probability* for human failure provides a reasonable *upper bound* when mathematically analyzing what *could happen* in an emergency. Furthermore, this *Axiom of Human Fallibility* is probably not far from the truth given that binary decisions randomized by stress and/or misinformation will, in the long run, go wrong about 50% of the time. Like the proverbial deer that makes an inappropriate decision when suddenly illuminated by the headlights of an oncoming car, even the most sincere, well-trained, well-rested, sober, and conscientious person will eventually break under pressure; as Leveson so aptly points out!

The dividing line between reliability and safety can sometimes be vague, especially when reliability concerns a safety-critical part, subsystem, or personnel. This argument has been going on for years. And, it doubtless will continue for many more years to come.

Assumption 2 Accidents are caused by chains of directly related events. We can understand accidents and assess risk by looking at the chain of events leading to the loss.

New Assumption 2 Accidents are complex processes involving the entire socio-technical system. Traditional event chain models cannot describe this process adequately.

Leveson's basic reason for rejecting the event chains (although her book is filled with them) is that "selecting some factors as relevant and others as irrelevant is, in most cases, arbitrary and entirely the choice of the modeler." This premise rests on two pillars. First, "the selection of an initiating event (first event of a chain, or 'root cause') is arbitrary and previous events and conditions could always be added." It is argued, convincingly, that companies will often attribute a mishap to initiating events that are politically acceptable and place the "blame" on low-level personnel, thereby avoiding the expense of hazard mitigation and shielding management from litigation. Several aviation accidents and the Bhopal disaster are cited as examples. And second, there is "subjectivity in selecting the chaining conditions"; meaning that there may be several reasons why one event of a chain triggers the next event. Leveson cites another aviation accident as an example, and says that "the selection of the linking conditions will greatly influence the cause ascribed to the accident". What Leveson is really talking about here is *Mathematical Induction*. If the first step is wrong, and there is no way to reliably go from the nth step to the (n + 1)th step, then the conclusion will be wrong. That is certainly true, any proof can be carried out incorrectly. However, logic is the best tool that humanity currently has. And, the selection of "initiating events" and "chaining conditions" is not as "arbitrary" as is claimed. First of all, common sense tells the analyst that information collected from one biased source will undermine the utility of an event chain. Serious accidents are all thoroughly investigated by *independent* parties. In fact, much of Leveson's information comes from such sources. In the end, a reliable event chain *can* be constructed. Event chains, called Mishap Chain Reactions (MCRs) in this book, can and should go back as far in time as necessary to ensure that *all* "root causes" have been identified. And, if an event chain branches at the nth step, all paths should be investigated to determine which ones played a part in a mishap. All active event paths can and should be included in an event chain, complexity is no limitation. If the same or similar accidents continue to occur, in spite of comprehensive careful analysis, because of unethical management behavior, that is not the fault of the event chain tool. It is a matter for an overseeing governmental authority, or the courts.

Unethical management behavior (cited by Leveson and directly observed by the author) includes cost-cutting efforts such as ignoring or delaying correction of design and engineering weaknesses, and replacing knowledgeable personnel who resign or retire with unskilled (or unqualified) workers, or eliminating a safety position altogether. Unless government safety requirements are issued as "*SHALL* statements", some corporations and contractors will do nothing! Although many organizations take system safety seriously, human behavior of the type just cited forces the author of this book to adopt an *Axiom of Human Noncompliance* for the *worst-case* calculations in Chap. 3 (*i.e., in the zero-sum game, between profits and safety, profits will always win at the expense of safety, so that the probability of future mishaps is unity*). However, it should be noted that this unfortunate situation may change due to the litigious environment of modern business. As settlements increase in cost, profits will decline to the point where safety becomes paramount. In truth, the cost of an engineering modification may be fairly trivial compared to the cost of a mishap. The same is true for the pay gap between a skilled engineer or chemist versus an unskilled replacement who is of little or no use during an emergency. Greed, like many diseases, is self-limiting! As a final remark, the author quotes the response of Carl Sandom to an audience question during his lecture of August 10, 2011 in Las Vegas [2], "Unless the system safety engineer has the full support of management, there is little that he can do to avert a mishap." Truer words were never spoken!

Assumption 3 Probabilistic risk analysis based on event chains is the best way to assess and communicate safety and risk information.

New Assumption 3 Risk and safety may be best understood and communicated in ways other than probabilistic risk analysis.

Leveson has several objections to Probabilistic Risk Assessment (PRA), which need to be discussed because they are, or can be, accommodated by the calculations associated with PRA. First, Leveson points out that in an accident involving multiple failures, the *independent failure assumption* (i.e., the assumption that the total probability of failure is simply the product of the individual failure probabilities) is often much too small. The author of this work agrees, and adds that such an estimate is *usually* too small by many orders of magnitude. The proper upper bound on the failure probability of complex accidents is calculated in Chap. 3 of this book.

Next, it is claimed that "design errors" and "component interaction" failures are "usually not considered" in PRA. Actually, this is not the case. First of all, when a component fails prematurely, it may do so for two reasons. It may cease operation because of a manufacturing defect, or it may be destroyed by an improper application (application design error). Both of these types of failures are captured by "historical" failure data during development and testing. Naturally, when such a failure occurs, a part will be replaced with a new part or a design will be corrected, often permanently, so that the failure rate will eventually decay. Probabilistic failure rate models (Chap. 2 of this book) can be built that mimic this life cycle behavior, sometimes quite well as will be demonstrated. So, the "Holy Grail" of PRA is not as unattainable as some might think. Furthermore, it should be noted that the tail of a probabilistic failure rate model never reaches zero until time t = ∞. That is to say, all probabilistic failure rate models take into account the reality that *every* complex system *always* has some latent flaws! This is contrary to the assertion that PRA considers "only immediate physical failures" and that "Latent design errors may be ignored and go uncorrected due to overconfidence in the risk assessment." As far as "dysfunctional interactions among non-failing (operational) components" are concerned, these too are considered failures (soft failures, see Assumption 1 above) and are captured in the historical test data.

Finally, Leveson asserts that PRA is "not appropriate for systems controlled by software and by humans making cognitive complex decisions, and there is no effective way to incorporate management and organizational factors, such as flaws in the safety culture, despite many well-intentioned efforts to do so." In Chap. 12 of this book, the PRA methodology will be extended to software. And, as far as unintentional human error and deliberate managerial and organizational safety flaws are concerned, these are captured by the Axiom of Human Fallibility and the Axiom of Human Noncompliance discussed above and used in Chap. 3 of this book.

In summary, most of Leveson's concerns about PRA have been, or can be, taken into account at some level. It must always be remembered that a submarine skipper cruising around in the deep will want some numerical assurance, and not just a qualitative "O.K." and a nod, that he will complete his mission and return home safely.

Assumption 4 Most accidents are caused by operator error. Rewarding safe behavior and punishing unsafe behavior will eliminate or reduce accidents significantly.

New Assumption 4 Operator behavior is a product of the environment in which it occurs. To reduce operator "error" we must change the environment in which the operator works.

This book does not really deal with operator behavior and the operator environment beyond the Axiom of Human Fallibility and the Axiom of Human Noncompliance.

And, it must always be understood that assuming unit probability for fallibility and noncompliance produces an *upper bound* for the probability of any given mishap. This approach has been taken because, as Leveson herself points out, "human behavior is much more complex than machines" and "the study of decision making cannot be separated from a simultaneous study of social context, the value system in which it takes place, and the dynamic work process it is intended to control." All these factors make detailed quantification of human behavior very difficult, if not impossible. Therefore, there is little for this author to say about Assumption 4, and Leveson's new counterpart, other than to emphasize a few of her comments. One suggestion made by Leveson is to "specify and train standard emergency responses." For example, "Operators may be told to sound the evacuation alarm any time an indicator reaches a certain level." This seems to be prudent advice when examined from a more mathematical point of view in Chap. 3 of this book. Finally, Leveson makes a general comment on the difficulty of "detecting and correcting potential errors before they have been made obvious by an accident". In fact, detection and correction of hazards will be the focus of much of this book.

Assumption 5 Highly reliable software is safe.

New Assumption 5 Highly reliable software is not necessarily safe. Increasing software reliability or reducing implementation errors will have little impact on safety.

There are two key statements in Leveson's argument. First, "Nearly all the serious accidents in which software has been involved in the past twenty years can be traced to requirements flaws, not coding errors. The requirements may reflect incomplete or wrong assumptions". Second, "It is possible and even quite easy to build software that we cannot understand in terms of being able to determine how it will behave under all conditions. We can construct software (and often do) that goes beyond human intellectual limits." It is certainly true that the software designer/ engineer must know what he or she wants to accomplish if safe software is to be produced. And, it is also true that "coding errors", in the narrowest sense of that term (i.e., trivial syntax errors), can easily be spotted during compilation so that they are seldom the source of software failure. In this sense, Leveson's first objection is strictly true. Leveson's second objection to Assumption 5 is also true. Code complexity can make software exhibit "<u>unintended</u> (and <u>unsafe</u>) behavior". However, the loss of designer/engineering *intent* is still considered a "coding (implementation) error" in the most general sense of that term. So, Leveson's New Assumption 5 is not strictly true. It may seem impossible to capture the notion of *intent* mathematically. Yet, that is precisely what will be done in Chaps. 13 and 14 of this book.

In the past, it has been a standard practice to assign a failure probability of unity to software. This practice is consistent with the Axiom of Human Fallibility since the success of even the most robust code ultimately relies on the fact that software designers/engineers really know what they want. Assigning a failure probability of

unity to software establishes a Safe Upper Bound. However, the science of estimating software normalized failure rates has become sufficiently advanced that it
now seems reasonable to also assign a Likely Failure Probability to software.
Calculation of the Likely Failure Probabilities will be the subject of Chap. 12.
Finally, Leveson suggests that one way to reduce "requirements flaws" is to use
"integrated product teams" where both design engineers and software engineers
meet together to discuss requirements. Another very prudent suggestion. Again,
*software designers/engineers must know exactly what they want in order to build
truly fail-safe software!* There is no way around this requirement, and it applies to
both hardware (especially electronic circuits) *and* software.

Assumption 6 Major accidents occur from the chance simultaneous occurrence of
random events,

New Assumption 6 Systems will tend to migrate toward states of higher risk. Such
migration is predictable and can be prevented by appropriate system design or
detected during operations using leading indicators of increasing risk.

What Leveson is talking about here is system "entropy". That is to say, any system
has many possible "disordered" states, but only one ideal state envisioned by
designers. Therefore, the disordered state is more probable than the ideal state given
an equal probability of occurrence for each state. If one models the life of a system
as a stochastic process consisting of a series of states s_i characterized by the system
safety engineer at the end of the ith inspection period, then it is clear that the system
as a whole will migrate to a very dangerous condition as disordered states are
further scrambled as $i \rightarrow \infty$. Furthermore, "energy" (in this case meaning "effort"
or "oversight") must be expended to maintain the ideal state. These ideas are neatly
captured in Leveson's New Assumption 6. More will be said about entropy within
the context of software in Chap. 13.

Assumption 7 Assigning blame is necessary to learn from and prevent accidents or
incidents.

New Assumption 7 Blame is the enemy of safety. Focus should be on understanding how the system behavior as a whole contributed to the loss and not on who
or what to blame for it.

What Leveson is talking about here is blaming individuals, usually the lowest
person in the authority chain, for *causing* a mishap. Contributions from "flawed
decision making, political and economic pressures…, poor problem reporting, lack
of trend analysis, a 'silent' or ineffective safety program, communications problems, etc." Leveson calls "reasons", as if there were no individuals behind these
problems to blame. However, this is a book on mathematical methods, so little will
be said about blame in the rest of this book, other than two brief related comments
at the end of Chaps. 15 and 17. Nevertheless, it seems appropriate here to make two
additional comments here concerning Assumption 7 and Leveson's new version.

Levison's "'silent' or ineffective safety program" is real, prevalent, and frightening. All too often a system safety engineer is shorn of all authority to change anything that design engineering does. In such cases, the system safety engineer can, at best, only make suggestions. The design team often views the system safety engineer as an annoyance rather than an integral member of the design team with the knowledge and experience to produce better, safer, and even more inexpensive products. Chapter 15 will hammer this point home with great force. Beyond the individual engineer lies the system safety organization. Tragically, these organizations are often little more than "dog and pony shows" for customers, especially government customers and agencies. Individual engineers are given explicit instructions not to interfere to strenuously in any program they are working with. And, the chairman of such organizations is sometimes just a "company man" who knows little or nothing about safety. The author knows of a particular case where the knowledge of a system safety supervisor was limited to the purely secretarial aspects of system safety. He knew almost nothing about mathematics, physics, chemistry, biology, medicine, geology/hydrology/environmental science, electronics, or software engineering. A serious mishap was bound to happen—and it did! Although there are many competent system safety engineers, cases of the type just described force the author to adopt an Axiom of Human Fallibility in a worst-case analysis. Finally, one last anecdote concerns what one might call the corporate safety culture and middle management's culpability in serious mishaps. Many years ago, at an aerospace company that manufactured rocket boosters, the author was told by a manager that, "they would rather have an occasional booster explode than get involved in non-destructive testing." And, this statement was made *after* a serious booster accident! It is uncertain whether this appalling view is just one man's opinion, or is a reflection of the larger safety culture (or nonexistence of a safety culture) at that company. And, it is also uncertain just how pervasive such a view is in the industry as a whole. In any case, it is because of attitudes of this kind that the author has adopted an Axiom of Human Noncompliance.

Summary: Leveson's seven assumptions highlight the similarities and differences between the heuristic and analytic approaches to system safety. They also help position this work within the context of system safety as a whole, and justify its basic assumptions. The heuristic and analytic approaches should not be thought of as the antitheses of one another, but rather as complimentary. Both approaches should be used in any safety program. The heuristic approach gives information that is general but imprecise. The analytic approach gives information that is precise but narrow. At the end of this book (Epilogue) Leveson's work will be revisited because hindsight yields an insight all its own. However, for now, the reader should note that analytic methods yield specific guidance as to how to detect and correct hardware and software hazards before they become mishaps. And, that is what will be discussed next. In the future, analytical methods will supplant heuristic methods for the analysis of faults not caused by human agency.

1.2   The Product Development Cycle-Hazard Prediction, Detection, and Correction

Typically, and perhaps classically, the prediction of reliability is based on statistics. Most textbooks that address the mathematical aspects of system safety, when it overlaps with reliability, do so in terms of statistics. The central question is simply this, "Is it possible to predict the future based on historical data?" Obviously, this book takes the position that it *is possible in many circumstances given a sufficient amount of information*, otherwise there would be nothing to write about. Although, some system safety practitioners maintain that there *is nothing to write about!* After all, how can you predict the behavior of a new system? But, is there really such a thing as a totally novel system? Isn't every *new* system just a collection of pre-existing parts and subsystems whose performance is known, perhaps well known, from past experience? This idea will be explored in the first three chapters of this book and in Chap. 12. However, beyond prediction of failure rates as a function of time lies the problem of early detection and correction of hazards *before* a product is deployed or reaches the market and the user. These last two *phases* of the product development *cycle* will occupy the remainder of this book. "Fail-safe", thorough, mathematical examination and correction procedures will be emphasized. The term "cycle" has been used here because prediction, detection, and correction may have to be repeated several times before a refined, "fail-safe" product is produced. Each "improvement", each change, may cause a new cascade of problems that require retesting, recalculation of performance predictions, reanalysis of failure modes, and re-correction. Finally, the product emerges, or reemerges, to become a finished product. Most of the problems encountered by system safety engineers fall into four basic types, which will be called the "Classical" branches of system safety. Therefore, the remaining chapters of this book (2 through 17) will be divided into four parts, each part concentrating on one branch.

1.3   The Four Classical Branches of System Safety

PART I—FUNDAMENTALS: In this first part of this text (Chaps. 2–4), general problems of system safety are considered. Here, the word "general" means that the concepts can be used to describe a variety of hardware systems (like electrical, mechanical, chemical, liquid, gas, or other types of systems), firmware systems, software systems, or any combination of these. However, as discussed above, the human element will only be modeled through the Axiom of Human Fallibility and the Axiom of Human Noncompliance, thereby establishing an upper bound to system behavior. All three of the chapters in Part I are "predictive" in nature.

PART II—ELECTRONICS: Chapters 5 through 11 deal with the detection and correction of electronic hazards. In particular, the Bent Pin, Sneak Circuit, and related electrical problems will be addressed. The solution to the Bent Pin Problem is interesting and useful in its own right. However, its real power is in the insight it lends to the solution of the Sneak Circuit Problem, a much more complex and important problem that turns out to be in some sense the compliment, or the mathematical *dual*, of the Bent Pin problem.

PART III—SOFTWARE: Chapters 12 through 14 are about software. Chapter 12 deals with predicting software failure rates, while Chaps. 13 and 14 deal with detecting and correcting deep software logical flaws (called *defects*). It will be shown that such flaws, or defects, are *Sneak Logic Paths* that are analogous to Sneak Circuit Paths in the analysis of electronic networks. So, the mathematicians old trick of using the solution of one problem to solve another will be applied here.

PART IV—DANGEROUS GOODS (Explosives, Propellants, Toxins, and Pollutants): Chapters 15 through 17 present solutions to three problems that are typical of chemical problems faced by the system safety engineer during the design, storage, and disposal phases of a system's life cycle. Handling dangerous goods problems requires extensive knowledge of mathematics, chemistry, materials science, biology, toxicology, medicine, geology, hydrology, and environmental science.

The Remainder: The working system safety engineer will typically be presented with a great variety of problems that will usually not exactly match the problems presented in the various chapters of this book. Therefore, the calculations presented should be thought of as models for how to do similar calculations. Furthermore, the system safety engineer will inevitably be presented with some technical problems that lie outside the scope of the four classical branches of system safety. In that case, the engineer will have to invent his/her own solution. But hopefully, the tools presented in this book will still be the engineer's most powerful resource. Therefore, the mathematical tools of our curious trade will be discussed next.

1.4   Tools of the Trade

Probability Density Function: The failure of complex systems is not purely random. On the contrary, complex system failures follow mathematical laws. A population of people offers a useful example of how a complex system fails. As soon as a person is born health problems begin. And, some deaths will occur at birth. A few people will die in infancy and childhood. But, most people will live into their 70s and 80s in the industrialized world. At that time the number of deaths per unit time interval (i.e., per year) will be at a maximum. Of course, a few people will live to be very old and, in the United States, about 25% of people will live to be 92. The number of deaths per year is now well below its peak. Not because old people are more robust than young people (in fact the surviving, aging, population as a whole is becoming *more hazardous*[†]), but because relatively few people will live to such

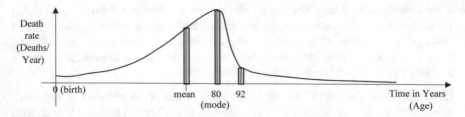

Fig. 1.1 Typical death rate versus age for people. Directly under the peak of the curve, along the age axis, is the *most common* age at which death occurs (called the *mode*). The *average* (or *mean*) age of death is a little smaller because the curve is not symmetrical. This curve supposes that all the people of this population were born at the same time. In reality, the members of a real population are born at different times and have different prospects for longevity due to advancing technology

an advanced age. The death rate (deaths/year) versus time graph for human beings has the general shape of Fig. 1.1. Death, of course, is a *permanent* "failure" of a member (person) of a complex system (population in this case). Its onset can be delayed by medical intervention but, once it occurs, it cannot be cured. So, death defines the *lifespan* of a member.

The curve of Fig. 1.1 displays a long tail as time approaches large values. In fact, this tail, theoretically, extends out to infinity. Of course, no one can live forever. The oldest known person for which a well-documented birth and death certificate exists is Jean Calment of France who reached 122 years of age [3]. She was born during the administration of President Ulysses S. Grant and died during the administration of William Clinton! So, perhaps, the death rate curve should be truncated at 122 years. But then, that would not admit the possibility that anyone could achieve an age greater than 122. And, that might be a mistake because Juana Rodriguez of Cuba celebrated her 126 birthday. Reports have also come in from other parts of the world of people reaching the ages of 128, 130, 150, and 152 years! Birth and death certificates apply to a minority of the world's population, especially 100 years ago, so the lack of such documents does not necessarily preclude that some very advanced ages have been reached. However, the *probability* of reaching such an advanced age is clearly *very* small. The point here is that the long tails of failure rate curves of complex systems, that in principle extend out to $t = \infty$, are necessary for the prediction of very rare events (called "Black Swans").

It is standard practice in statistics to *normalize* the Failure Rate (in this case death rate) curve by dividing the instantaneous number of failures per unit time by the total number of failures. Again, with respect to people, normalization of Fig. 1.1 would involve dividing the number of deaths recorded for each age by the total number of people in the population being studied. The resulting curve, of course, has the same shape as in Fig. 1.1, but a different vertical scale applies. The ith data point of the normalized curve would then be the fraction f_i (greater than 0 but less than 1) of people in the entire population that died in the ith year (assuming a 0th data point for birth at $t = 0$).

When all these fractions (probabilities per year) are multiplied by the width of their associated age bins ($\Delta t = 1$ year), and added up for all years in which there was even a single living member of the population, the result is unity (i.e., 1; the normalization property).

$$\text{Probability of death} = f_1\Delta t + f_2\Delta t + f_3\Delta t + \cdots = f_1(1) + f_2(1) + f_3(1) + \cdots$$
$$= f_1 + f_2 + f_3 + \cdots \rightarrow 1.$$

$$(1.1)$$

The *normalized failure rate* curve is called a Probability Density Function (or pdf), or *failure probability rate* curve. The pdf is just the instantaneous probability of failure per unit time. The area under the entire curve is unity because death is an eventual certainty for all of us.

Cumulative Probability Distribution: There is an ancient Hindu saying, "All compound objects must eventually decay." Quite true! But, what is the probability P that a complex system will fail between any two times t_1 and t_2? It is simply the area trapped under the pdf between t_1 and t_2. A more precise method of calculating this area than manually adding up the area of the rectangles under the pdf (see Fig. 1.1 and Eq. 1.1) is to *integrate* the pdf (denoted by ρ) from t_1 and t_2. Symbolically,

$$P = \int_{t_1}^{t_2} \rho dt.$$

$$(1.2)$$

The *integral* is an infinite sum (Riemann sum) of an infinite number of rectangles whose width is infinitesimal (i.e., in the limit as $\Delta t \rightarrow 0$). Naturally, if $t_1 = 0$ and $t_2 = \infty$ then $P = 1$, as it must in the limit of Eq. 1.1, with a pdf like that of Fig. 1.1. However, mathematical integration will depend on having an *equation* for the pdf. And, these pdfs typically involves one or more free parameters that can be chosen by the system safety engineer to best fit the finite experimental data. And, there are many models available to choose from [4]. The most common are discussed in the next subsection.

A Few Popular and Useful Model pdfs:

The Gaussian (Normal) Distribution Model

The *Gaussian*, or *normal*, pdf is probably the oldest failure probability rate model. Its utility is usually limited to fairly simple systems, such as a collection of light bulbs. The equation for the normal pdf (or normal distribution) is given by Pearson [4] as

$$\rho = (1/\sqrt{2\pi})(1/\sigma)\exp\left(-\tfrac{1}{2}[(t-\mu)/\sigma]^2\right),$$

$$(1.3)$$

where μ and σ are the *mean* and *standard deviation* of this "bell-shaped" distribution, respectively. An important shortcoming of the normal distribution as a model is that its tail at t = 0 does not equal zero for large positive μ. So, there is always an unphysical nonzero probability of failure even before a system is created! However, so long as the width of the distribution (standard deviation σ) is much smaller than the location of its peak (mean μ) from the origin (i.e., σ << μ, or σ/μ → 0), the probability trapped under the tail for negative time is so small that it can usually be ignored. A simple pdf that is very similar to the Gaussian, but does not have a nonzero value in the lower tail at t = 0 is the *Rayleigh* pdf that will be discussed next.

The Rayleigh Distribution Model

The *Rayleigh* pdf was another early equation used to model the failure probability rate of complex systems, and is described by Eq. 1.4 [5].

$$\begin{aligned} \rho &= (t/\sigma^2)\exp\left(-\tfrac{1}{2}[t/\sigma]^2\right) \quad &\text{for } 0 \le t < \infty \text{ and} \\ \rho &= 0 &\text{for } -\infty < t < 0 \end{aligned} \tag{1.4}$$

where σ is the mode, or peak, of the distribution. At t = σ, failures are most likely to occur. Or, in the case of software, defects are most likely to be detected. Comparison of Eqs. 1.3 and 1.4 shows that the Gaussian and Rayleigh models are very similar. However, the pre-factor of t in Eq. 1.4 forces the pdf to have zero value at t = 0. This seemingly minor detail, together with the condition on ρ for −∞ < t < 0, insures that there is a zero failure probability rate before the system is created, thereby eliminating the contradiction inherent in the Gaussian model. In Fig. 1.2 a Rayleigh failure probability rate distribution is plotted with σ = 1. This

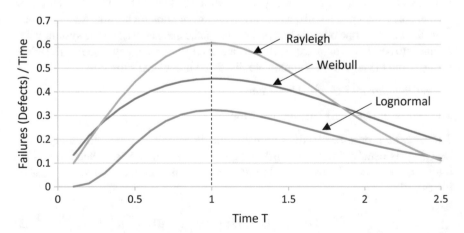

Fig. 1.2 Comparison between the Rayleigh, Weibull, and Log-normal distributions. All models have their mode (peak) set to unity in time

will be the standard model against which the other models described below will be compared. The principal problem of the Rayleigh model is that the position of the distribution's mode and the width of the distribution cannot be controlled independently. The two-parameter normalized failure rate models considered next will remove this difficulty.

The Weibull Distribution Model

Probably the single most used failure probability rate (defect probability rate) model for complex systems is the *Weibull Model* (Eq. 1.5) [5],

$$\rho = (\alpha\beta)t^{\beta-1}\exp(-\alpha t^{\beta}) \quad t > 0; \alpha, \beta > 0.$$
$$\rho = 0 \qquad\qquad\qquad \text{for} -\infty < t < 0$$

$$(1.5)$$

The *arbitrary parameters* α and β are chosen by the analyst to best fit the normalized failure (defect) probability rate data. When $\alpha = \frac{1}{2}$ and $\beta = 2$, the Weibull distribution collapses identically to the Rayleigh distribution. But, if α is lowered to 0.4, and β is lowered to 1.7, the two curves can be separated (Fig. 1.2). The mode (peak) of the Weibull curve has now been depressed while maintaining its position at $t = 1$. Probability has now been pushed into the tail to maintain unit total probability under the distribution. The long, *slowly decaying*, tail that has developed is of the greatest importance because most complex systems display a slowly decaying normalized failure rate curve for *large* values of time t ($t \rightarrow \infty$) [6]. The term "slowly decaying" is in comparison to the Rayleigh distribution in standard position ($\sigma = 1$). And, the word "large" here refers to times much greater than the time at which the mode of the distribution appears (in this case $t \gg 1$). The slope of the tail can be controlled at the expense of the height of the mode. However, both parameters α and β affect the location of the Weibull distribution's mode, as well as the slope of its tail (and therefore the height of its peak), in a complex way. This interdependence makes searching α, β space for the optimal Weibull fit to any given data set more difficult. Finally, it should be noted that a Weibull-like model has been developed with system *parameters* that can be related to subsystem properties. This breakthrough allows a priori computation of a system failure rate model even before construction and testing of a complex system begins; provided the appropriate subsystem properties are known from past experience. Such potentially powerful design tools are topics of current research [7].

The Log-normal Distribution Model

The *Log-normal* distribution is given by Pearson [4]

$$\rho = [\alpha/(\sqrt{2\pi}t)]\exp\left[-\frac{1}{2}(\alpha\ln t + \beta)^2\right], \quad \text{where } 0 < t < \infty \text{ and } \alpha > 0, -\infty < \beta < \infty$$
$$\rho = 0 \qquad\qquad\qquad\qquad\qquad \text{for} -\infty < t \leq 0$$

$$(1.6)$$

As before, the arbitrary parameters α and β are chosen by the analyst to best fit the normalized failure rate (defect rate) data. The Log-normal distribution produces a curve that is very similar in shape to the Weibull distribution (c.f. Fig. 1.2), but now the location of the peak is *primarily* (but not exclusively) controlled by β, while the distribution's width is *primarily* controlled by α. It is helpful to set $\beta = -\beta'$. In that case, the peak moves out along the t-axis as β' increases. And, the distribution becomes more *peaked* as α increases (or flatter as α decreases). These simple intuitive relationships make the Log-normal distribution a little easier to fit to data than the Weibull distribution. The Log-normal curve in Fig. 1.2 was generated by setting $\alpha = 1.0$ and $\beta' = 0.42$.

The Gamma Distribution Model

The name of this distribution is a little misleading since the Gamma function is only used to evaluate the leading numerical coefficient. The required Rayleigh-like shape is actually furnished by a power of t and its exponential; where the power of t brings the distribution function down to zero at the origin. The Gamma Distribution is another common two parameter fit to data given by Pearson [4]

$$\rho = \{1/[\Gamma(\alpha)\beta^{\alpha}]\}\{t^{\alpha-1}\}\{\exp(-t/\beta)\}, \quad \text{where } t > 0, \text{and } \alpha, \beta > 0.$$
$$\rho = 0 \qquad\qquad\qquad\qquad\qquad \text{for} -\infty < t \leq 0 \tag{1.7}$$

The inverse of the Gamma Function [i.e., $1/\Gamma(\alpha)$] is given by Euler's infinite product [8]

$$1/\Gamma(\alpha)] = \alpha\, e^{\gamma\alpha} \prod_{n=1}^{\infty} \left[(1+\alpha/n)e^{-\alpha/n}\right] \tag{1.8}$$
$$\text{where } |\alpha| < \infty \text{ and } \gamma = 0.57721\ldots\text{(the Euler-Mascheroni Constant)}$$

The Exponential Distribution Model

The *Exponential Distribution* is a good failure probability (defect probability) rate model for well tested systems that are already in the tail of a Rayleigh-like distribution, and has the form [4]

$$\rho = \lambda\exp(-\lambda t) \quad \text{for } t \geq 0$$
$$\rho = 0 \qquad\qquad\quad \text{for } t < 0, \tag{1.9}$$

where λ is the decay constant of the distribution. Since ρ is a *normalized failure probability rate*, the number of failures per unit time, dF/dt, is just given by

$$dF/dt = N\lambda\exp(-\lambda t) \quad \text{for } t \geq 0, \tag{1.10}$$

where N is the total number of latent defects at delivery ($t = 0$). And, the observed defect rate at delivery ($t = 0$) is just $dF/dt \mid_{t = 0} = N \lambda$. To calculate the mean time between failures as a function of time, MTBF(t), start by taking the inverse of Eq. 1.10 and then set $dF \approx \Delta F = 1$. In that case $dt \approx$ MTBF(t) so that

$$\text{MTBF(t)} = 1/[N\lambda \exp(-\lambda t)] \quad \text{for } t \geq 0. \tag{1.11}$$

And, the initial MTBF at delivery is just given by MTBF ($t = 0$) = $1/N\lambda$ [6].

The Uniform Distribution Model

When all the distributions above fail to mimic the system failure probability rate behavior, it is often due to the fact that none of them decays fast enough in the tail of the distribution as $t \to \infty$. After some time t, many complex systems display an almost steady-state normalized failure rate behavior. There are several reasons for this. The introduction of new software defects, or future hardware failures, after the repair of old defects/failures, as well as defects and future failures introduced by upgrades and modifications, insures that defects and failures never completely vanish.

Of course, the area under a constant function approaches infinity as integration proceeds over the whole timeline from 0 to +∞, integration over negative time being disallowed as usual because a system cannot fail before it has been created. Clearly, the uniform distribution is only valuable for modeling a small finite time interval centered about some "large" value of time t. The word "large" in this context means a system so mature that its failure (defect) probability rate no longer seems to drop as a function of time. If t_1 and t_2 are the endpoints of an interval along the positive time axis of a normalized failure probability rate curve, and if $k < 1$ is the mean expected failure probability rate over this whole interval, then the uniform distribution is given by Pearson [4]

$$\begin{aligned} \rho &= k/(t_2 - t_1) \quad &\text{for } t \in (t_2 - t_1) \text{ and } t_2, t_1 \gg 0 \\ \rho &= 0 \quad &\text{for } t \notin (t_2 - t_1). \end{aligned} \tag{1.12}$$

In reality, a truly constant failure (defect) probability rate never occurs. Instead, the real-world failure (defect) rate has a "lumpy" appearance with an average value that may match the constant (uniform) level quite well.

So far, the discussion of "Tools" has been confined to "Probability". It is now time to briefly discuss "Statistics". And, the statistic that will be of most value, but requires some explanation, is the Chi-squared (χ^2) statistic. So, that is what will be discussed next.

"Goodness-of-Fit" and the Chi-square Test: There are many tests for "Goodness of Fit" between a data set and its analytic model. However, the most straightforward and useful procedure is to use a standard χ^2 test [5]. This simple test is intuitively appealing because it implies that the smaller the difference between the data and the model, the smaller the value of χ^2. Let $\mathcal{D}(t_i)$ represent the raw failure rate data at time t_i, and let $\mathcal{N}$ represent the total number of raw data points. If $\rho(t_i)$ is the probability density function used to model the failure characteristics of a part, subsystem, or system, then

$$\chi^2 \approx X^2 \equiv \sum_{i=1}^{n} \left\{ [\mathcal{D}(t_i) - \mathcal{N}\rho(t_i)]^2 / \mathcal{N}\rho(t_i) \right\}, \qquad (1.13)$$

where the index i runs from the first to the nth time bin. The model that yields the lowest value of χ^2 (or more precisely X^2) is the best model for the system. For example, consider 100 light bulbs ($\mathcal{N} = 100$). And, suppose that the failure rate law for the light bulbs is expected to be exponential as a function of time. Then the (normalized) probability density function has the form $\rho(t) = \alpha \exp(-\alpha t)$. Suppose the analyst checks at unit time intervals t_i, (say, once a week) to see how many bulbs are still working. According to the model, there should be $\mathcal{N}\rho(t_i)$ bulbs alive. If the value of χ^2 calculated by Eq. 1.13 for the exponential model is less than that calculated for some other model, then the exponential model would be preferred.

However, there are problems associated with curve fitting via the χ^2 statistic. First of all, the defect count during the early stages of any software project tends to be very noisy, so that minimizing χ^2 emphasizes minimizing the numerator for the early data since these terms are the biggest contributors to the sum. Furthermore, by the time $t = 2t_{max}$, where t_{max} is the time at which the maximum of the data curve occurs, the earliest data is too obsolete to be useful for future prediction of system performance. Both of these difficulties can be overcome by only fitting the data after $t = t_{max}$. There is also a deeper problem that is harder to understand. If the frequency of the data noise is very high, so that it "averages out" within a given time bin, the χ^2 test will not be able to detect this type of deviation from the model. In this book, calculations will employ time bins that are sufficiently small to unmask this type of deviation of the data from its analytic model.

Analysis: Analysis is the study of functions and their calculus. In particular, this edition will only employ functions that map real numbers into real numbers, so that its scope will be limited to *Real* Analysis. Chapter 2 will focus on expanding the *normalized failure rate* curve (a probability density function) for a *complex system* (call it ρ_{global}) in terms of the *normalized failure rate* curves (probability density functions) of its i constituent subsystems (call them ρ_i). Typically, the ρ_{global} curve has certain shape features that are not modeled by the standard probability density functions that are available in the literature. For example, ρ_{global} may have a very slowly decaying tail as $t \to \infty$. By contrast, the subsystem Failure Rate curve is usually quite simple having a *Rayleigh shape*. The ρ_i form a set of *independent* functions, in the sense that no nonzero element of the set of functions ρ_i can be expressed as a linear combination of other elements of that set. However, it is not known for certain if this set truly *spans* the ρ_{global} function space, although it seems to. That is to say, it is not certain if every possible *global* (entire complex system) normalized failure rate curve can be expressed as an expansion of Rayleigh functions. Therefore, one can only speak of the ρ_i as being a *"basis set"* for the ρ_{global} function space in a colloquial sense, and with great caution. Of course, it is not *necessary* to choose *Rayleigh* "basis functions". Some other set of functions may prove to be a better "basis set". However, the Rayleigh functions seem to be the most useful and intuitively appealing at this time. It should also be noted that the ρ_i are not *orthogonal functions*. That is to say,

$$\int_0^\infty \rho_i^2 dt = \text{a constant} \neq 1, \text{ and } \int_0^\infty \rho_i \rho_j dt = \text{a constant} \neq 0. \tag{1.14}$$

Fortunately, these special properties are not needed in the developments that follow. Clearly, there is a great deal of theoretical work left to be done in this area. Each Rayleigh ρ_i is distinguished by its own *mode* (σ_i), or the time at which the peak failure rate occurs. And, when the number of subsystems i is very large, it is best to describe the set of modes σ_i by a distribution function all their own (called $f(\sigma)$). Therefore, the global probability density function will simply be the sum (or more properly the integral) of the Rayleigh distribution weighted by the distribution of Rayleigh modes. One might even say that ρ_{global} is a distribution of distributions of the form

$$\rho_{global} = \int \rho_i f(\sigma) d\sigma, \tag{1.15}$$

where *both* functions within the integral sign are distribution functions; a fact that will prove very useful in the next chapter where some important integrals of this kind will be evaluated by expanding them as a series solution via the method of Papoulis [9], while others will be evaluated in terms of the solutions to a certain differential equation.

System safety engineers are, however, not just interested in predicting failure rates as a function of time. They also want to know where the greatest system *Risk* lies in severity-probability space (S, P space; where severity is measured in dollars and probability is a dimensionless number between 0 and 1). As a practical matter, S and P range over so many orders of magnitude that it is easier to work with their logarithms to the base 10; where $\mathcal{S} \equiv \text{Log } S$ and $\mathcal{P} \equiv -\text{Log } P$. Let $F(\mathcal{S}, \mathcal{P})$ be a function that describes the fraction of hazards from a hazard analysis (preliminary or final) that fall within a differential neighborhood $d\mathcal{S}\, d\mathcal{P}$ around a point $(\mathcal{S}, \mathcal{P})$ in logarithmic severity-probability space, then the contribution to a system's average risk per hazard from area A of $(\mathcal{S}, \mathcal{P})$ space is defined as

$$\underline{R}_A = \iint_A S\,P\,F(\mathbb{S}, \mathcal{P})\,d\mathbb{S}\,d\mathcal{P}. \qquad (1.16)$$

Notice that the integrand contains both S and P (explicitly) as well as $\mathbb{S}$ and $\mathcal{P}$. This last double integral will prove to be very important, but also very difficult to evaluate. It will be a formidable exercise in the calculus of several variables.

No attempt will be made here to review all of calculus. A 2-year (four semester) course of study (limits, infinite series and differential calculus, integral calculus, the calculus of several variables, and ordinary differential equations) is a prerequisite for reading this book, which was intended to be a first-year graduate-level course in engineering. In this subsection, only some general information has been covered to provide some continuity between a university curriculum and the practical problems encountered by working system safety engineers.

Matrix Algebra: A *Matrix* is nothing more than a collection of numbers arranged in rows and columns. In this book, these numbers will usually be integers, and the number of rows and columns will be equal so as to form a "square" matrix. However, when Linear Programming is discussed, any real number will be possible, and the number of rows and columns will generally be unequal so as to form a "rectangular" matrix called a "tableau". Furthermore, an *Algebra* is a set of mathematical objects (in this case matrices; call any three of them A, B and C) for which multiplication by a number, and distributive laws, are defined. If c is some constant (e.g., a real number), and if "∘" represents some mathematical operation between two matrices (call it "composition"), then an algebra requires

$$(cA) \circ B = c(A \circ B) = A \circ (cB) \ (\text{"Scalar" multiplication}), \text{ and}$$
$$(A + B) \circ C = A \circ C + B \circ C \text{ and } C \circ (A + B) = C \circ A + C \circ B \text{ (Distributive Laws)}.$$
$$(1.17)$$

Notice that a Commutative Law for composition is not required of an algebra. This is an important point that will be revisited. If "∘" represents *element by element multiplication*, that would certainly be allowed according to Eq. 1.17. For example,

$$\left(5\begin{pmatrix}1&2&3\\4&5&6\\7&8&9\end{pmatrix}\right)\circ\begin{pmatrix}1&1&1\\2&2&2\\3&3&3\end{pmatrix}=\begin{pmatrix}5&10&15\\20&25&30\\35&40&45\end{pmatrix}\circ\begin{pmatrix}1&1&1\\2&2&2\\3&3&3\end{pmatrix}=\begin{pmatrix}5&10&15\\40&50&60\\105&120&135\end{pmatrix}.$$

$$\|$$

$$\text{But, }\left(5\begin{pmatrix}1&2&3\\4&5&6\\7&8&9\end{pmatrix}\right)\circ\begin{pmatrix}1&1&1\\2&2&2\\3&3&3\end{pmatrix}=5\begin{pmatrix}1&2&3\\8&10&12\\21&24&27\end{pmatrix}=\begin{pmatrix}5&10&15\\40&50&60\\105&120&135\end{pmatrix}.$$

$$\|$$

$$\text{And, }\begin{pmatrix}1&2&3\\4&5&6\\7&8&9\end{pmatrix}\circ\left(5\begin{pmatrix}1&1&1\\2&2&2\\3&3&3\end{pmatrix}\right)=\begin{pmatrix}1&2&3\\4&5&6\\7&8&9\end{pmatrix}\circ\begin{pmatrix}5&5&5\\10&10&10\\15&15&15\end{pmatrix}=\begin{pmatrix}5&10&15\\40&50&60\\105&120&135\end{pmatrix}.$$

$$(1.18)$$

So, element-by-element multiplication certainly satisfies the rules of scalar multi-plication, and the Distributive Law is satisfied as well. Provided, of course, that matrix addition (and subtraction—the addition of a negative number) is also defined to be "element by corresponding element" for matrices of equivalent dimensions. To summarize, *matrices of equivalent dimensions (size), together with element-by-element addition and multiplication, form a "matrix algebra"*. However, there is another way to define "multiplication" called *matrix multipli-cation* that is tremendously useful, indeed essential, and that also satisfies the rules of an "algebra" (Eq. 1.17). If $A(i, j)$ are the elements of an m x n matrix A (where m is the number of rows and n is the number of columns of A), and $B(i, j)$ are the matrix elements of the n × p matrix B, then the matrix elements of the matrix product AB = C are given by

$$C(i,j) = \sum_{k=1}^{n} A(i, k)B(k, j). \qquad (1.19)$$

For example,

$$AB = \begin{pmatrix} 3 & 4 & 2 \\ 2 & 3 & -1 \end{pmatrix} \begin{pmatrix} 1 & -2 & -4 \\ 0 & -1 & 2 \\ 6 & -3 & 9 \end{pmatrix} = \begin{pmatrix} 15 & -16 & 14 \\ -4 & -4 & -11 \end{pmatrix} = C,$$

$$(1.20)$$

$$\text{and, } C(2,3) = \begin{pmatrix} 2 & 3 & -1 \end{pmatrix} \begin{pmatrix} -4 \\ 2 \\ 9 \end{pmatrix} = 2(-4) + 3(2) + (-1)(9) = -11.$$

The other elements of C can be computed in a similar fashion. It should be noted that matrix multiplication is noncommutative. That is to say, $A \circ B \neq B \circ A$ in general. However, matrix multiplication *is* associative (i.e., $(A \circ B) \circ C = A \circ (B \circ C)$). When a matrix A is matrix multiplied by itself it is called a power of A, so that $A \circ A$ is denoted by A^2. Powers of matrices will prove extremely useful in diagnosing problems in electronic and software systems. In the next section matrix algebra will be employed in solving systems of linear equations because this technique will be central to the solution of the *Linear Programming* problems associated with the *minimization* or removal of dangerous goods from newly designed systems. In all the matrix calculations that follow, the usual customary practice of either replacing the "o" symbol with a "dot" ("·"), or ignoring it completely, will be followed so that $A \circ B \equiv A \cdot B \equiv AB$.

Linear Programming: The Linear Programming problem is an *optimization* prob-lem that involves finding the maximum, or the minimum, of some quantity. For example, an analyst may wish to *maximize* heat conduction, or the strength of a particular structural member(s). Or, he/she may be interested in *minimizing* cost, or the use of some hazardous material. In either case, a set of simultaneous linear equations will have to be solved. And, the best way to do that is by matrix methods. So, that is what will be reviewed here. Consider the following system of equations:

$$2x - y + z = 1$$
$$3x + 2z = -1 \qquad\qquad (1.21)$$
$$4x + y + 2z = 2.$$

This system of equations can be recast, using the definition of matrix multiplication from the previous section, to read

$$\begin{pmatrix} 2 & -1 & 1 \\ 3 & 0 & 2 \\ 4 & 1 & 2 \end{pmatrix} \begin{pmatrix} x \\ y \\ z \end{pmatrix} = \begin{pmatrix} 1 \\ -1 \\ 2 \end{pmatrix}. \qquad (1.22)$$

At this point, it is customary to join the 3×3 matrix on the left to the 1×3 matrix on the right of Eq. 1.22. The vertical (x, y, z) vector and the equal sign may now be omitted and replaced by a dashed line to yield the so-called *augmented matrix*.

$$\begin{array}{cccc} x & y & z & c \end{array}$$
$$\left(\begin{array}{ccc|c} 2 & -1 & 1 & 1 \\ 3 & 0 & 2 & -1 \\ 4 & 1 & 2 & 2 \end{array} \right) \qquad (1.23)$$

Now, this augmented matrix looks very similar to the *tableau* that will be used in Chap. 15 to solve certain Dangerous Goods problems. The chief difference being that a horizontal dashed line will be introduced to separate the bottom row of figures from those above. The reason for this is that the bottom row of figures in a tableau will be seen to have special significance in Chap. 15. For now, the horizontal line can be dispensed with because the primary reason for the current exercise is to demonstrate the *row reduction* procedure as a method to solve the linear system of equations in 1.21 above. Notice that the variables x, y, and z, and the constant c have been placed above the columns of the augmented matrix 23. This is usually not done. However, these labels will serve as a reminder to the reader, connecting the coefficients below with the appropriate variable above, and the c column contains the constants on the far right of Eq. 1.21. The labeling will also preserve continuity between this discussion and the discussion in Chap. 15.

Since the augmented matrix 23 just represents the system of linear equations in 1.21, it is always possible to perform the following three operations without altering the solution of the system 22: (1) any two rows can be interchanged, (2) any row can be multiplied by a constant, and (3) any row can be added to another row. Of course, the second and third operations can be combined so that any row can be multiplied by a constant (positive or negative) and added to another row. Suppose the following operations are performed on the augmented matrix 23:

(1) multiply row 2 by 1/3,
(2) this new row 2 is now multiplied by –2 and added to row 1,
(3) the new row 2 is now multiplied by –4 and added to row 3,
(4) interchange rows 1 and 2,
(5) add row 3 to the current row 2,
(6) interchange rows 2 and 3,
(7) add 2/3 of the current row 3 to the current row 1,
(8) add –2/3 of the current row 3 to the current row 2,
(9) multiply the current row 3 by –1.

The result of all these operations is the augmented matrix below.

$$\begin{matrix} x & y & z & c \end{matrix}$$
$$\begin{pmatrix} 1 & 0 & 0 & 3 \\ 0 & 1 & 0 & 0 \\ 0 & 0 & 1 & -5 \end{pmatrix} \tag{1.24}$$

The solution to the original system 21 can now be read off as follows. In the column labeled x, look for the "1" below. Then move to the right. In the "c" column read 3. Therefore, x = 3. Similarly, y = 0 and z = –5. Why is this so? Simply return the augmented matrix 24 to its matrix equation form to get

$$\begin{pmatrix} 1 & 0 & 0 \\ 0 & 1 & 0 \\ 0 & 0 & 1 \end{pmatrix} \begin{pmatrix} x \\ y \\ z \end{pmatrix} = \begin{pmatrix} 1 \\ 0 \\ -5 \end{pmatrix}. \tag{1.25}$$

Employing matrix multiplication on the left side of this equation yields the set of equations x = 3, y = 0, and z = –5.

The discussion of matrix methods in the last two subsections has been necessarily brief, and is not intended to be a substitute for a basic course in Linear Algebra or Linear Programming, but only a refresher for those who have taken such courses [10, 11].

Game Theory and Graph Theory: There are two basic types of games in this world. There are "*games of chance*", and there are games of *strategy* (also called "*games of no chance*"). Between the two exist an infinite number of mixed possibilities. Both basic types of games play a role in system safety. The first because future outcomes are often uncertain, so that the system safety engineer must make decisions based on what is most probable, or has the greatest *chance* of happening. The second because the system safety engineer may seek a known, or desired, outcome (a winning *strategy*).

Games of chance can, themselves, be divided into two types. There are games that can be characterized purely numerically, like calculating the probabilities of throwing various sums in dice. And, there are games that display their randomness (or chance) in a geometrical way, like Buffon's needle [12]. In the next chapter, two numerical games of chance will be encountered that are based on the "coin toss" game. A coin will either come up heads or tails when dropped. Similarly, a light bulb is either live (heads) or dead (tails) after a time T typical of its lifespan, and an electronic binary random number generator (Schrodinger Cat Experiment) will either output a "1" (heads) or a "0" (tails). Geometrical games of chance are also represented in the text (Chap. 9). Georges-Louis Leclerc, Comte de Buffon (1777), a French mathematical genius born into a generation of geniuses, proposed the first such "game". Consider dropping a needle of length L onto a floor that is ruled with two straight lines also separated by L. If the needle is dropped a large number of times, Buffon proved that multiplying the number of random drops by 2, and dividing by the number of times the needle touched a line, was equal to π [12]! In Chap. 9, a very similar children's game called "pick up sticks" will be discussed, whose random geometrical behavior may be used as a model for the "Bent Pin Problem" in which the long thin pins of an aerospace connector undergo random bending during a faulty insertion. Another geometrical game of chance involves random walks, or the random turning of the moving parts of a Rubik's Cube (1974). This game has some bearing on the complexity of geometrically solving Linear Programing problems (Chap. 15) in n variables (n dimensions). Of course, Rubik's Cube can also be solved (unscrambled) by strategy, so those are the types of games that will be discussed next.

The most common strategy games are board games which, like games of chance, go back all the way to ancient Egypt (*Senet*). And, almost every culture has them. Europe has *Chess*, a game that probably has its origins in India. Africa has *Warri*, that may be as old as Senet. And China has *Go*. In the Americas, certain stone tablets called *Yupanas* (typically about 27L × 23.5W × 7.5H cm) have been found with 16 incised (identical) square, and 5 rectangular, depressions. At first, these were thought to be floor plans of Recuay culture (Peru, 200–600 AD) buildings (i.e., architectural models). However, recent research recognizes them as game boards [13], and a dozen or more have now been excavated. The board's reflection symmetry (or in some cases 180° rotational symmetry about a perpendicular axis) possibly suggesting the Yupanas' use by two players (opponents). It may be that board games are an almost universal hallmark of human culture. Strategy games of this kind are of great interest, and will be discussed in more detail in Chaps. 5 and 11, because the moves of a piece mark out *paths*, *circuits*, and *network* on a game board. And, paths, circuits, and networks are of central importance when trying to understand electronic circuits and programming logic. For example, a system safety engineer may want to know if a network contains a circuit possessing a certain safety-critical part. The engineer may also want to know how many such circuits exist, and what is the longest circuit of this kind that does not contain any loops or switchbacks? Game theory helps to answer these questions. The theory of games is intimately connected to a branch of mathematics called *Graph Theory*. A graph summarizes the possible moves that a game piece can make, or the connections in an electrical network, or the possible connections that a set of bent pins can make in an aerospace connector, etc. So, much of this book will involve studying graphs and their properties. In particular, the numerical representation of a graph called an Adjacency matrix will be central, which brings the engineer back to matrix algebra. So, system safety ties all of these mathematical ideas together closely.

Summary: This book is primarily a book on mathematical techniques for the system safety engineer. Therefore, the mathematics presented in this "Tools" section of the introductory chapter has been a brief review of necessary ideas, results, and methods with a view toward filling in any gaps in the text that might require extensive digressions that would break up the flow of subsequent chapters and this book as a whole. This review was never intended to be a replacement for rigorous course work in Probability and Statistics, Calculus, Linear Algebra, and Graph Theory. However, as previously mentioned, an attempt has been made to bridge the gap between classroom instruction and the practical problems faced by the safety engineer in the field. Furthermore, the undergraduate education of the system safety engineer should not just be limited to mathematics either. System safety is very much an interdisciplinary field. Consequently, extensive knowledge of physics, chemistry (organic, inorganic, physical, and analytic), biochemistry, biology/toxicology/medicine, geology/hydrology/environmental science, and materials science are all valuable to the safety engineer. And, information from these allied disciplines will be introduced into the text where it is required for modeling and analytical calculations.

1.5 Populations, Hardware, Software, and Firmware

The mathematical discussion of the previous section of this chapter began with actuarial probabilities (probability of death) as a model for the nonliving complex systems (hardware, software, firmware, or chimeras formed from combinations of these) dealt with by the system safety engineer. However, there is a key difference. The members (parts) of most expensive complex systems (hardware, software, etc.) are not discarded and removed from the initial population like the deceased after failure. Instead, they are continually repaired (hopefully permanently), upgraded, or modified, until the user decides to retire the system for reasons of obsolescence. In this case, "failure"[‡] is a *temporary* condition that does not define the lifespan of a member. So, the population of members remains essentially constant. This behavior has important consequences for the failure rate versus time curve. At first, few failures are observed in new systems. But then, as testing proceeds, weaknesses become evident, and the failure rate rises. As repairs or modifications are made, the failure rate begins to fall, and a "tail" begins to form that seems to asymptotically approach zero. The decay of the tail (first derivative of the failure rate, d[failure rate]/dt, or the second derivative of failures with respect to time, d^2[failures]/dt^2) for various times past the peak failure rate will depend on system details. The tail can be very flat, or it can decrease very rapidly. In fact, the failure rate versus time curve can be nothing more than a rapidly decaying exponential with its peak at the zero of time. This monotonic decrease of failure rate (and the normalized failure rate) for large times t is not due to depletion of the number of system members, but is due to depletion of the pool of possible failure modes. That is to say, the system as a whole is becoming *less and less hazardous as it ages*! This is especially true of software, and quite opposite to what is observed for people or light bulbs (where no repair is possible and members are just discarded after failure).

The probability equations of the previous section on "Tools" are general, and apply to either hardware, software, or firmware. However, there are features of these classes of complex systems that are unique to each. Software, for example, doesn't fail in the usual sense of that word. That is to say, it doesn't "wear out" or "break" the way hardware does. For this reason, software can be thought of as being *immortal*. However, software can display undesirable, or unexpected, behavior because of some *defect* in the code. A defect can be as simple as a typographical error in an equation (quickly fixed), or it can be much deeper, involving the underlying logical structure of the software (difficult to identify). Usually, correction of a defect (especially the simpler ones) is permanent. Or, at least, that is what the programmer is striving for during *debugging* procedures. Occasionally, it may happen that during the removal of some defects, other new problems are introduced that do not manifest themselves until much later. The net result is that after an initial rapid rise in defects per unit time, a long and very slowly decaying tail can result for large times t [6]. Software under development gradually depletes the pool of possible defects and, in contrast to human populations, becomes *less hazardous* as it ages. Theoretically, a point will be reached (t → ∞) at which all possible defects are depleted for a given application, and maintenance is no longer needed. Copies of such "finished" software are identical, and also finished.

By contrast, the parts (members) of complex hardware systems do break. When a prototype submarine is built, it will at first display many abnormalities of varying severity. Failed parts of a new, unreliable, boat are not just removed from the system (boat), instead, they are repaired or replaced. If a bolt in a particular location fails because of a crack (*recurrent* bolt mass production flaw), it may be replaced with a new bolt in perfect condition. Or perhaps the bolt failed because it was too weak for that particular application (*one-time design flaw*), in which case it can be replaced by one of much higher tensile strength. Hopefully, this repair will be permanent in either case. As initial problems are removed, the system becomes *less hazardous*, again in contrast to human populations. Fewer failures/time are observed, until finally, the boat is ready for a deep dive. Of course, after very long times, parts will begin to wear out, failures/time will increase, and the cycle of repairs will repeat again. This is the "bathtub curve" (Fig. 1.3). To simplify calculations, this book will assume that repairs are either permanent, or that the system is between "bathtub" peaks (recurrent peaks in the normalized failure rate curve). So, hardware is not quite immortal in the same sense as software. Hardware immortality depends on periodic *maintenance*! Finally, it is important to note that

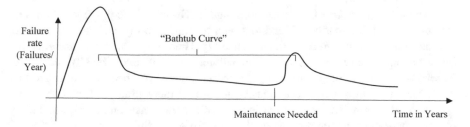

Fig. 1.3 Typical failure rate versus time for complex hardware – long-term view

when a copy is made of the prototype boat, it will have a failure rate curve that looks like the prototype failure rate curve[#] *minus* all one-time design flaws that have been permanently removed (corrected) from the prototype design. So, hardware copies are never "finished" in the sense of software copies because flaws will always be present that can result in *recurrent failures*.

Firmware occupies a position halfway between hardware and software. Firmware consists of electronic hardware into which code (software) has been programmed. Sometimes the *state* of a *bit* can be upset by an electrical disturbance. This disturbance can be internal to the circuitry (e.g., noise), or external to the entire system (such as a cosmic ray induced bit upset). And, it can be *temporary*, leaving no lasting effect on the firmware device after reset. The frequency of such induced events will depend on how well the device is shielded from noise or cosmic rays, the amount of charge that is required to cause a bit state change, and on whether the state of a bit changes from a 0 to a 1 or vice versa. In a fixed radiation environment these *soft failures* are distributed more or less *uniformly* in time until the device physically fails (called a *hard failure*) and needs to be replaced. However, in this case, unlike both hardware and software, replacement does not preclude, or even reduce, further soft failures.

Therefore, complex hardware/firmware/software systems can display permanent (hard) failures, repairable hard failures, recurrent repairable hard failures, one-time repairable hard failures, firmware soft failures, and software defects. These types can be further divided into *severity categories* such as *Catastrophic, Critical, Marginal*, and *Negligible*, as defined by the U.S. government safety document 882 D and E [14, 15]. The rate of any particular type and severity category of failure or defect can be normalized and plotted versus time to yield a pdf. And, it is essential for the analyst to understand the type/severity category, or mix thereof, that is being counted. For any hardware project, *maintenance records* catalogue the various failure types as a function of time. Whereas, software defects are recorded in the *change control* records and are typically plotted against time in defects per thousand lines of code (Defects/KLOC).

* The word "chimera" refers to a "monster made of body parts from a variety of animals". In biochemistry, it refers to a macromolecule made up of smaller macromolecules of diverse origin. Here the term will be used to refer to a system composed of some combination of software, firmware, hardware, and personnel.

† The term *hazardous* has many meanings. In Systems Safety, the word *hazard* refers to a condition that could lead to a *failure*. That is to say, a hazard is a failure that has not yet manifested itself. In Reliability Theory, the *hazard function* is the instantaneous failure rate (measured in failures per unit time). Finally, the word *hazardous* has a colloquial meaning. Here, an attempt has been made to use the word hazardous in a way that is consistent with all these usages.

‡ Here, the term "failure" is used in its most general sense, and applies to both a physical hardware malfunction as well as a software coding defect.

For an in-depth philosophical discussion of probabilities for individual systems versus ensembles of systems, see Ref. [16].

1.6 Problems

(1) The following problem is a modification of one proposed by Pickover (2009) [12]. Suppose there is a population of black and white butterflies. Genetics tells a biologist that these two colors should occur with equal frequency. But field netting reveals that out of 100 butterflies captured only 10 are black and the rest are white. Calculate χ^2 for this experiment. (Hint: there are only two terms in the sum for χ^2. Therefore, n = 2 and $\mathcal{N}$ = 100.)

Note: The large value of χ^2 for this experiment suggests that the initial hypothesis (i.e., that the sample was drawn from a population with an equal number of the two colors) is probably wrong.

(2) What is the matrix product

$$\begin{pmatrix} 1 & 2 & 3 \\ 4 & 5 & 6 \\ 7 & 8 & 9 \end{pmatrix} \begin{pmatrix} 1 & 0 & 0 \\ 0 & 1 & 0 \\ 0 & 0 & 1 \end{pmatrix} = ?$$

The matrix with 1's on the main diagonal is an *identity matrix* for matrix multiplication.

(3) What is the matrix product

$$\begin{pmatrix} 1 & 2 & 3 \\ 4 & 5 & 6 \\ 7 & 8 & 9 \end{pmatrix} \begin{pmatrix} 1 & 1 & 1 \\ 1 & 1 & 1 \\ 1 & 1 & 1 \end{pmatrix} = ?$$

The unit matrix (all ones) is an *identity matrix* for element-by-element multiplication.

(4) Given a square matrix A that has zeros on its main diagonal, is symmetric across its main diagonal, and all of whose off-diagonal elements are either 0 or

1, prove that the matrix $S \equiv A \, X \, A^T = A$; where the "T" indicates the transpose matrix formed by the interchange of the rows and columns of A, and "X" indicates element by element multiplication.

(5) Given a square matrix A that has zeros on its main diagonal, prove that the matrix

$$R_3 = A \cdot d(A^2) + d(A^2) \cdot A - A \, X \, A^T$$

will always have zeros on its main diagonal. The "d" operation on any square matrix keeps the entries on the main diagonal, but replaces all off-diagonal entries with zeros. The "dot" indicates ordinary matrix multiplication, while the "cross" indicates element by element multiplication. For any *square* matrix A, there is usually no problem with mixing both definitions of multiplication in the same expression since both conform to the rules of an associative matrix algebra by themselves. However, the two types of multiplication are not associative with each other [i.e., in general, $(A \cdot B) \, X \, C \neq A \cdot (B \, X \, C)$ and $(A \, X \, B) \cdot C \neq A \, X \, (B \cdot C)$]. Therefore, in such cases, brackets are necessary to specify the order of operations. In Chap. 7 some very complex calculations will be performed using these operations.

(6) Prove Buffon's Theorem:

$$\pi = \frac{2 \, (\text{number of needle drops})}{(\text{number of line touches})},$$

provided the number of drops approaches infinity.

(7) Show that an upper bound on the number of positions of a $3 \times 3 \times 3$ Rubik's cube (with 6 colors) is about 1.7×10^{20}. The actual number of positions is 43,252,003,274,489,856,000 (or about 0.4×10^{20}) [12].

References

1. Leveson, N. G. (2011). *Enginering a safer world: Systems thinking applied to safety* (pp. 7–60). Cambridge, MA: MIT Press.
2. Sandom, C. (2011). Lecture/Address of August 10, 2011, Las Vegas, NV, *International System Safety Conference 2011*, Las Vegas, NV.
3. Anon. (2013). The Oldest Woman is 126 Years old! http://youtube.com/watch?v=TveIDQxt8uU. Accessed 23 Jan 2013.
4. Pearson, C. E. (1983). *Handbook of applied mathematics: Selected results and methods* (2nd ed., pp. 1226–1230). New York: Van Nostrand.
5. Meyer, P. L. (1970). *Introductory probability and statistical applications* (2nd ed., pp. 220, 234–5, 316–37, 350–1). Reading, MA: Addison-Wesley.
6. Peterson, J., & Arellano, R. (2002, March). Modeling software reliability for a widely distributed, safety-critical system. *Reliability Review, 22*(1), 5–26.

7. Zito, R. R. (2013, August). How complex systems fail-I: Decomposition of the failure histogram. In *Proceedings of the International Systems Safety Conference, 2013*, Boston, MA (pp. 12–16).
8. Abramowitz, M., & Stegun, I. A. (1965). *Handbook of mathematical functions* (p. 255). New York: Dover.
9. Papoulis, A. (1965). *Probability, random variables and stochastic processes* (pp. 151–152). New York: McGraw-Hill.
10. Shields, P. C. (1968). *Elementary linear algebra*. New York, N.Y.: Worth Publication.
11. Trustrum, K. (1971). *Linear programming* (p. 88). London: Routledge & Kegan Paul.
12. Pickover, C. A. (2009). *The math book* (pp. 194–5, 300–1, 452–3). New York: Sterling.
13. Pillsbury, J., Sarro, P. J., Doyle, J., & Wiersema, J. (2015). *Design for eternity: Architectural models from the ancient Americas* (pp. 18, 77–79). New York: Metropolitan Museum of Art, 2015.
14. MIL-STD-882D (2000, February 10). *Standard Practice for System Safety* (pp. 18–20).
15. MIL-STD-882E (2002, May 11) *Standard Practice for System Safety* (pp. 11–12).
16. Jammer, M. (1974). *The philosophy of quantum mechanics* (pp. 440–441). New York: Wiley.

Part I
Fundamentals

Chapter 2
Decomposition of the Failure Histogram

2.1 Background and Definitions

It is customary in the system safety community to think of a "system" in terms of "subsystems". The "system" executes the designer's goal, such as the intercept of a ballistic missile. The "subsystem" is simply a collection of related parts that perform some function that contributes to the system goal. The possible failures of individual subsystems are described in the worksheets of the Subsystem Hazard Analysis (SSHA). The possible failures of the system are described in the worksheets of the System Hazard Analysis (SHA). The SHA failures include interface failures, such as the failure of a conducting ribbon or its connectors, that joins two subsystems. Interface components are separate and distinct from the subsystems they join at either end. So, the union of all subsystems (S_i) is not necessarily equal to the system (S) itself. There might be some embarrassing interface parts left over. That is to say,

$$\bigcup_i S_i \subseteq S \tag{2.1}$$

In this book, interface parts will not be allowed, so that a ribbon conductor and its connectors will have to belong to the subsystem at either one end or the other. Otherwise, the ribbon conductor and its connectors will have to be considered a subsystem in and of itself. In that case

$$\bigcup_i S_i \equiv S. \tag{2.2}$$

In fact, *Eq. 2.2 may be thought of as the definition of a system.*

Next, the notion that a subsystem is a collection of parts that performs a function related to the system goal must be modified. In the context of this chapter, a

© Springer Nature Switzerland AG 2020
R. R. Zito, *Mathematical Foundations of System Safety Engineering*,
https://doi.org/10.1007/978-3-030-26241-9_2

subsystem is a related collection of parts that fails according to some "simple" law as a function of time. Furthermore, *it will be assumed in this chapter that the normalized subsystem failure rate law as a function of time t [called ρ(t)] is a Rayleigh distribution, where the cumulative probability of failure by time T [called P(T)] is just the time integral of the normalized rate law, that is to say,*

$$\rho_i(t) = \left(t/\sigma_i^2\right)\exp\left(-\tfrac{1}{2}[t/\sigma_i]^2\right) \quad \text{and} \quad P_i(T) = \int_0^T \rho_i(t)dt. \quad (2.3)$$

Normalization means that the instantaneous number of failures per unit time at any time t (instantaneous failure rate) has been divided by the total number of observed failures. Therefore, the normalized failure rate curve ρ(t) (also called the failure probability density curve; P(t)/Δt) is just the observed (un-normalized) failure rate curve divided by the total number of observed failures. The subscripts "i" indicate that the failure rate law and the cumulative probability of failure applies to subsystem i. The mode σ_i is the value of t at the peak of the ith Rayleigh distribution (i.e., the value of t corresponding to the highest failure rate; normalized or un-normalized). Note that the units of ρ_i are 1/sec to measure a *rate*, while the probability P_i is dimensionless, as it should be. Notice also that the limits of integration in the calculation of P_i run from zero to T because $\rho_i(t) = 0$ for t $\leq$ 0, and $P_i(T) \to 1$ for T $\to \infty$ as it must. Why should the normalized failure rate law be a Rayleigh distribution? It is because a failure cannot occur in negative time. That is to say, a subsystem cannot fail before it is switched on. So, the failure rate (normalized or un-normalized) at t = 0 must be zero. On the other hand, as time progresses, the natural lifetime of a given (*non-repairable*) subsystem will be reached. When this happens, its failure rate will be a maximum. However, like people, a few such subsystems will go on to outlive their fellows. As the population of survivors thins out, the failure *rate* drops. And, as t $\to \infty$, the failure rate approaches zero because there are simply no more subsystems (parts) left to fail per unit time. The oldest known person with a documented birth and death certificate is Jeanne Calment of France who died at age 122! However, it is *possible* (though unlikely) that a person can live even longer. Hence, the death (failure) rate of actuarial models is not *identically* zero when t = 122 years, but it is *very* small—an important point that will be revisited in the next chapter. The shape of the Rayleigh distribution is shown in Fig. 2.1-left, and the corresponding cumulative probability is shown in Fig. 2.1-right.

A discrete example will help bring the ideas of the previous paragraph into sharper focus. Consider a subsystem consisting of a single light bulb. The bulb has a probability of 0.2 that it will fail before reaching its 500 h required life. If 20 such subsystems are tested, what is the probability that exactly k of these bulbs fail prematurely? So, k is a measure of total illumination time for these *defective* parts; which should have all lasted 500 k hours ideally. Each bulb can exist in one of only two states; either it's alive or dead. And, the probability that k bulbs are dead is just

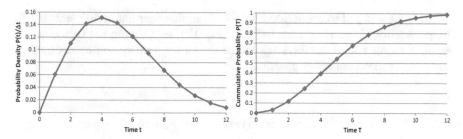

Fig. 2.1 The Rayleigh probability density function, or simply the Rayleigh distribution (left). The Rayleigh cumulative (probability) distribution function (right)

given by the binomial term $P(k) = C(20, k)(0.2)^k(1 - 0.2)^{20-k}$, where $C(20, k)$ is the binomial coefficient 20 choose k, or $20!/(k![20 - k]!)$. Now, referring to Fig. 2.2-left, at each value of k on the k-axis ($\sim$ time axis), construct a rectangular box one k-unit wide and P(k) tall. Both the height and the area of this box is P(k) for some fixed k value, because the rectangle is only one unit wide. Therefore, the vertical axis can be thought of as $\rho(k)$ or $P(k)$ because according to Eq. 2.3, $P(k) = \rho(k)\Delta k = \rho(k) \times 1 = \rho(k)$ for each rectangle. Of course, at $t = 0$ and before, P must equal zero, but $P(k = 0)$ is not necessarily zero because it is possible (but unlikely) for all 20 bulbs to survive 500 h. Evaluation of P(k) for all k from 0 to 10 yields 0.012, 0.058, 0.137, 0.205, 0.218, 0.175, 0.109, 0.055, 0.022, 0.007, and 0.002. The values of P(k) for $k \geq 11$ are very small and have been ignored. That is to say, the reservoir of possible defective parts is effectively depleted at $k = 10$ (i.e., it is very unlikely that a sample of 20 bulbs has more than 10 that are defective). Finally, connect straight line segments between the origin and the upper right-hand corners of each rectangle. The result is Fig. 2.2-left. Notice the approximately Rayleigh shape (long tail), and the fact that the total area under the curve (i.e., the cumulative probability P(T) as $T \to \infty$) is unity (i.e., $0.012 + 0.058 + 0.137 + \cdots \approx 1$) (see Fig. 2.2-right). Also note that the peak (mode) of the distribution lies at $k = 4$, meaning that it is most likely that four out of

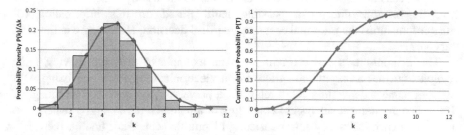

Fig. 2.2 The Rayleigh-like binomial distribution for a system of binary subsystems (left). The Rayleigh-like binomial cumulative (probability) distribution function (right)

the 20 bulbs will fail to reach 500 h of life. And, that four out of 20 yields 0.2 for the probability of failure of each bulb (4/20 = 0.2), as it should be! Hence, as this simple example shows the Rayleigh failure model for a subsystem makes intuitive sense and is mathematically plausible.

2.2 Schrödinger's Cat and Systems of Finite Complexity

The Rayleigh subsystems above (Eq. 2.3) can be combined to form a system. Consider the following *gedanken experiment* (thought experiment) called Schrödinger's Cat Experiment; originally devised by quantum mechanicians to visualize the peculiarities of that probabilistic science. First, consider an integrated circuit supplying eight random outputs for each unit of time (say, once a second). Each output is either a 1 or a 0. If all eight outputs are 1, then a relay IC allows a current to pass from a power supply to an electromagnetic hammer that smashes a vial of prussic acid in a box containing a cat. After a time t, the interior of the box is checked by an outside observer (safety engineer). The question is, "Will the cat still be alive after time T?" At first, the probability of the cat's death is very small—just $1/2^8$. But, after a time, say after n operations of the random number generator, the probability of death, P(n), begins to grow (P(n) $\rightarrow$ $n/2^8$). After 2^8 s death is probable. Beyond this point in time, the cat's survival is unlikely. So, the *cumulative probability distribution* for the cat's death approaches unity (P(n) $\rightarrow$ 1) as time progresses. This behavior is exhibited in Fig. 2.1-right and Fig. 2.2-right. Notice that these cumulative probability distributions have a point of inflection, where they become concave down and asymptotically approach unity, at abscissa values corresponding to the peak of the Rayleigh (or Rayleigh-like) probability density functions in Fig. 2.1-left and Fig. 2.2-left, respectively. Therefore, *the probability density function is the derivative of the cumulative probability distribution, and the cumulative probability distribution is the integral of the probability density function* as described in Eq. 2.3. The Schrödinger Cat Experiment may seem esoteric, but many real-world safety problems fall into this category. For example, consider a pilot who reads (or miss-reads) a display, which may, or may not, be accurate due to a possible fault. Based on this information (or miss-information) the pilot makes a safety-critical decision. How long can the pilot survive?

The Schrödinger's Cat Experiment is an example of a system having *finite complexity* (four subsystems). There is an electronic subsystem consisting of a random number generating IC, a relay IC, a power supply, and an electromagnet. There is a mechanical subsystem consisting of a hammer, a vial, and a sealed box. There is a chemical subsystem consisting of prussic acid. And, finally, there is a living subsystem consisting of the cat. Each of these four simple subsystems has its own cumulative failure probability yielding a probability P_i, where i = {1, 2, 3, 4}. So, by Eq. 2.2, the composite failure law for the whole system is P = $\sum P_i$ for all i.

If, in addition, the subsystems all have normalized failure rates described by a Rayleigh probability density functions, then from Eq. 2.3

$$P = N \sum_{i=1}^{n} \int_0^T (t/\sigma_i^2) \exp\left(-\tfrac{1}{2}[t/\sigma_i]^2\right) dt = N \sum_{i=1}^{n} \left\{ 1 - \exp\left(-\tfrac{1}{2}[T/\sigma_i]^2\right) \right\},$$

(2.4)

where $n = 4$ in this case. Note that an overall normalization factor N ($=1/n$) is required because each subsystem probability density function is, itself, normalized. The integrated result on the right side of Eq. 2.4 can always be used to compute the overall cumulative probability of failure as a function of time T, even if it has to be applied on a part by part basis. Furthermore, it is clear that the time T_{50}, when system has a 50% chance of failure, can be calculated by setting $P = 0.5$, substituting in the values for the σ_i, and numerically solving for T.

Equation 2.4 is a major result because it allows computation of the system P from the properties of the Rayleigh subsystems, or even individual Rayleigh parts. Furthermore, the failure properties of subsystems may be known from legacy usage, and the failure properties of individual parts may be known from their manufacturers. However, although Eq. 2.4 is practical, it can be tedious and messy, especially for large n. And, it says nothing profound about the nature of complex systems. These shortcomings will be remedied in subsequent sections where distributions of σ_i are considered for complex systems containing an infinite number, or at least a very large number, of Rayleigh subsystems. Examples of complex systems include submarines, modern rockets, nuclear power plants, and jets, just to mention a few.

2.3 A Distribution of Distributions for Systems of Infinite Complexity

Suppose a system is so complex, and has so many subsystems, that the Rayleigh modes (σ_i values) of all the subsystems are themselves most conveniently characterized by a smooth distribution function. In that case, the *global* probability density function for the *system*, $\rho_{global}(t)$,[†] is given by the integral of the Rayleigh distribution weighted by the probability density function of the modes $f(\sigma)$. That is to say,

$$\rho_{global}(t) = \int_0^{\infty} (t/\sigma^2) \exp\left(-\tfrac{1}{2}[t/\sigma]^2\right) \ f(\sigma) d\sigma.$$

(2.5)

The integration proceeds from 0 to ∞, since σ is always positive, and is analogous to the summation in Eq. 2.4. Equation 2.5 can be integrated exactly in a few special cases of great importance. However, the probability density function f that is most useful in describing the "tail" of a normalized system failure rate versus time histogram will require approximate methods to integrate Eq. 2.5. The technique described by Papoulis [2] is most suited to integrals where one of the two functions in the integrand is a probability density function. This method also gives an answer that is theoretically useful since it will provide a measure of the deviation from Rayleigh-like behavior. It can be shown [2] that given an integral I such that

$$I = \int_0^\infty g(x)f(x)dx, \text{ where } f(x) \text{ is a probability density function, that}$$

$$I \cong g(\eta) + g''(\eta)(\mu_2/2) + \cdots g^{(n)}(\eta)(\mu_n/n!), \text{ where } \eta = \int_{-\infty}^\infty xf(x)\,dx \quad \text{and}$$

$$\mu_n = \int_{-\infty}^\infty (x - \eta)^n f(x)dx.$$

$$(2.6)$$

These formulas apply provided that g is "smooth." In this context "smooth" means that the function g does not vary suddenly or discontinuously over the region where f takes on significant values. In the next two sections both exact and approximate evaluations of Eq. 2.5 will be considered.

2.4 Exact Methods

Dirac Delta Function distribution on σ: The very simplest distribution of modes is the Dirac Delta Function [3] distribution, δ. The delta function has the property that it is infinitesimally narrow, and it is also infinitely tall, with the net result that the area under this function is unity (like any probability density function). Furthermore, if a delta function is located at position σ_m, and multiplied by a second function, and then integrated over all nonzero values of the second function, the result is just the second function evaluated at location σ_m. If the Dirac Delta Function is substituted for $f(\sigma)$ in Eq. 2.5, only a single global value (σ_m) for the mode σ will survive, and Eq. 2.5 becomes

$$\rho_{\text{global}}(t) = \int_0^\infty (t/\sigma^2) \exp\left(-\frac{1}{2}[t/\sigma]^2\right) \delta(\sigma - \sigma_m) d\sigma$$

$$= (t/\sigma_m^2) \exp\left(-\frac{1}{2}[t/\sigma_m]^2\right).$$

$$(2.7)$$

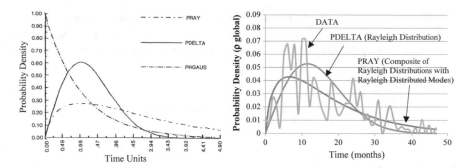

Fig. 2.3 (Left) The probability density (per KSLOC), $\rho_{global}(t)$, plotted as a function of t for the three exactly integrable distributions on σ; **a** the Dirac Delta function (PDELTA), **b** the Rayleigh distribution (PRAY), and **c** the half-Gaussian distribution (PHGAUS). (Right) RMS best fit of PDELTA and PRAY to normalized smoothed experimental data; module C of Peterson and Arellano [1] (DATA). The data involved analysis of 74.4 KSLOC and was originally expressed in defects/KSLOC/month prior to normalization

That is to say, $\rho_{global}(t)$ is just the Rayleigh distribution; as expected for a system with only one value of σ. Such a "system" is just composed of, and is synonymous with, a single subsystem. In Fig. 2.3-left, the Rayleigh distribution (right side of Eq. 2.7) has been plotted with $\sigma_m = 1$. All other global probability distributions, $\rho_{global}(t)$, will be compared to this Rayleigh curve.

In the study by Peterson and Arelliano [1], the "failure histories" of four software modules (denoted by A, B, C, and D) were examined. Of course, software doesn't usually "fail" in the usual sense of that word. Here, the word "failure" refers to a software "bug". That is to say, a coding error at some level that causes the software to do something that the designer did not intend. The failure history, measured by failures per thousand software lines of code (KSLOC), from development testing to field use, was plotted as a function of time for each module. Failure data was taken at 1-month intervals. When this data is properly normalized to produce a probability density function, modules A, B, and D can be fit by a Rayleigh function, *so long as one is not too concerned about underestimating the probability in the tails*. This statement is equivalent to saying that modules A, B, and D are "simple enough" to be considered subsystems in the mathematical sense of the word used above. Again, with the caveat that modeling the long term behavior of these modules is of little concern. However, module C is quite different. It is too complex to ever be considered a subsystem, and it will be discussed in great detail in the next subsection. In general, when a subsystem is too complex to be modeled by a Rayleigh failure rate law, it may have to be cut up into even smaller pieces (perhaps functional sub-subsystems) before Rayleigh failure laws can be realized.

Finally, it should be clear that if the distribution of modes consists of several Delta functions, each with its own σ_m, then a cumulative distribution function given by Eq. 2.4 will arise after proper normalization.

Rayleigh Distribution on σ: Suppose that the modes of Rayleigh subsystems are themselves Rayleigh distributed with mode M. This model has intuitive appeal since a subsystem mode can never have a negative value, but may be very large and positive. Therefore,

$$p_{global}(t) = \int_0^\infty (t/\sigma^2) \exp\left(-\tfrac{1}{2}[t/\sigma]^2\right)(\sigma/M^2)\exp\left(-\tfrac{1}{2}[\sigma/M]^2\right)d\sigma. \qquad (2.8)$$

Setting $\alpha \equiv \tfrac{1}{2}(\sigma/M)^2$ yields

$$p_{globad}(t) = \int_0^\infty (t/M^2)(1/2\alpha)\exp\left\{-\left[\alpha + (t/2M)^2(1/\alpha)\right]\right\}2\sqrt{\alpha}\,d\sqrt{\alpha}. \qquad (2.9)$$

Since $d\alpha = d(\sqrt{\alpha})^2 = 2\sqrt{\alpha}\,d\sqrt{\alpha}$, Eq. 2.9 becomes

$$p_{global}(t) = (t/2M^2)\int_0^\infty (1/\alpha)\exp\left\{-\left[\alpha + (t/2M)^2(1/\alpha)\right]\right\}d\alpha. \qquad (2.10)$$

Define $\kappa = \tfrac{1}{2}(t/M^2)$ and $\beta = (t/2M)^2$ such that $\kappa = \sqrt{\beta}/M$. Then

$$p_{global}(t) = \kappa\int_0^\infty (1/\alpha)\exp(-\alpha - \beta/\alpha)d\alpha \qquad (2.11)$$

The integral of Eq. 2.11 is given by [4, 5] $p_{global}(t) = 2\kappa\,K_0(2\sqrt{\beta})$, where K_0 is a function whose properties are well known since it is the solution of the second-order differential equation [6]

$$u'' + (1/z)u' - \left(1 + [v/z]^2\right)u = 0, \qquad (2.12)$$

where u is a function of z and v is a constant. The series solution of Eq. 2.12 yields [7]

$$K_0(2\sqrt{\beta}) = -(\ln\sqrt{\beta})\sum_{j=0}^\infty \left[(\sqrt{\beta})^{2j}/(j!)^2\right] + \sum_{j=0}^\infty \left[(2\sqrt{\beta})^{2j}/(2^{2j}j!^2)\right]\psi(j+1).$$

$$(2.13)$$

Therefore,

$$\rho_{global}(t) = [(2\sqrt{\beta})/M] \sum_{j=0}^{\infty} \left\{ \left[\beta^j/(j!)^2 \right] [\psi(j+1) - \ln\sqrt{\beta}] \right\}. \tag{2.14}$$

It is convenient to expand $\psi(j+1)$ for integers j as follows [8, 9]:

$$\psi(j+1) = -\gamma + \sum_{n=1}^{j} (1/n), \tag{2.15}$$

where γ is the Euler–Mascheroni constant whose approximate value is 0.57721… [10, 11] and the summation in Eq. 2.15 is not evaluated for $j = 0$. Typically, j need not exceed a value of 9. In Fig. 2.3-left, Eq. 2.14 has been plotted so that its peak aligns with the peak of the Rayleigh curve. After its maximum value has been reached, Eq. 2.14 displays a slow decrease of probability density as a function of time (PRAY in Fig. 2.3-left). Appendix A gives numerical values of $\rho_{global}(t)$ as a function of t for a few typical integer values of M. Values of the Cumulative Probability Distribution $P = \int \rho_{global}(t)dt$, integrated from 0 to T, are also given in Appendix A. Figure 2.3-right shows the RMS best fit of Eq. 2.14 to module C of Peterson and Arellano [1]. Module C data has been reduced to a smoothed, normalized, probability density function. For comparison, the RMS best fit simple Rayleigh Distribution is also plotted on Fig. 2.3-right. Visual inspection shows that Eq. 2.14 offers the superior fit. After its peak, the Rayleigh Distribution spends most of its time either above or below the data set. However, the minimum RMS errors for these two models differs by only 10.5% (i.e., 0.0152 for PDELTA with $\sigma_m = 11.5$ vs. 0.0136 for PRAY with $M = 10.9$). Obviously, the RMS error is not a very sensitive measure of "Goodness-of Fit". The Chi-square test is much better [3]. For PDELT with $\sigma_m = 11.5$, $\chi^2 = 453.63$, and for PRAY with $M = 10.9$, $\chi^2 = 117.08$. Therefore, PRAY offers about a factor of 4 improvement (reduction) in the value of χ^2 over a simple Rayleigh model. Curiously, if one tries to optimize fitting by minimizing χ^2, the result is almost the same as that achieved by minimizing the RMS error. The result for PRAY is a curve with $M = 11.2$ and $\chi^2 = 116.79$; a trivial improvement in Chi-square of only 0.29. Finally, it should be noted that that the critical value tabulated for Chi-square at the 1% confidence level is 69.957 for 45 degrees of freedom (i.e., 47 data points minus 1, minus an additional 1 because M was estimated from the data) (Appendix B). In other words, there is only a 1% chance that the calculated χ^2 will exceed ~ 70. However, 117 is slightly larger than this critical value. Why? It is because the data of Module C (Fig. 2.3 above) is *very* noisy! That is to say, there is enough scatter in the data to admit other possible models (a point that will be revisited in Chap. 12). This insight is part of the power of the χ^2 method! In summary, module C is *not* a simple subsystem. It is a system in itself, composed of Rayleigh subsystems whose modes are themselves Rayleigh distributed.

Half-Gaussian Distribution on σ: The half-Gaussian distribution function is defined by

$$f(\sigma) = \left(2/\sqrt{[2\pi]}\right)(1/s)\exp\left(-\tfrac{1}{2}[\sigma/s]^2\right) \quad \sigma \geq 0$$
$$\text{and} \quad f(\sigma) = 0 \qquad\qquad\qquad\qquad\qquad \sigma < 0,$$

(2.16)

where s is the standard deviation. This distribution of modes is useful in modeling the normalized software failure rate *after* the development process [1]. By contrast, the Delta function and Rayleigh distribution of subsystem modes above results in normalized failure rate laws that includes *all* faults, of a given type and severity, discovered throughout the development process [1]. Notice also that a half-Gaussian distribution of modes looks a lot like the "tail half" (from M to +∞) of the previous Rayleigh distribution of modes (sec. previous subsection). In other words, software system "development" fixes the modules (subsystems) with "early" failure rate distribution modes so that only a half-Gaussian remains, with a new time zero at the end of development. In that case, as will be shown, the detection of faults follows the classic exponential decay law as a function of time [1]. The factor of 2 in Eq. 2.16 (top) is there to insure that $f(\sigma)$ possesses a unit integral over t. The global probability density function is given by

$$\rho_{\text{global}}(t) = \int_0^\infty (t/\sigma^2)\exp\left(-\tfrac{1}{2}[t/\sigma]^2\right)(2/\sqrt{[2\pi]})(1/s)$$
$$\exp\left(-\tfrac{1}{2}[\sigma/s]^2\right)d\sigma,$$

(2.17)

Defining $a \equiv \tfrac{1}{2}\,t^2$, and $b \equiv \tfrac{1}{2}(1/s^2)$, yields

$$\rho_{\text{global}}(t) = (2t/\sqrt{[2\pi]})(1/s)\int_0^\infty (1/\sigma^2)\exp(-a[1/\sigma^2] - b\sigma^2)d\sigma.$$

(2.18)

This last equation integrates to

$$\rho_{\text{global}}(t) = (1/s)\exp(-t/s).$$

(2.19)

Setting s = 1 yields the curve PHGAUS in Fig. 2.3-left. The cumulative failure (defect) probability from t = 0 to maturity time T is just

$$P(T) = \int_0^T (1/s)\exp(-t/s)dt = 1 - \exp(-T/s).$$

(2.20)

If italic N is the total number of defects at delivery, then the cumulative number of defects (faults F) that have occurred by maturity time T is given by

$$F(T) = N\,P(T) = N[1 - \exp(-T/s)] = N[1 - \exp(-\lambda T)], \qquad (2.21a)$$

where $\lambda \equiv 1/s$. And, the number of faults remaining in the system is

$$F(\text{remaining}) = N - F(T) = N\exp(-\lambda T). \qquad (2.21b)$$

This is just Eq. 1.1-1 of Peterson and Arellano [1]. *The fact that the decay constant λ is just the reciprocal of the standard deviation of the underlying half-Gaussian distribution of Rayleigh subsystem modes is a profound result.* The instantaneous fault rate is the derivative of Eq. 2.21a with time.

$$dF(T)/dT = N\,\lambda\exp(-\lambda T), \quad \text{and at } t = 0 \quad dF(T)/dT|_{T=0} = N\lambda. \qquad (2.22)$$

Since the mean time between failures (MTBF) is just the time per unit fault, or $dT/dF(T)$, it is clear that

$$MTBF(T) = 1/[N\,\lambda\exp(-\lambda T)], \quad \text{and at } t = 0 \quad MTBF(t = 0) = 1/N\lambda. \qquad (2.23)$$

These, of course, are Eqs. 1.1-3 and 1.1-4 of Peterson and Arellano [1]. Therefore, the exponential fault laws (Eqs. 2.21b, 2.22, and 2.23) are a natural consequence of a half-Gaussian distribution of σ values for the component Rayleigh subsystems of a complex software system.

Summary for Exact Methods: So, each of the exactly integrable functions $f(\sigma)$ explains some commonly occurring software failure behavior. There is one last problem that may seem minor at first, but is actually of great importance in understanding the failure of both software and hardware. The tails of the Rayleigh distributions that Peterson and Arellano used to model modules A, B, and D are too low. That is to say, they significantly underestimate the true observed normalized failure rate for large t. The same has been observed for hardware, as will be discussed in Sect. 2.6 on "Parametric Methods" below. Now, such an observation may seem trivial because there is so little probability trapped under such a tail. But, in fact, it is the rare events, whose probability falls in this "tail" region, that result in the unexpected, and extraordinarily destructive, "Black Swan" events. "Black Swan" events are so very destructive because there is generally no contingency for dealing with them! In the next section, approximate methods will be used to model the tails of normalized failure rate curves more accurately.

2.5 Approximate Methods

Narrow Gaussian Distribution on σ: In this section it will be supposed that σ is distributed by a narrow Gaussian about some nonzero value. That is to say, the set of parts or software lines of code are "almost" a simple subsystem with a Rayleigh distribution. Whether the analyst models this set as a system, or a subsystem, will depend on the level of detail desired. If the tails are important, then the analyst is dealing with a system. If not, then a Rayleigh subsystem description may suffice. Now, there is a certain amount of inconsistency with this "narrow Gaussian" assumption since a Gaussian will have a tail that eventually extends into the negative region of the σ-axis on the f(σ) versus σ graph. And, as previously mentioned, σ can never be negative. However, if the Gaussian is narrow, and possesses a mean which is far from the origin, little probability will be trapped under the tail in the region of negative σ. In that case the probability of achieving a negative σ will be very close to zero. For similar reasons, it also makes sense to extend the integration of Eq. 2.5 to the interval $(-\infty, +\infty)$. Now, the global probability density function is given by

$$\rho_{\text{global}}(t) = \int_{-\infty}^{+\infty} (t/\sigma^2) \quad \exp(-\tfrac{1}{2}[t/\sigma]^2) \quad (1/\sqrt{[2\pi]})(1/s)$$
$$\exp(-\tfrac{1}{2}[\{\sigma - \sigma_m\}/s]^2)d\sigma, \tag{2.24}$$

where, as in the case of the δ-function, σ_m represents the displacement of the mode of the distribution f from the origin of the σ axis in the f versus σ graph, and as in the half-Gaussian case, s is the standard deviation of f(σ). As a practical matter, $\sigma_m = (\Sigma_i \, \sigma_i)/n$ and $s = \sqrt{[\Sigma_i \, (\sigma_i - \sigma_m)^2/n]}$, where the sums are over all i subsystems and $1 \leq i \leq n$ (large n). Since

$$f(\sigma) = (1/\sqrt{[2\pi]})(1/s)\exp(-\tfrac{1}{2}[\{\sigma - \sigma_m\}/s]^2), \quad \text{and}$$
$$g(\sigma) = (t/\sigma^2)\exp(-\tfrac{1}{2}[t/\sigma]^2), \tag{2.25}$$

it is easy to show from Eq. 2.6 that $\eta = \sigma_m$ and $\mu_2 = s^2$. Therefore, to second order, Eq. 2.6 yields

$$\boxed{\rho_{\text{global}}(t) \cong (t/\sigma_m^2) \exp\left(-\tfrac{1}{2}[t/\sigma_m]^2\right)\left\{1 + \tfrac{1}{2}(s/\sigma_m)^2[6 - 7(t/\sigma_m)^2 + (t/\sigma_m)^4]\right\}} \tag{2.26}$$

In this form, it is clear that Eq. 2.26 has a Rayleigh part and a part in curly brackets that represents the departure from Rayleigh behavior. Equation 2.26 for $\rho_{\text{global}}(t)$ is graphed in Fig. 2.4-left together with the standard Rayleigh curve for comparison. Appendix C gives numerical values of $\rho_{\text{global}}(t)$ as a function of t for $\sigma_m = 10$ and a

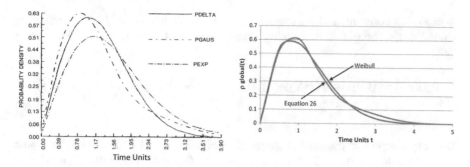

Fig. 2.4 (Left) The probability density $\rho_{global}(t)$ plotted as a function of t for the two approximately integrable distributions on σ; **a** the narrow Gaussian distribution (PGAUS), and **b** the exponential distribution (PEXP). The exactly integrable $f(\sigma) = \delta$, with $\sigma_m = 1$, is also included for comparison (a Rayleigh curve denoted by PDELTA). Note that PGAUS and PEXP can always be made to coincide exactly. Hence, the peaks of PGAUS and PEXP have been deliberately displaced from the peak of PDELTA to make all three curves more visible. (Right) Comparison between Eq. 2.26 and the Weibull parametric distribution. The two curves interlace

few typical integer values of s. Values of the Cumulative Probability Distribution $P = \int \rho_{global}(t)dt$, integrated from 0 to T, are also given in Appendix C. Generally speaking, the global probability density of Eq. 2.26 shows Rayleigh-like behavior with a tail that decays more slowly than that of the Rayleigh distribution, but faster than Eq. 2.14. This behavior is very typical of the normalized failure rate for both software and hardware [1, 13].

Equation 2.26 is an important result because it will ultimately allow the a priori (beforehand) computation of the "Black Swan" probability in terms of the system fundamental constants s and σ_m. This will be accomplished in the next chapter on the computation of the "Black Swan" probability. Because of its central position in the mathematical developments that follow, *a system having an infinite number of Rayleigh subsystems whose modes are Gaussian distributed such that $s \ll \sigma_m$, and whose normalized failure rate is governed by Eqs. 2.24 and 2.26, will define a Model Infinite System (MIS).* As will be discussed in the last two sections of this chapter, the failure behavior of real-world complex systems usually approaches the behavior of the MIS, especially when t is large (i.e., as $t \rightarrow \infty$). The definition of the MIS is in analogy to the definition of the Model Infinite Connector (MIC) that will be used in Part 2 of this book [14].

Displaced Exponential Distribution on σ: The last distribution for f that will be examined by approximate methods is the displaced exponential below,

$$
\begin{aligned}
f(\sigma) &= \lambda \exp(-\lambda[\sigma - \sigma_m]) \quad &\sigma \geq \sigma_m \\
f(\sigma) &= 0 \quad &\sigma < \sigma_m,
\end{aligned}
\tag{2.27}
$$

where λ is a constant, and the moments $\eta = \sigma_m + 1/\lambda$ and $\mu_2 = 1/\lambda^2$. Therefore, to second order, the global probability density function (error rate) $\rho_{global}(t)$ is given by

$$\rho_{global}(t) \cong (t/\eta^2) \exp(-\tfrac{1}{2}[t/\eta]^2)\{1 + \tfrac{1}{2}(1/\lambda\eta)^2[6 - 7(t/\eta)^2 + (t/\eta)^4]\}.$$

(2.28)

This expression for $\rho_{global}(t)$ is graphed in Fig. 2.4-left (PEXP). It has the same basic character as $\rho_{global}(t)$ generated from the Gaussian distribution on σ. It is difficult to distinguish between the two models by fitting data. However, the Gaussian distribution of modes is more intuitively appealing. It is interesting to note that Eq. 2.28 again has the form of a Rayleigh term times a term which represents the deviation of $\rho_{global}(t)$ from Rayleigh behavior.

2.6 Parametric Methods

In this section, two- and three-parameter functions are considered for fitting (mimicking) the normalized failure rate data for hardware and software. Examples are the log-normal distribution [15–17] and the Weibull distribution [13]. The lognormal distribution takes the form

$$\rho_{global}(t) = 0 \qquad\qquad\qquad\qquad\qquad\qquad\qquad\qquad t < c$$
$$\rho_{global}(t) = \{1/[(t - c)\sigma\sqrt{(2\pi)}]\}\{\exp[-(\ln(t - c) - m)^2/2\sigma^2]\} \quad t > c,$$

(2.29)

where c, m, and σ are adjustable constants. Note that Eq. 2.29 has the same form as Eq. 2.6 of Chap. 1 if $c = 0$, $1/\sigma \equiv \alpha$ and $m/\sigma \equiv \beta$. The Weibull distribution has been widely used to model failure rates since its detailed description in 1951. These studies will not be repeated here, only referenced [18–22]. The Weibull distribution takes the form

$$\rho_{global}(t) = \alpha\beta\, t^{\beta-1} \exp(-\alpha t^\beta),$$

(2.30)

where α and β are adjustable. In general, good fits may be obtained between these parametric distributions and experimental data; with the Weibull distribution being the better of the two. However little insight is gained about the physics of failure from such techniques, since the parameters are not easily connected to physically meaningful quantities. Furthermore, values of α and β cannot be established unless the system has been in existence for some time and undergone a statistically meaningful number of failures. So, the determination of Weibull parameters depends on a posteriori (historical or past) knowledge. Figure 2.4-right shows the match between the Weibull distribution and the all-important Eq. 2.26. The mode and the standard deviation of Eq. 2.26 were arbitrarily set to 1.0 and 0.3, respectively. Note that s $\ll$ σ_m, as it must be to satisfy the requirements of the derivation of Eq. 2.26. The RMS best fit Weibull distribution was then plotted on the same time scale. A global best fit was achieved when $\alpha = 0.631$ and $\beta = 1.71$. The two curves match nicely. Notice how they interweave. In fact, the curves cross each other four times. If $\sigma_m = 1$ and $s = 0.1$ in Eq. 2.26, with $\alpha = 0.51$ and $\beta = 2.0$ in the Weibull distribution, the two curves would be virtually coincident. That is to say, as $s/\sigma_m \rightarrow 0$, $\alpha \rightarrow 0.5$ and $\beta \rightarrow 2.0$, and both curves approach a Rayleigh distribution as they should. However, Eq. 2.26 involves fundamental system parameters that can be known to the analyst *before* fielding a system! Finally, it should be recalled that Eq. 2.26 is the second term truncation of a full Papoulis expansion of the integral in Eq. 2.24. It is partly for this reason that Eq. 2.26 integrates to slightly less than unity (~ 0.979 over $0 \leq t \leq 5$). As the number of terms approaches infinity, the area under the expansion approaches unity. Equations 2.24 and 2.26 will be called a *Pseudo Weibull Distribution (or PWD)*, and a *Truncated Pseudo Weibull Distribution (TPWD)* respectively; where the terms "distribution", "probability density function", and "normalized failure rate law" are being used interchangeably. Furthermore, since the TPWD in Eq. 2.26 is truncated after 2nd order terms in μ_n, it is more properly described as a *Truncated Pseudo Weibull Distribution of second order (TPWD2)*, where TPWD∞ $\equiv$ PWD. The family of curves TPWDn (where n = 2, ... ∞), form a new set of mathematical objects essential for a full understanding of system safety and reliability in general.

Many normalized failure rate data sets are Weibull-like. That is to say, their tails decay slower than a Rayleigh distribution but faster than the global probability density function with Rayleigh distributed modes (Eq. 2.14). The Peterson–Arellano modules A, B, and D are typical examples of this behavior. Although the simple Rayleigh distribution is generally a good fit to these data, its value is much too small in the tails. And, the tails are where the probability of rare events are to be found. Figure 2.5-left shows the match between the smoothed normalized data set of module B [1] and the RMS best fit TPWD2; where $\sigma_m = 9.47$ and $s = 3.09$. For this fit $\chi^2 = 215.32$. By comparison, $\chi^2 = 58{,}128.05$ for the RMS best fit simple Rayleigh distribution. Therefore, PGAUS (TPWD2) does a much better job of fitting the experimental data globally than PDELTA (simple Rayleigh model). The fit is especially good for large t (Fig. 2.5-right). Between months 10–39, the contribution to the PGAUS Chi-square is only 26.87. If the fit over the entire time interval had had such a low Chi-square, the analyst would have to accept the narrow Gaussian distribution of modes hypothesis at the 97.5% confidence level for the 44 degrees of freedom available for module B (Appendix B). Alas, early development noise ruined the Chi-square figure of merit. It should also be noted that the area under the TPWD2 is 0.999 over the $0 \leq t \leq 50$ interval (Appendix C). Therefore, the all-important normalization property of the model probability density function, needed for accurate χ^2 calculations, is almost perfect. Examination of Fig. 2.5-right also shows that the experimental data lies far above the best fit Rayleigh curve by an order of magnitude in the tail of the distribution. It is precisely these small Rayleigh probabilities for large t that cause the last terms of the χ^2 sum

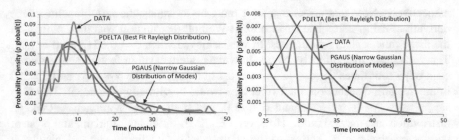

Fig. 2.5 (Left) Comparison between the global probability density function (per KSLOC) for a narrow Gaussian distribution of Rayleigh modes versus Peterson–Arellano module B. The RMS best fit Delta function model (pure Rayleigh fit) has also been plotted on the same graph for comparison. (Right) Comparison for large t between the narrow Gaussian model, the Delta Function model (pure Rayleigh fit), and the smoothed normalized data of module B. The module B data involved analysis of 43.8 KSLOC and was originally expressed in defects/KSLOC/month prior to normalization

(Eq. 1.13, Chap. 1) to "blow-up" into the thousands. Ultimately, this type of fit failure results in gross errors when trying to evaluate the probability of rare events. Therefore, the Rayleigh model can be safely eliminated for the Module B data set.

2.7   Summary and Conclusion

The observed normalized failure rate laws of complex systems are simply due to a distribution of Rayleigh distributions, where each Rayleigh distribution applies to the failure of a particular subsystem. This simple theorem provides a unified framework for understanding the failures of both hardware and software. Many types of distributions on σ have been described in this chapter, and all the resulting expressions for $\rho_{global}(t)$ involve only fundamental system constants that can be estimated by the analyst. The choice of model for any given system will depend on the level of modeling accuracy required (especially in the tails), and the value of system parameters. Table 2.1 summarizes the choices and conditions.

Table 2.1 Distribution of modes versus system/subsystem type

Distribution on σ	Conditions
Dirac Delta Function	Small χ^2 in tail not required; simple subsystems
Multiple Dirac Delta functions, each with unique σ_m	System with a small number of simple subsystems
Half-Gaussian	Modeling after development; complex systems
Rayleigh Distribution of modes	Small χ^2 required in tail; modes widely spread or uncertain; poorly tested complex systems
Narrow Gaussian	Small χ^2 required in tail; s $\ll$ σ_m; well tested complex systems

One last point deserves special mention. It has been noticed that some failure rate curves (both normalized and un-normalized) for complex systems, especially complex software systems, seem to have a steady-state number of failures per unit time for very large t [23]. These systems do not appear to be described by any of the models in Table 2.1. Apparently, new defects are being added at approximately the same rate at which they are being removed. Peterson has suggested that eventually correction of simple Rayleigh modules of such complex systems creates new Rayleigh failure curves, thereby maintaining the more or less steady state level of the normalized global failure rate curve [23].

The stage is now set for the next chapter on the a priori (i.e., predictive) computation of the "Black Swan" probability (probability of occurrence of "rare events") and the mean "Black Swan" time to failure as a function of system parameters.

† Note that, as a matter of nomenclature, $\rho_{global}(t)$ could have been called $\rho_{sys}(t)$ in this context. However, the expressions that will be developed for $\rho_{global}(t)$ have far-reaching applications beyond system safety, and beyond systems analysis in general. For example, $\rho_{global}(t)$ can be applied to the analysis of Synthetic Aperture Radar (SAR) imagery histograms [24]. Therefore, the more generic nomenclature will be retained.

2.8 Problems

(1) Compute the probability trapped under the tail of the PDELT and PGAUS probability density functions in Fig. 2.5 (right) from month 25 to month 50.
(2) A system safety engineer is confronted with a complex system consisting of 10 subsystems. Each Rayleigh subsystem i is known to have a peak failure rate at the time given in the table below when first installed. This data is known from legacy (previous) applications, manufacturers specifications, and accelerated bench testing. Assume that all repairs are permanent, so that each subsystem becomes more reliable with time after it reaches its maximum failure rate.

Subsystem Number (i)	σ_i (months)
1	7
2	6
3	8
4	6
5	12
6	11
7	12.5
8	13.5
9	14
10	10

(a) What is the value of σ_m for the whole system?
(b) What is the value of s for the whole system?
(c) What is the ratio s/σ_m? Is it reasonable to use a TPWD2 function to model the normalized failure rate curve for this complex system?
(d) How long does the system safety engineer have to wait before he is sure that 90% of his problems are past? Hint: consult the Cumulative Distribution Function graph and table in Appendix C.
(e) How long does the system safety engineer have to wait to be sure that 99% of the system failures are past?
(f) Suppose project requirement demand fielding the system after 2 years with an assurance that 94% of all failures have occurred and are now in the past. What can the system safety engineer demand of the subsystem suppliers to meet that deadline?

Note: This problem shows how the modeling in this chapter can be used to design a system with safety and reliability goals in mind a priori. Such a procedure is in contrast to system construction, followed by a posteriori documentation of safety and reliability history, followed by extrapolation into the future.

(3) A complex system is to be fielded after 26 months with 90% of all its "bugs" worked out. The Rayleigh subsystem failure rate peaks are themselves Rayleigh distributed in time with a peak at M; a fact known from legacy (previous) applications, manufacturers specifications, and accelerated bench testing. Using the graph in Appendix A, what demands must the system safety engineer make on subsystem manufacturers to meet project requirements? What value of M must be achieved?

References

1. Peterson, J., & Arellano, R. (2002, March). Modeling software reliability for a widely distributed, safety-critical system. *Reliability Review, 22*(1).
2. Papoulis, A. (1965). *Probability, random variables and stochastic processes* (pp. 151–152). New York: McGraw-Hill.
3. Pearson, C. E. (Ed.) (1983). *Handbook of applied mathematics* (pp. 163, 1228, 1256–1259). New York: Van Nostrand.
4. Gradshteyn, I. S., & Ryzhik, I. M. (1965). *Table of integrals, series, and products* (4th ed., p. 340). New York: Academic Press.
5. DeHaan, D. B. (1867). *Nouvelles tables D'Integrales definies* (p. 144). New York: Hafner.
6. Gradshteyn, I. S., & Ryzhik, I. M. (1965). *Table of integrals, series, and products* (4th ed., p. 972). New York: Academic Press.
7. Pearson, C. E. (Ed.) (1983). *Handbook of applied mathematics* (p. 961). New York: Van Nostrand.
8. Jahnke, E., & Emde, F. (1945). *Tables of functions* (4th ed., p. 19). New York: Dover.
9. Abramowitz, M., & Stegun, I. A. (1972). *Handbook of mathematical functions* (p. 258). Washington: NBS.
10. Jahnke, E., & Emde, F. (1945). *Tables of functions* (4th ed., p. 2). New York: Dover.

11. Abramowitz, M., & Stegun, I. A. (1972). *Handbook of mathematical functions* (p. 255). Washington: NBS.
12. Hoel, P. G., Port, S. C., & Stone, C. J. (1971). *Introduction to statistical theory* (p. 226). Boston: Houghton Mifflin.
13. Anon. Weibull distribution. *Wikipedia*. http://en.wikipedia.org/wiki/Weibull_distribution. Accessed 29 Dec 2012.
14. Zito, R. R. (2012). The curious bent pin problem-IV: Limit methods. In *30th International System Safety Conference*, Atlanta, GA.
15. Jakeman, E., & Pusey, P. N. (1976). A model for non-rayleigh sea echo. *IEEE Transactions Antennas and Propagation, AP-24*(6), 806–814.
16. Trunk, G. V., & George, S. F. (1970). Detection of targets in non-Gaussian sea clutter. *IEEE Transactions Aerospace Electron Systems, AES-6*, 620–628.
17. Fisz, M. (1963). *Probability theory and mathematical statistics* (p. 171). New York: Wiley.
18. Montgomery, D. C. (2012). *Statistical quality control* (p. 95). New York: Wiley. ISBN 9781118146811.
19. Anon. (2013). System evolution and reliability of systems. http://www.sys-ev.com/reliability01.htm. Accessed 4 Jan 2013.
20. Crevecoeur, G. U. (1993). A model for the integrity assessment of ageing repairable systems. *IEEE Transactions Reliability, 42*(1), 148–155.
21. Crevecoeur, G. U. (1994). Reliability assessment of ageing operating systems. *European Journal of Mechanical and Environmental Engineering, 39*(4), 219–228.
22. Nelson, W. B. (2003). *Applied life data analysis* (pp. 1–662). New York: Wiley.
23. Peterson, J. R. (2013, 7 Feb). Raytheon network concentric systems. Fullerton, CA, 714-446-2927, Private Communications.
24. Zito, R. R. (1988). The shape of SAR histograms. *Computer Vision, Graphics, and Image Processing, 43*, 281–293.

Chapter 3
Bounding the "Black Swan" Probability

3.1 The Anatomy of a Black Swan: Four Case Histories

(a) *The USS Thresher*: On April 9, 1963, the submarine USS *Thresher* headed to sea from the Portsmouth Naval Shipyard. The next morning at 9:18 A.M., during a deep dive exercise off Boston, the *Thresher* broke apart and was lost at sea with all hands [1, 2]. Subsequent investigation revealed the following sequence of events [2]. First, trouble began during construction, testing, and qualification. Fourteen percent of the piping that moves high-pressure seawater within the boat failed ultrasonic inspection testing. This, in itself, is not harmful since known risks can be corrected. However, in this case, hundreds of joints were not tested. Therefore, the mean lifetime of this subsystem was uncertain, and potentially zero. During the first deep dive, an engine room leak from this subsystem caused a fuse to be blown, which, in turn, caused a reactor shut down. Although an emergency reactor shut down would normally be a rare event, it was precipitated with certainty by the proceeding events. Finally, when ballast was blown, there was insufficient buoyancy to float the *Thresher*, due to an older ballast subsystem design, the presence of seawater within the boat, and no reactor output to "power out" of a dangerous situation. The ability of the system to compensate for failures was overwhelmed as the *Thresher* slowly sank below the 1500 ft. level and imploded in the deep. Figure 3.1 summarizes the MCR.

(b) *The USS Scorpion*: In May 1968, the *Scorpion* was returning from navy exercises in the Mediterranean. The last message received from the boat on May 21 indicated that the *Scorpion* would reach Norfolk at about 1 PM EST on the 27. Instead, the boat never arrived. It had simply disappeared somewhere in the North Atlantic. Eventually, the wreckage of the boat was located on the seafloor via hydrophone data and, curiously, it was found that the *Scorpion* was heading east; away from Norfolk. This lead to the first theory of how the

© Springer Nature Switzerland AG 2020
R. R. Zito, *Mathematical Foundations of System Safety Engineering*,
https://doi.org/10.1007/978-3-030-26241-9_3

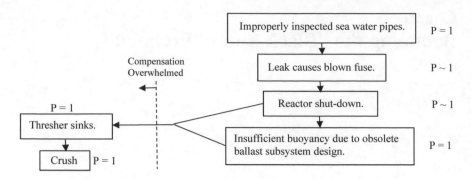

Fig. 3.1 The *Thresher* MCR has no "rate limiting" step. That is to say, there is no step in this sequence of events with a small probability of occurrence such that the number of failures per unit lifetime is limited. The known design and inspection shortcomings doomed this boat to failure from the start

Scorpion failed [1, 2]. It was thought that perhaps an electrical short (a rare event) may have initiated the launch of a torpedo. In that case, a 180° turn of the Scorpion should have disarmed its own torpedo; a safety mechanism built into every torpedo to prevent it from striking its own ship. Only this time, it didn't work (another rare event). So, in this scenario, the Scorpion would have been struck by its own torpedo. However, mathematically speaking, this is a very unlikely theory because its probability is the product of the small probabilities of two independent and unlikely events.

The second theory is based on more information, is mathematically more probable, and follows a familiar pattern of Black Swan mishaps. Torpedo batteries, of the type deployed on the Scorpion, contained a small foil component that controlled the internal flow of electrolyte. It had been discovered during laboratory testing that this thin foil component could rupture during intense vibration, causing initiation of the battery, an abnormal overheating condition, and a battery explosion that was capable of causing a "cook-off" (low-level) warhead explosion. In fact, a severe vibration had been noted at least once before on the Scorpion, without any resolution as to its cause. Examination of the Scorpion on the seafloor revealed that the battery well was almost completely destroyed. A battery initiation event would have immediately convinced the commanding officer to turn the boat 180° in accordance with his training. An ensuing blast may have opened external hatches in the torpedo room causing flooding of the submarine. Flooding would have caused the boat to dive deeper until it was crushed by the subsurface pressure of the sea. The ability of the system to compensate for failures was simply overwhelmed.

In the final analysis, it seems likely that a rare vibration event-triggered breakage of the foil component within a torpedo battery with high probability. This event caused battery overheating/explosion and warhead cook-off with near unit probability, followed by the inevitable flooding, further submersion and, ultimately, the

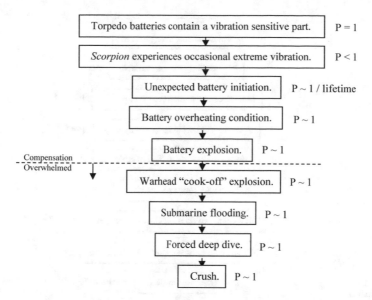

Fig. 3.2 The *Scorpion* MCR. The rate-limiting step of this MCR is the frequency of occurrence of extreme submarine vibration events

implosion that was detected by the hydrophones. The MCR seems clear (Fig. 3.2), and has a much higher probability than might have been expected because it depends on only a *single* rare event.

(c) *Three Mile Island*: Volumes have been written about the Three Mile Island Nuclear Accident of March 28, 1979 that rated a five on the seven-point Nuclear Event Scale [3, 4]. In part, the extensive nature of the literature is due to the complexity of the accident. And, all the details cannot be covered here except by reference. However, the "flow chart" of Fig. 3.3 summarizes the salient events of the MCR. Before discussing each block of Fig. 3.3, it is helpful to review the design and operation of a *pressurized water reactor*. The basic design of the Three Mile Island reactor involves two basic working fluid loops, both containing water. The *primary loop* contains *pressurized water* that is used to cool the reactor core. It is for this reason that the Three Mile Island reactor design is referred to as a "pressurized water reactor". If the pressure in the primary loop gets too high due to heating, or overheating, a valve (*Pilot-Operated Relief Valve* or *PORV*) will relieve the excess pressure by releasing some of the primary loop water into a *relief tank*. Hot pressurized water from the reactor core is then feed into a *steam generator*, which is basically a boiler. It is not the primary loop water that is boiled (that water remains liquid because it is pressurized), but water contained within a *secondary loop* that also passes through the steam generator. So, the secondary loop is a water/steam loop; water is pumped ("feed") into the steam generator

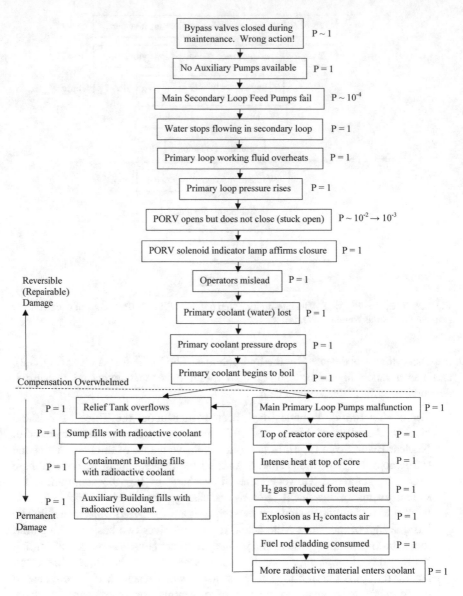

Fig. 3.3 The MCR for the Three Mile Island nuclear accident

and steam comes out. And, this exiting steam is used to drive a turbine that turns a generator to make electricity.

On March 28, 1979, at 4 A.M., a chain reaction of events started that precipitated a Level 5 incident. The trouble started with the failure of *Main Secondary Loop Feed Pumps*, for reasons which are still unknown, during maintenance operations (blockage removal) on one filter ("*polisher*") that was used to clean the secondary loop water. It is, of course, tempting to suspect that the two events were in some way connected. In the author's experience, water pump failures are usually progressive, releasing debris that can be trapped in filters somewhere else in the system prior to total failure. By the time noticeable filter blockage occurs, pump damage is so severe that replacement is the only option. So, perhaps progressive pump failure precipitated filter maintenance. However, no such theory has ever been proven to the author's knowledge. In any case, this mishap by itself would not have precipitated a Level 5 disaster because *Auxiliary Secondary Loop Feed Pumps* are activated automatically to bypass the Main Secondary Loop Feed Pumps in the unlikely event of a failure. However, something *very* irregular had been done. The *bypass valves* were shut during this "routine" maintenance. And, *this was a violation of the Nuclear Regulatory Commission (NRC) rules!* So, a fatal judgment error had been made at the very first step of the MCR. The binary decision, *"turn valves off or turn valves on"*, was incorrectly made. Because human beings often behave irrationally, the system safety engineer must always assume that the wrong decision will be made with unit probability.

At this point, water stopped flowing in the secondary loop. Without heat removal from the primary working fluid through the steam generator, acting as a heat exchanger, the pressure in the primary loop began to rise (with unit probability). The reactor was "scrammed" but the problem of decay heat remained. During such an emergency, the PORV is capable of bleeding off some of the primary water, thereby lowering the primary loop pressure. And, in fact, this did happen. However, this valve never closed due to a mechanical fault. It is for this reason that manufacturers of such equipment advise regular testing at pressure [5]. A probability of 10^{-2}–10^{-3} failures per year per valve [6] might be assigned to a sticking bypass valve, especially when there is working fluid contact and a long time interval between cycles. Worse still, the indicator light, which was keyed to the valve solenoid and not the valve itself, indicated that the PORV had returned to its closed position. This misinformation confused operators into thinking that all was well. So, they did not seek alternative confirmation of PORV closure. Again, this was a wrong human decision whose probability must be assumed to be close to unity. In fact, primary working fluid was leaving the primary loop at a prodigious rate and was filling an overflow reservoir (*relief tank*).

Next, a series of events took place in which one event triggers the next with near unit probability. As pressure dropped in the primary loop, the primary working fluid began to boil causing two other mishaps. First, steam started to displace water in the reactor core, so the relief tank began to overflow causing the containment building sump to fill. Eventually, radioactive coolant began to leak out into the general containment building, and some was even pumped from the sump into an auxiliary building outside the main containment building until these pumps were eventually stopped. In parallel with this overflow, steam in the primary working fluid caused the *Primary Loop Main Pumps* to malfunction ("*cavitate*") [7] and shake violently [4] as a steam/water mixture passed through them instead of liquid water. The core temperature continued to rise as boiling continued until finally the top of the reactor core was exposed. The intense heat now produced a chemical reaction between the Zircalloy (zirconium alloy) fuel rod cladding and the steam in the reactor that released even more heat.

$$Zr^0 + 2H_2O \rightarrow ZrO_2 + 2H_2 + \Delta H. \tag{3.1}$$

The hot hydrogen gas released then exploded on contact with air. Furthermore, as cladding was consumed, the damaged fuel pellets released even more radioactivity into the reactor coolant.

This was the state of affairs at Three Mile Island when a new shift of personnel arrived at 6 A.M. on the 28. During much of this accident, the old crew of operators had assumed that the reactor's problem was *too much coolant*. After all, coolant levels were apparently high. They didn't realize that these levels were due to percolation – like coffee in a coffee pot. But, a fresh set of eyes soon realized that the reactor's *true* problem was a *loss of coolant* through the PORV. A backup valve that was in line with the PORV was then shut to stop the loss of coolant. However, by this time 32,000 U.S. gallons of coolant was lost, and the reactor core had partially melted. Slowly, the power plant was brought under control. Nevertheless, the damage was done, and it took about 14 years to clean up the mess!

Each step of the MCR, a rare event by itself, occurred with unit probability, or nearly so. Like a row of falling dominos, each subsystem or part failure guaranteed failure of the next. Mechanically speaking, the MCR had only two rate-limiting steps; i.e., the failure of the Main Secondary Loop Feed Pump and the failure of the PORV. Each of these two events is unlikely, but the failure of the Main Secondary Feed Pump during a maintenance operation seems suspicious. In any case, the probability of the Three Mile Island accident was clearly much higher than the designers had suspected when combined with a series of human misjudgments. Given a sufficient amount of time all water pumps will fail but, judging from automotive water pumps where extensive repair records and experience exists, regular maintenance and/or replacement may reduce the failure probability to $\sim 10^{-4}$ per year [8]. Since, as stated above, the failure probability of the PORV lies in the 10^{-2}–10^{-3} range per year per valve, and since the failed TMI-2 reactor was in operation for about 1 year, the overall failure probability is on the order of $10^{-6} \rightarrow 10^{-7}$ per year. The true overall failure probability may be higher given the fact that TMI-2 had already experienced a coolant pump failure one year earlier on March 28, 1978 [4] that should have served as a "heads-up" that this plant had to be watched *very* carefully! The calculated overall failure probability is not a large number, but it isn't negligibly small either. And, it's certainly not small enough for a system as dangerous as a nuclear reactor.

(d) *The Fukushima Daiichi Nuclear Disaster*: The Fukushima Daiichi Nuclear Disaster [9] is one of the only two Level 7 mishaps on the Nuclear Event Scale; the other being Chernobyl. On March 11, 2011, the Tōhoku earthquake shook the Fukushima Daiichi nuclear power plant, knocking out power to the plant electronics and coolant systems. Immediately, reactors 1, 2, and 3 (of the 6) shut down automatically. This first step in the MCR would not have been disastrous by itself, but the tsunami that followed greatly complicated the situation because it flooded low-lying rooms that housed emergency generators. These generators supplied power to pumps that circulate coolant water even during a power outage. This circulation must be maintained for several days even after reactor shut down, due to the heat generated by radioactive decay of short-lived fission products. The flooded generators failed. Now, comes a very critical step. If the reactors had been immediately flooded with seawater as an emergency measure, much of the disaster would have been averted. But, the decision to do this was delayed because it would have ruined the reactors permanently. By the time an affirmative decision was made, it was already too late! What followed next was inevitable, and beyond the compensation capability of the system. As water in the reactors boiled away, the intense heat allowed a chemical reaction to take place between the Zircalloy fuel element cladding and steam (see Eq. 3.1). This reaction produced hydrogen gas and still more heat. The reactors' fuel rods began to melt. Eventually, the cores of all three reactors completely melted. Several hydrogen–air explosions also occurred, followed by venting of radioactive gas to the atmosphere. A circular area with a radius of 20 km around the Fukushima plant had to be evacuated. However, the damage did not stop there. A large volume of seawater that had been exposed to the melting fuel rods became radioactive and was returned to the sea for several months.

In the end, both sea and groundwater had become contaminated. And, cesium-137 levels of concern were measured 30–50 km from the plant causing the Japanese government to ban the sale of food grown in that area. There were also 37 injuries, including radiation burns.

Each step in the MCR triggered the next with near unit probability. And, the true mishap probability for this Black Swan mishap was really just the probability of the first step (the probability of a tsunami-producing earthquake)—which is much too high for a dangerous system like a nuclear power plant. An underwater earthquake must rank above 6.75 on the Richter scale to produce a tsunami. There are about six such seismic events per century, and 90% of these occur in the Pacific Ocean [10]. Conservatively estimating the lifetime of a Nuclear Reactor to be 50 years [11]. A seaside plant in Japan is likely to experience three tsunami events during its lifetime, on the average. The MCR is summarized in Fig. 3.4. The progress of the chain could have been arrested with timely action in step 5, but this did not happen. And, it is impossible to assign an exact probability to this step because it involves human response, evaluation, and unfortunately, emotion. The only conservative position is to assume that the wrong decision will be made under pressure with unit probability. It would have been better to demand that affirmative action is always, and immediately, taken if step 5 is reached—*with no equivocation involved!*

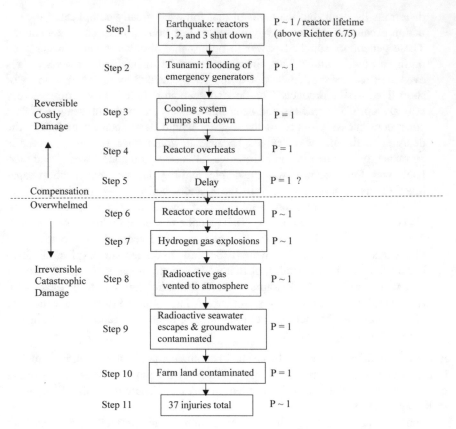

Fig. 3.4 The MCR for the Fukushima Daiichi Nuclear Disaster. The rate-limiting step is the probability of a tsunami-producing earthquake within an effective radius around the reactor complex, whose lifetime is taken to be 50 years

3.2 Calculations

The previous case studies, and especially the "fault trees" [12] of Figs. 3.1 through 3.4, clearly indicate that the probability of a Black Swan event is much higher than analysts might at first expect. Events which are rare by themselves might occur with nearly unit probability when they are part of an MCR. Furthermore, an MCR can start with the failure of almost any subsystem or part, even a seemingly trivial one whose cost is only a few dollars. Of course, not every failure will manifest itself in a Black Swan disaster. Therefore, the probability of a Black Swan event P_{BS} is greater than or equal to the product, but less than or equal to the sum of individual subsystem failure probabilities. That is to say,

$$\prod_{i=1}^{N} P_i \leq P_{BS} \leq \sum_{i=1}^{N} P_i. \tag{3.2}$$

P_i is the probability of failure for the ith subsystem or part, and i runs from 1 to N, where N is the total number of subsystems or parts. Note that P_{BS} calculated from a sum is much larger than P_{BS} calculated from a product. In the previous chapter [13], several explicit models were put forward for the overall system normalized failure rate (or instantaneous probability of failure per unit time) of large complex systems. Each model is an appropriate description of a particular type of system. For example, is the system "well-tested", or is it still in the early stages of development? Is it a mechanical system, or is it software? Or, is the system very mature, having undergone many upgrades? All of these factors influence the selection of the appropriate normalized failure rate model; which will simply be denoted by $\rho_{global}(t)$. What is important, however, is that *the various expressions for $\rho_{global}(t)$ were developed on the assumption that each system is composed of a distribution of Rayleigh subsystems. That is to say, the normalized failure rate of each subsystem follows a Rayleigh law as a function of time. Each Rayleigh law is defined by its own statistical "mode". Therefore, each complex system has a normalized failure rate law that is composed of a distribution of distributions.* The function $\rho_{global}(t)$ also tacitly assumes that the repair of each failure is permanent, so that the system being modeled becomes more and more reliable as $t \rightarrow \infty$.

Suppose a complex system has been in operation for a time T, what is the upper limit on P_{BS}? It is the probability that a failure will occur in any safety-critical subsystem. And, that is just the area under the tail of $\rho_{global}(t)$ from T to ∞.

$$P_{BS} \leq \int_{T}^{\infty} \rho_{glebal}(t)dt \tag{3.3}$$

Equation 3.3 is the continuous analogue of the summation Eq. 3.2. The value of P_{BS} will very much depend on the shape and magnitude of the tail of $\rho_{global}(t)$ from T to ∞. If the maximum value of $\rho_{global}(t)$ lies over a value t_{max} on the time axis, and if $T \gg t_{max}$, and if the tail of $\rho_{global}(t)$ is decreasing rapidly, then the value of P_{BS} will be very small. On the other hand, if the normalized system failure model has a slowly decaying tail, and/or if T is not $\gg t_{max}$, then P_{BS} will be large. In the previous chapter it was shown how historical (a posteriori) failure rate information from the manufacturers of individual parts, or legacy subsystems (i.e., subsystems that have been previously employed in other systems), can lead to the a priori prediction of a normalized failure rate law for a newly designed system even before construction, testing, and deployment. All that is required is a knowledge of the distribution of modes for all Rayleigh subsystems. Once a system is constructed, tested, and deployed, the design phase (normalized) failure rate model can be refined. The longer a complex system is field-tested, the greater the confidence in the normalized failure rate model, and the greater the certainty that P_{BS} will be truly small. Observation will confirm, or correct, the value of t_{max} and the rate of decay of $\rho_{global}(t)$ for $T > t_{max}$. Once a model has been selected during design, or confirmed or corrected during deployment, Eq. 3.3 can then be integrated out to ∞.

As a numerical example, consider a complex system with an exponential normalized failure rate curve of the form

$$\rho_{global}(t) = \alpha \exp(-\alpha t), \tag{3.4}$$

where α is the decay constant, and time t is measured in years, so that integration of Eq. 3.4 from $t = 0$ to ∞ yields 1. If observation indicates a decay constant of 1 year^{-1} (i.e., $\alpha = 1$), then the normalized failure rate is reduced by $1/e$ (1/2.72, or 0.368) for each year of operation. After 10 years of operation, what is the upper bound on the Black Swan probability? It is just

$$P_{BS} \leq \int_{10}^{\infty} e^{-t}dt = -\int_{10}^{\infty} (-e^{-t})dt = -\int_{10}^{\infty} d(e^{-t}) = -\{e^{-t}\,|_{10}^{\infty}\} = -\{0 - (e^{-10})\}$$
$$= e^{-10} = 4.68 \times 10^{-5}.$$

$$\tag{3.5}$$

This is a small probability, but not a negligible one for a dangerous system. Other models with other decay rates in the tail of the normalized system failure rate curve, as well as other observation times (other limits), will yield a different result; possibly much higher or lower. One way to lower the value of P_{BS} is to design a system in such a way that a catastrophic event requires the *independent* failure of at least *two* equivalent parallel subsystems. In that case, if independence can truly be guaranteed, $P_{BS} = (4.68 \times 10^{-5})^2 = 2.19 \times 10^{-9}$ for the figures used in this example. A probability of 2.19×10^{-9} is much more acceptable for a Black Swan mishap because it means that a single such system will only experience two such events in a billion years—a truly geologic time scale! This technique might have saved the Three Mile Island reactor if the individual failure probabilities associated with the Main Secondary Loop Feed Pump and the PORV had been lower.

3.3 Conclusion

Periodically, spectacular and catastrophic mishaps attract the attention of the popular press and the public. And, like the proverbial 100-year floods that seem to occur once a decade, such high-profile mishaps occur far too frequently. One has to conclude that the probability of catastrophic events is, perhaps, orders of magnitude larger than the minute probabilities and the enormous "mean-time-to-failure" values predicted by many analyses. These first two chapters of Part I have attempted to provide a "rare event" probability correction by (1) noting that MCRs have a probability of catastrophic failure bounded by the *sum* of the probabilities of all possible safety-critical failures, and by (2) noting that the "tail" (large t values) of normalized failure rate histograms are much longer, flatter, and higher, than the tail predicted by simple Rayleigh models.

Notice that no attempt has been made to anticipate the *cause* of rare Black Swan mishaps. It is only assumed that some *known* safety-critical part, line of code, subsystem, etc. will fail if the system safety or reliability engineer observes a system for a long enough period of time. As in quantum mechanics, if an event *can* happen then, eventually, it *will* happen. The *cause* of the failure is, in this sense, irrelevant! And, to discount the potential failure of a part because its *cause* cannot be clearly identified, or even imagined, is a mistake in reasoning. Mathematically, only the existence of the tail of a failure law is important. This philosophy contradicts "common sense" lectures on Black Swan events that concentrate on *trying* to anticipate the unexpected *cause* of events without explaining exactly how this is to be done. By definition, it makes no more sense to *try to anticipate an unknown, or unexpected, cause* than it does to ask, "What is north of the north pole?" The question itself makes no sense!

3.4 Problem

What is the upper bound on the Black Swan probability of failure for a Rayleigh system after 10 years of operation if its mode $\sigma_m = 1$ year?

References

1. Dunmore, S. (2002). *Lost Subs* (pp. 138–139, 140–151). Da Capo/Madison Press, Toronto, Canada.
2. Anon. (1999). *Submarines, secrets and spies*. PBS.
3. Anon. *Three mile island accident, 1/7/1013*. http://en.wikipedia.org/Three_Mile_Island_accident. p. 18.
4. Walker, S. J. (2004). *Three mile island*. Berkeley, CA: University of CA Press.
5. Anon. (1994). *Swagelok Catalogue MS-01-14*, Willoughby, OH 44094.
6. McElhaney, K. L., & Staunton, R. H. (1996). Reliability estimation for check valves and other components. In *Paper presented at the 1996 ASME Pressure Vessels & Piping Conference*. Montreal, Quebec, Canada, 7/21-26/1996.
7. Markel, A. (2016). *Water pump cavitation: What cavitation in pumps means*. Babcox Media. Retrieved March 30, 2016, from, http://www.underhoodservice.com/water-pump-cavitation-vehicle/.
8. Anon. (2016). Private communications, automotive service, 6420 E. 22nd St., Tucson AZ 85710, (520) 747-2000.
9. Anon. (2013). *Fukushima Daiichi nuclear disaster*, p. 47. Retrieved July 1, 2013, from http://en.wikipedia.org/wiki/Fukushima_Daiichi_nuclear_disaster.

10. Anon. (2016). How often do Tsunamis occur? *GEOL 105 Natural Hazards*. Retrieved March 30, 2016, from http://geol105naturalhazards.voices.wooster.edu/should-the-u-s-expect-a-tsunami/.
11. Behr, P. (2016). How long can nuclear reactors last? *Scientific American*. Retrieved March 30, 2016, from http://www.scientificamerican.Com/article/how-long-can-nuclear-reactors-last/.
12. Ericson II, C. A. (2011). *Fault tree analysis primer*. CreateSpace Charleston NC.
13. Zito, R. R. (2013). How complex systems fail- I: Decomposition of the failure histogram. In *Proceedings of the International Systems Safety Conference* August 12–16, 2013. Boston, Mass.

Chapter 4
The Risk Surface

4.1 Introduction

Risk (R), or the *expectation value of failure cost*, is the product of *mishap severity* (S), or cost, by the *probability* (P) *of failure per lifetime* (i.e., Risk R = severity × probability = S × P) [1, 2]. Risk is defined for a particular hazard. And, all hazards are taken together from the *system risk* ($\mathcal{R}$). Therefore,

$$\mathcal{R} = \sum_{i=1}^{N} R_i = \sum_{i=1}^{N} S_i \times P_i, \tag{4.1}$$

where the index i for the individual hazards runs from 1 to the total number of hazards, N. The individual hazards, themselves, are simply the entries from the worksheets of the Preliminary Hazard Analysis (PHA), or the Final Hazard Analysis (FHA). Therefore, the procedures and calculations presented here do not replace, but are built upon, well-established hazard analysis techniques.

System risk may, in principle, be represented by a three-dimensional bar graph, where the height of each parallelepiped bar represents the number, n_I, of hazards that fall within a particular severity-probability bin (labeled by index capital I) of the S-P chart (plane) discussed in the next section. And, $N = \Sigma\, n_I$, where the summation extends over all I from 1 to N, and N is the total number of severity-probability bins (usually 20). Let $F_I = n_I/N$. Then F_I represents the fraction of hazards in bin I, and ΣF_I, over all I, equals unity. In that case, Eq. 4.1 becomes

$$\mathcal{R} \approx \sum_{I=1}^{N} n_I \times S_I \times P_I = N \sum_{I=1}^{N} F_I \times S_I \times P_I, \tag{4.2}$$

© Springer Nature Switzerland AG 2020
R. R. Zito, *Mathematical Foundations of System Safety Engineering*,
https://doi.org/10.1007/978-3-030-26241-9_4

where S_I and P_I are just the severity and probability of occurrence associated with the centroid of the Ith S-P bin in this discrete approximation to $\mathcal{R}$. The product N $(F_I \times S_I \times P_I)$ is just that part of the system risk associated with bin I. The sum of these products for several contiguous bars is that part of the system risk associated with hazards of the corresponding S-P bins below. The sum of all bars is just the total system risk measured in the average number of dollars per lifetime that the system owner expects to payout for the operation of a given system. Finally, the summation $\sum F_I \times S_I \times P_I$ over all I, from 1 to N, is just the average risk per hazard $(\mathcal{R}/N)$, and will be denoted by $\underline{R}$. While each term of this summation $(F_I \times S_I \times P_I)$, a "risk function", is just the contribution to the average risk per hazard from bin I; denoted by $\underline{R}_I$. Therefore, $\underline{R} = \sum \underline{R}_I$ overall I from 1 to N.

4.2   Qualitative Risk Characterization Chart (S-P Plane)

Following MIL-STD-882D [2], *Severity* is divided into four discrete "categories". *Catastrophic* mishaps correspond to category I failures, while *Critical*, *Marginal*, and *Negligible* mishaps correspond to categories II, III, and IV, respectively. These categories are not just defined by numerical cost. They also admit nonnumerical costs. For example, what is the "cost" of a human life? What is the "cost" of irreversible severe environmental damage? In this mathematical presentation, a numerical cost will always be required. Therefore, the onset of a category I (catastrophic) mishap, for example, will begin at one million dollars, regardless of whether that cost is paid in compensation to the family of deceased personnel or represents the total loss of property due to environmental damage. Also, following MIL-STD-882D, the *Mishap Probability* per planned system life expectancy will be divided into five "levels"; *Frequent* $(P > 10^{-1})$, *Probable* $(10^{-1} > P > 10^{-2})$, *Occasional* $(10^{-2} > P > 10^{-3})$, *Remote* $(10^{-3} > P > 10^{-6})$, and *Improbable* $(<10^{-6})$. Both the Severity and the Probability can be organized into a single chart that summarizes the *risk* associated with a particular hazard. This Qualitative Risk Characterization Chart is displayed in Fig. 4.1. All of this standard formalism will be kept intact in this presentation, provided that an *origin* is placed in the upper right-hand corner of the chart as shown below. Extending from this origin are a Severity axis (S) and a Probability axis (P). This S-P plane can now serve as a base upon which the number of hazards (n_I), or the fraction of hazards (F_I), can be plotted as bars above each Severity-Probability bin. In fact, the later of these presentations will prove the more useful since it is easier to extend to the infinite systems discussed in the next section.

Mishap Probability per Life (P)	Mishap Severity, S ($)			
	Catastrophic (>1M)	Critical (1M-200K)	Marginal (200K – 10K)	Negligible (<10K)
Frequent ($>10^{-1}$)	HIGH	HIGH	SERIOUS	MEDIUM
Probable (10^{-2} - 10^{-1})	HIGH	HIGH	SERIOUS	MEDIUM
Occasional (10^{-3} - 10^{-2})	HIGH	SERIOUS	MEDIUM	LOW
Remote (10^{-6} - 10^{-3})	SERIOUS	MEDIUM	MEDIUM	LOW
Improbable ($<10^{-6}$)	MEDIUM	MEDIUM	MEDIUM	LOW

$\infty \leftarrow \text{Log}_{10} \, S$

O

$-\text{Log}_{10} \, P \rightarrow +\infty$

Fig. 4.1 Standard qualitative risk characterization S-P chart. MIL-STD-882D has been used because mishap probability P is numerically defined

4.3   The Model Infinite System (MIS)

As defined in Chap. 2 [3], an infinite system has an infinite number of subsystems and, therefore, an infinite (or very large) number of possible failure modes (hazards). How do the discrete system concepts above translate into infinite systems? First, imagine that the bins of the S-P plane are shrunk to infinitesimal size. In that case, the F_I bar graph can be replaced by a smooth curved surface defined by the tops of an infinite number of infinitesimally narrow bars. This is the *Hazard Distribution Function (HDF)* F. An example of a *normalized* HDF is shown in Fig. 4.2.

It is important to note that the scales of the S and P axes of the Qualitative Risk Characterization Chart (Fig. 4.1) are expressed as positive and negative powers of 10, respectively. For that reason, it is best to redefine these axes as $\mathbb{S} = \text{Log}_{10} \, S$ and $\mathcal{P} = -\text{Log}_{10} \, P$. This has been done in Fig. 4.2. Notice that all three axes, F, $\mathbb{S}$, and $\mathcal{P}$, now form a standard right-handed coordinate system when F is plotted above the $\mathbb{S}$, $\mathcal{P}$ plane so that integrations may be conveniently performed. The $\mathbb{S}$ and $\mathcal{P}$ scales of Fig. 4.3 now run from 0 to 6 on both scales. Of course, higher values can always

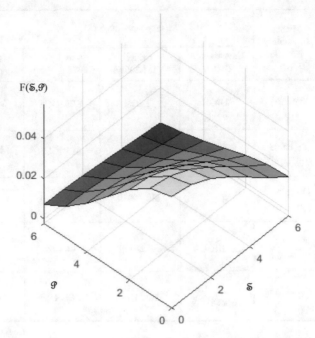

Fig. 4.2 A Bivariate Normal (Gaussian) Hazard Distribution Function for an ideal system with $\rho = 0$, $\sigma_{\mathbb{S}} = 4$, and $\sigma_{\mathcal{P}} = 3$. The surface peaks at the origin, decays far from (0, 0), and has unit volume trapped under it

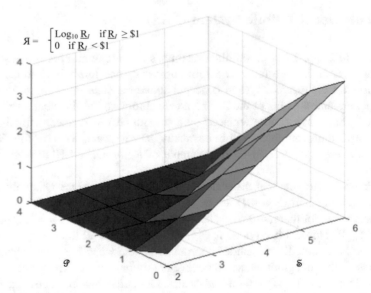

Fig. 4.3 MIS *Modified Risk Surface* corresponding to the BND F($\mathbb{S}$, $\mathcal{P}$) of Fig. 4.2 with $\sigma_{\mathbb{S}} = 4$, $\sigma_{\mathcal{P}} = 3$, and $\rho = 0$

be achieved on these axes. However, many system safety issues will be confined to this zero to six on a side square $S \times \mathcal{P}$ area. This is, especially, true of the more inexpensive systems.

The next question is, "What function should be used to model this "smooth distribution" over $S \times \mathcal{P}$ space?" The answer to this question must be based on experience. In the military context, WSESRB would have to decide, based on the past performance of weapon systems, what would be an acceptable distribution of "High", "Serious", "Medium", and "Low" risk hazards. In this research, a simple Gaussian *Bivariate Normal Distribution* (BND) has been selected since, in the author's experience, *the BND is typical of safe system performance* (c.f. Figs. 4.1 and 4.2). For example, if the severity and probability of a hazard are both large (i.e., $S \rightarrow \infty$ while $P \rightarrow 1$, or $S \rightarrow \infty$ while $\mathcal{P} \rightarrow 0$), then the integer n_I (or fraction F_I) must approach zero (i.e., $F \rightarrow 0$). In fact, it may be reasonably argued that F should identically equal zero if the severity exceeds some threshold value. In any case, F should decrease *very* rapidly along the S-axis as S increases. Much more will be said about this phenomenon later. What about hazards that are frequent (or even *certain* to occur) but have a negligible severity? There may be very many of these, since they are typically maintenance items like a blown fuse. The cost of the fuse is negligible, as is the temporary denial of usage. So, a maximum for F is expected as $S \rightarrow 0$ and $\mathcal{P} \rightarrow 0$. What about hazards that are *both* rare and have negligible consequence? It's hard to even imagine what such a hazard might be. But, the very fact that it is a rare event eliminates even the inconvenience of temporary denial of usage. And, it is questionable whether such "hazards" should be considered hazards at all. So, there should be very few of these, and $F \rightarrow 0$ as $S, \mathcal{P} \rightarrow \infty$. In any case, detailed modeling of such unimportant hazards is unnecessary. Finally, there are hazards that are both catastrophic and rare. These are the "Black Swan" events. A designer/analyst tries to eliminate as many of these as possible, but there are always a few because "Black Swan" events are usually unexpected. Therefore, $F \rightarrow 0$, but never equal zero, as $S \rightarrow \infty$ and $\mathcal{P} \rightarrow \infty$. The surface in Fig. 4.2 has all these general properties. *Therefore, the BND will be the model used to describe the instantaneous fraction of hazards per unit $S \times \mathcal{P}$ area for an ideal system in the remainder of this chapter.*

The BND also has the advantage of well-known, and tractable, mathematical properties. Since the peak of the $F(S, \mathcal{P})$ BND, along both the S and $\mathcal{P}$ axes, lies at the origin, $F(S, \mathcal{P})$ can be written as [3]

$$F(S, \mathcal{P}) = \left[4/\left(2\pi \, \sigma_S \sigma_{\mathcal{P}} \sqrt{(1 - \rho^2)}\right)\right]$$
$$\exp\left\{\left[-1/\left(2(1 - \rho^2)\right)\right]\left[(S/\sigma_S)^2 - (2\rho \, S\mathcal{P}/\sigma_S \sigma_{\mathcal{P}}) + (\mathcal{P}/\sigma_{\mathcal{P}})^2\right]\right\},$$

$$(4.3)$$

where σ_S and $\sigma_{\mathcal{P}}$ are the standard deviations of the BND in the S and $\mathcal{P}$ directions. The factor of 4 in the numerator of the pre-factor is necessary to insure that $\iint F(S, \mathcal{P}) \, dS d\mathcal{P} = 1$ when S and $\mathcal{P}$ are both integrated from 0 to $+\infty$ (recall the condition $\sum F_I = 1$, above Eq. 4.2, for discrete systems). In other words, the factor of 4 insures the normalization property on F when only the positive quadrant ($\mathcal{S}$ and $\mathcal{P}$ positive only) of this 3D bell-shaped distribution is used. The parameter ρ is the correlation coefficient between $\mathcal{S}$ and $\mathcal{P}$; where $-1 < \rho < +1$. If the severity and probability axes are uncorrelated (i.e., the severity of a hazard does not a priori imply its probability), then $\rho = 0$ and Eq. 4.3 simplifies to

$$F(\mathcal{S}, \mathcal{P}) = [4/(2\pi\sigma_{\mathcal{S}}\sigma_{\mathcal{P}})] \exp\left\{-\tfrac{1}{2}\left[(\mathcal{S}/\sigma_{\mathcal{S}})^2 + (\mathcal{P}/\sigma_{\mathcal{P}})^2\right]\right\}. \tag{4.4}$$

This last model is particularly useful for comparison with the PHA worksheets, before mitigations have taken place. Mitigations *anti-correlate* severity (S) and probability of occurrence (P). That is to say, the more severe the hazard (i.e., the greater its value of S and $\mathcal{S}$), the more engineering will try to reduce its P (anti-correlation). But, a decrease in P means an increase in $\mathcal{P}$. Therefore, $\mathcal{S}$ and $\mathcal{P}$ will be *correlated*, and that means $0 < \rho < 1$. Equation 4.3 might be a better model for comparison with the FHA worksheets.

The contribution to a system's average risk per hazard from area A in the $\mathcal{S}$, $\mathcal{P}$ plane may now be written as the double integral

$$\underline{R}_A = \iint_A d\underline{R} = \iint_A S\,P\,F(\mathcal{S},\mathcal{P})d\mathcal{S}d\mathcal{P}. \tag{4.5}$$

Since $\mathcal{S} = \mathrm{Log}_{10}\,S = (0.43429....) \ln S$, it is clear that $S = \exp(\mathcal{S}/0.43429)$ to five significant figures. Similarly, $P = \exp(-\mathcal{P}/0.43429)$. Equation 4.3 or 4.4 can then be used to evaluate $F(\mathcal{S}, \mathcal{P})$, whichever is more appropriate. If Eq. 4.4 is used, then Eq. 4.5 can be written as

$$\underline{R}_A = \iint_A \{\exp(\mathcal{S}/0.43429)\}\{\exp(-\mathcal{P}/0.43429)\}\left\{[2/(\pi\,\sigma_{\mathcal{S}}\sigma_{\mathcal{P}})]\exp\left(-\tfrac{1}{2}\left[(\mathcal{S}/\sigma_{\mathcal{S}})^2 + (\mathcal{P}/\sigma_{\mathcal{P}})^2\right]\right)\right\}d\mathcal{S}d\mathcal{P}.$$
$$\tag{4.6}$$

The double integral of Eq. 4.6 can be replaced by the product of two single integrals after separation of variables.

$$\underline{R}_A = [2/(\pi\,\sigma_{\mathcal{S}}\sigma_{\mathcal{P}})]\left[\int_a^b \{\exp(\mathcal{S}/0.43429)\}\{\exp[-\tfrac{1}{2}(\mathcal{S}/\sigma_S)^2]\}d\mathcal{S}\right]$$
$$\left[\int_c^d \{\exp(-\mathcal{P}/0.43429)\}\{\exp[-\tfrac{1}{2}(\mathcal{P}/\sigma_{\mathcal{P}})^2]\}d\mathcal{P}\right]. \tag{4.7}$$

Notice that the limits of integration are now from "a to b" along the $\mathcal{S}$-axis, and from "c to d" along the $\mathcal{P}$-axis. Equation 4.7 for $\underline{R}_A$ is of central importance because its integration over the physically meaningful part of the $\mathcal{S}$, $\mathcal{P}$ plane (i.e., from 0 to ∞

along both axes) yields the *system average risk per hazard*. It is most instructive to perform this integration explicitly, taking one integral at a time. Using a standard integral table [4] it can be shown that

$$\int_u^\infty \{\exp(\mathcal{S}/0.43429)\}\{\exp[-\tfrac{1}{2}(\mathcal{S}/\sigma_\mathcal{S})^2]\}d\mathcal{S}$$

$$= \{\sigma_\mathcal{S}\sqrt{(\pi/2)}\}\{\exp[\tfrac{1}{2}(\sigma_\mathcal{S}/0.43429)^2]\}\{1 - \Phi(-\sigma_\mathcal{S}/[0.43429\ \sqrt{2}] + u/[\sigma_\mathcal{S}\sqrt{2}])\},$$

(4.8)

$$\int_u^\infty \{\exp(-\mathcal{P}/0.43429)\}\{\exp[-\tfrac{1}{2}(\mathcal{P}/\sigma_\mathcal{P})^2]\}d\mathcal{P}$$

$$= \{\sigma_\mathcal{P}\sqrt{(\pi/2)}\}\{\exp[\tfrac{1}{2}(\sigma_\mathcal{P}/0.43429)^2]\}\{1 - \Phi(\sigma_\mathcal{P}/[0.43429\ \sqrt{2}] + u/[\sigma_\mathcal{P}\sqrt{2}])\},$$

(4.9)

where u $\geq$ 0, and where

$$\Phi(-\sigma_\mathcal{S}/[0.43429\ \sqrt{2}] + u/[\sigma_\mathcal{S}\sqrt{2}]) = (2/\sqrt{\pi}) \int_0^{-\sigma_\mathcal{S}/[0.43429\ \sqrt{2}]\,+\,u/[\sigma_\mathcal{S}\sqrt{2}]} \exp(-t^2)dt$$

$$= (2/\sqrt{\pi}) \sum_{k=0}^\infty \{[(-1)^k(-\sigma_\mathcal{S}/[0.43429\ \sqrt{2}] + u/[\sigma_\mathcal{S}\sqrt{2}])^{2k+1}]/[k!(2k+1)]\}$$

(4.10)

and

$$\Phi(\sigma_\mathcal{P}/[0.43429\ \sqrt{2}] + u/[\sigma_\mathcal{P}\sqrt{2}]) = (2/\sqrt{\pi}) \int_0^{\sigma_\mathcal{P}/[0.43429\ \sqrt{2}]\,+\,u/[\sigma_\mathcal{P}\sqrt{2}]} \exp(-t^2)dt$$

$$= (2/\sqrt{\pi}) \sum_{k=0}^\infty \{[(-1)^k(\sigma_\mathcal{P}/[0.43429\ \sqrt{2}] + u/[\sigma_\mathcal{P}\sqrt{2}])^{2k+1}]/[k!(2k+1)]\}.$$

(4.11)

The last two Φ integrals (Eqs. 4.10 and 4.11) are called "Probability Integrals" and their value is well tabulated. All that is required are the values of $\sigma_\mathcal{S}$ and $\sigma_\mathcal{P}$ so that the limits of integration can be determined. Alternatively, the summations on the right in Eqs. 4.10 and 4.11 may be used. Equations 4.7 through 4.11 provide a complete solution for the evaluation of the contribution to a system's average risk per hazard from *any* area A in the $\mathcal{S}$, $\mathcal{P}$ plane (R_A); even though the integrals in Eqs. 4.8 and 4.9 only run from u to ∞. To understand why this is so, consider a rectangular area in the $\mathcal{S}$, $\mathcal{P}$ plane defined by "a and b" along the $\mathcal{S}$ axis such that b > a, and "c to d" along the $\mathcal{P}$ axis such that d > c; as in Eq. 4.7. Furthermore, let the integral over $\mathcal{S}$ be called I_1 and the integral over $\mathcal{P}$ be called I_2. Then, I_1 between a and b is just given by

$$I_1 \bigg|_a^b = I_1 \bigg|_a^\infty - I_1 \bigg|_b^\infty . \tag{4.12}$$

Similarly,

$$I_2 \bigg|_c^d = I_2 \bigg|_c^\infty - I_2 \bigg|_d^\infty . \tag{4.13}$$

The comprehensive analytic capability of Eqs. 4.7 *through* 4.13 *provides the second reason for selecting the BND as a model for F.* In fact, the integration can be extended over the entire positive quadrant of the $\mathbb{S}$, $\mathscr{P}$ plane from the origin all the way out to infinity on both axes to yield a system's (total) average risk per hazard ($\underline{R}$). In that case, $a = 0$, $c = 0$, $b = \infty$, and $d = \infty$ in Eq. 4.7. And, the integrals in Eqs. 4.8 and 4.9 can be used directly by setting $u = 0$, so that Eqs. 4.12 and 4.13 are not needed. When these calculations are performed, a finite number results. *The very fact that a numerical solution for $\underline{R}$ always exist means that $\underline{R}$ is bounded for the MIS. A system with a bounded value of $\underline{R}$ is naturally, normally, or Gaussianally (recall Fig.* 4.2*), safe in the sense that, on the average, the cost of a mishap cannot be infinite. That is to say, owner/operator liability (in the legal sense of that word) is limited. Furthermore, this property holds for any bounded system Hazard Distribution Function having a tail that decays faster than the MIS. These powerful statements provide the third reason why the BND was selected as a model for $F(\mathbb{S}, \mathscr{P})$ of the MIS.*

Now, one may reasonably ask, "Can you give me an example of a realistic distribution F for a system that would yield an unbounded (∞) value of $\underline{R}$?" Consider the "exponential distribution" given by

$$F(\mathbb{S},\mathscr{P}) = \mathfrak{N} \exp(-\alpha\, \mathbb{S} + \beta\mathscr{P}), \tag{4.14}$$

where $\mathfrak{N}$ is a normalization factor and α and β are constants. If $\alpha < 1/0.43429\ldots \approx 2.3026$ then the integral I_1 will become infinite, and that means $\underline{R}$ will be unbounded. Such a system is naturally, normally, or Gaussianally unsafe in the sense that the average risk per hazard can be infinite (infinite liability)! This might seem counterintuitive. After all, the Hazard Distribution Function in Eq. 4.14 approximates the shape of Fig. 4.2 qualitatively. Where's the problem that causes such a huge difference in $\underline{R}$. The difference is in the details. The exponential distribution has a long tail that does not decay as rapidly as the BND. This seemingly harmless difference eventually causes $\underline{R}$ to become infinite. Everyone has heard of a car that is a "lemon". What does that actually mean? It means that the average risk (cost) per hazard is so great that the owner/operator will eventually experience a failure that forces him/her to sell that vehicle. In other words, the repair costs have increased without bound, at least for that owner's pocketbook. In fact, the colloquial term "lemon" can be defined precisely in a mathematical sense. *A system is a "mathematical lemon" if $\underline{R}$ is undefined (i.e., ∞).* Notice the use of the word *mathematical*, because in practice, an owner's threshold for defining a lemon might be considerably less than infinity. That is to say, there is a difference between a *mathematical lemon* and a *practical lemon*! *The BND separates the set of all possible $F(\mathbb{S},\mathcal{P})$ functions into those that apply to Gaussianally safe systems, and those that apply to Gaussianally unsafe systems (mathematical lemons and the worst of the practical lemons). This is the fourth reason for selecting the BND as a model for F.* Finally, it must be recalled that all the calculations in this section depend on the fact that $\sigma_{\mathbb{S}}$ and $\sigma_{\mathcal{P}}$ can be determined for the MIS. Therefore, these two quantities and their effect on $\underline{R}_A$ will be discussed next.

4.4   Explicit Evaluation of $\sigma_{\mathbb{S}}$, $\sigma_{\mathcal{P}}$, and $\underline{R}_A$

Based on TABLE A-IV ("Example mishap risk categories and mishap risk acceptance levels") of 882D [2], WSESRB seems to feel that the "Serious" risk level forms a soft boundary between what is acceptable and what is unacceptable for any hazard. That is to say, "serious" and "High" risks are subject to special approval by customer executives. With this thought in mind, one might scale the BND by setting $\sigma_{\mathbb{S}} = \mathrm{Log}_{10}\ S = \mathrm{Log}_{10}\ (\$10\ K) = 4$, since even this small severity will define the lower edge of "Serious" risk if it happens often enough. Similarly, one might define $\sigma_{\mathcal{P}} = -\mathrm{Log}_{10}\ P = -\mathrm{Log}_{10}\ (10^{-3}) = 3$, since even this "Remote" probability can result in a "Serious" risk if the severity is high enough. Using these parameters in Eqs. 4.7 through 4.13 yields Table 4.1. Each entry ($\underline{R}_I$) in Table 4.1 is a digital contribution to $\underline{R}$. The italic index *I* runs overall "cells" (from 1 to 36) that are one unit wide on the $\mathbb{S}$ axis (one order of magnitude wide on the S axis), and one unit wide on the $\mathcal{P}$ axis (one order of magnitude wide on the P axis). The word "cell" has been used to distinguish each box in the $\mathbb{S}$-$\mathcal{P}$ plane of Table 4.1 from a "bin" in Fig. 4.1. Cells are smaller, and more precisely and uniformly defined, than bins. For example, the cell at $(\mathbb{S}, \mathcal{P}) = (6, 6)$ has a Severity $\mathbb{S}$ that centers on 6 (i.e., $S = 10^6$), but runs from 5.5 to 6.5 (i.e., $\$3.16 \times 10^5 < S < \$3.16 \times 10^6$). Similarly, $\mathcal{P}$ also centers on 6 (i.e., $P = 10^{-6}$), and runs from 5.5 to 6.5 (i.e., $3.16 \times 10^{-6} > P > 3.16 \times 10^{-7}$).[†] Calculations were truncated at a severity center value of $\$10^6$ and a probability center value of $P = 10^{-6}$. Shade coding (unshaded, light gray, medium gray, and dark gray) has been used to mark off WSESRB zones of "High", "Serious", "Medium", and "Low" risk, respectively, for *individual* hazards. And, it is instructive to compare the individual and system risks in each cell. First, medium gray and dark gray areas for $\mathbb{S} \leq 3$ also correspond to low *system* risk contributions (small numerical entries). The light gray area also seems intuitively to be about where it should be (i.e., intermediate individual hazard risk translates into intermediate system risk contributions). But, the light gray zone below $\mathcal{P} = 4$ seems to have system risk contributions that are lower than what might be expected from the individual hazard risk. There is nothing abnormal about this situation. It just means that the underlying function F is decaying *very fast* for large values of $\mathcal{P}$. Unshaded cells $(\mathbb{S}, \mathcal{P}) = (5, 0)$, and $(6, 0)$ have system risk values that seem abnormally high. This happens because the BND has a hazard count (or fraction of total hazards) that tapers off *too slowly* as $\mathcal{S} \to \infty$ to depress the value of $\underline{R}_I$ in cells (5, 0), and (6, 0). The problem is exacerbated by a large value for the standard deviation $\sigma_{\mathcal{S}}$, which will slow decay. The MIS system's average risk per hazard (sum of all cell entries) is \$12,078. If Table 4.1 were continued out to a Severity center value of $\mathcal{S} = 7$ (i.e., $6.5 < \mathcal{S} < 7.5$ so that the maximum total system loss value S is \$3.16 $\times$ 10^7) then the MIS system's average risk per hazard would be \$81,302. In fact, these costs per hazard (costs per failure as judged from maintenance records) are of the correct order of magnitude, in 2016 U.S. dollars, for aircraft systems whose total loss is 3.16 and 31.6 million dollars, respectively (e.g., a small turboprop airplane and a jet, respectively—[5]).

The entries in Table 4.1 can be plotted in three dimensions over the $\mathcal{S}$, $\mathcal{P}$ plane to form a piecewise continuous tessellated *Risk Surface* embedded in a 3D right-handed coordinate system space. However, the range of risk values $\underline{R}_I$ varies over so many orders of magnitude that it is *visually* best to plot Log $\underline{R}_I$ on the vertical axis. Furthermore, any risk less than, or equal to, \$1 should be mapped into zero, since such risks are of no importance to the safety analyst. Finally, if only the interesting region bounded by $2 \leq \mathcal{S} \leq 6$ and $0 \leq \mathcal{P} \leq 4$ is plotted, the *Modified Risk Surface* ($\mathcal{A}$) of Fig. 4.3 emerges.

Normally, risk $\underline{R}_I$ (or $\mathcal{A}$) would be a function of three variables F($\mathcal{S}$, $\mathcal{P}$), $\mathcal{S}$, and $\mathcal{P}$. But, there are obvious display difficulties associated with building a four-dimensional graph. However, since F is also a function of $\mathcal{S}$ and $\mathcal{P}$, it is possible to plot risk directly above the $\mathcal{S}$-$\mathcal{P}$ plane and avoid the 4D trap. Therefore, two 3D graphs (i.e., Figs. 4.2 and 4.3) suffice to completely describe the risk status of the MIS. Notice how $\mathcal{A}$ rises to a maximum at $\mathcal{S}$, $\mathcal{P} = (6, 0)$ in Fig. 4.3. What this graph is trying to say is that even though the number of severe hazards with a high probability (see the tail along the $\mathcal{S}$-axis of Fig. 4.2) is very small, it is not small enough (and it is certainly not zero in the BND model). The tragic result is that most of the cost associated with operating a MIS system will come from this small pool of hazards (defects). The great importance of $\underline{R}_I$ will become even clearer in the next section.

4.5   The Norm

IS IT SAFE? That is the central question. But, "safe" is a relative term. One must ask, "Safe relative to what?" In this section that question will be answered and, as the reader has probably guessed, *safety will be measured relative to the MIS* via *a "Norm"* that is defined discretely by

$$Norm = \sum_{I=1}^{\mathcal{N}} [\underline{R}_I^{MIS}(\mathbb{S}_I, \mathscr{P}_I) - F_I(\mathbb{S}_I, \mathscr{P}_I)\, S_I\, P_I] = \sum_{I=1}^{\mathcal{N}} [\underline{R}_I^{MIS}(\mathbb{S}_I, \mathscr{P}_I) - F_I(\mathbb{S}_I, \mathscr{P}_I)\, 10^{\mathbb{S}_I}\, 10^{-\mathscr{P}_I}],$$

$$(4.15)$$

where the summation is over all $\mathcal{N}$ risk cells that the analyst wishes to consider (36 in Table 4.1). $F_I(\mathbb{S}_I, \mathscr{P}_I)\, S_I\, P_I$ is the contribution from the Ith risk cell to the real system's average risk per hazard, and $\underline{R}_I^{MIS}(\mathbb{S}_I, \mathscr{P}_I)$ is the contribution from the Ith risk cell to the MIS $\underline{R}$. The hazards used to evaluate $F_I(\mathbb{S}_I, \mathscr{P}_I)\, S_I\, P_I$ come from the real system's worksheets and define a new table with entries $R_I^{sys}(\mathbb{S}_I, \mathscr{P}_I)$ that are analogous to those of Table 4.1 for the MIS. Furthermore, a piecewise continuous tessellated real system *Risk Surface* (or *Modified Risk Surface*), analogous to that in Fig. 4.3, could also be generated. This surface could then be replaced by a smooth best-fit function, if the analyst so desires. This last step would greatly simplify any *analytical* integration of the real system risk surface. However, a smooth best-fit function is not necessary if a *digital* integration is going to be performed. In either case

$$Norm = \sum_{I=1}^{\mathcal{N}} \left[\underline{R}_I^{MIS}(\mathbb{S}_I, \mathscr{P}_I) - \underline{R}_I^{sys}(\mathbb{S}_I, \mathscr{P}_I) \right]. \qquad (4.16)$$

Notice that even if $\underline{R}_I^{MIS}(\mathbb{S}_I, \mathscr{P}_I)$ and $\underline{R}_I^{sys}(\mathbb{S}_I, \mathscr{P}_I)$ are smooth functions, the *Norm* is still only being evaluated from the difference of the MIS and real system surfaces at a finite number of $\mathcal{N}$ points.

The properties of the *Norm* are clear. If the *Norm* is positive, then the real system's average risk per hazard $\underline{R}$ is less than that of the MIS over all the risk cells of interest to the analyst. Such a system is Gaussianally Safe. If the *Norm* is negative, then the real system is Gaussianally unsafe.

Table 4.1 MIS average risk contributions (in dollars) as a function of severity and probability; rounded to two decimal places

← +∞

6	5	4	3	2	1	0	$\mathbb{S}$	$\mathscr{P}$
8612.91	1214.59	160.91	20.02	2.34	0.26	0.03	0	
1629.45	229.78	30.44	3.79	0.44	0.05	0.01	1	
137.93	19.45	2.58	0.32	0.04	0	0	2	
10.45	1.47	0.20	0.02	0	0	0	3	
0.71	0.10	0.01	0	0	0	0	4	
0.04	0.01	0	0	0	0	0	5	
0	0	0	0	0	0	0	6	

+∞ ↓

4.6 Summary

A great many questions about the nature of risk in complex systems have been answered in this chapter. Although the calculations may seem involved, the basic principles are easy to understand.

(1) The basic Qualitative Risk Characterization Chart has been only slightly modified so as to create a plane over which analytic computations may be performed.
(2) A *Bivariate Normal Distribution* (BND) has been used to model the number of hazards occurring over different parts of this plane. A system with a BND *Hazard Distribution Function* is called a *Model Infinite System (MIS)*.
(3) The *system risk* for a BND MIS system is *bounded* (i.e., it is a limited liability system).
(4) If the sum of the differences of cell risks (*Norm*) for the MIS, and a real system to be analyzed, is positive, then the system is *Gaussianally safe*. If the *Norm* is negative, then the real system is *Gaussianally unsafe*.

The purpose of this procedure is to make the process of system safety designation as objective as possible. And, it has the appealing advantage of being built upon previous standard methods so that nothing has to be changed to reevaluate older systems. The calculations presented here can simply be added on to previous analyses.

All the results calculated in this chapter depend on the selection of the underlying Hazard Distribution Function. And, in the case of the BND, the values of $\sigma_{\mathbb{S}}$, $\sigma_{\mathbb{P}}$, and ρ. The BND seems to decay too slowly in the $\mathbb{S}$ direction to be realistic. And, the value of $\sigma_{\mathbb{S}}$ that was selected seems to contribute to this problem. So, the BND has its problems. Nevertheless, the calculations presented above produce an average risk per hazard for the MIS system that is not incompatible with the average risk per hazard actually observed for the two real-world test systems examined. And, other recent research supports this view [6, 7]. On the other hand, the MIS criterion for separating safe from dangerous systems is not trivial to satisfy either, as will be demonstrated by the two problems at the end of this chapter where limited liability systems fail to be Gaussianally safe. So, the MIS criterion is just about what it should be, demanding but not impossible to reach. It makes manufacturers stretch without exhausting their technical expertise, resources, and money. Whatever decisions WSESRB experts eventually make concerning these issues, the calculations presented here will serve as a model for future work in this area.

† Cell sizes along the axes require extra clarification. Consider the cell (0, 3) along the $\mathbb{P}$ axis. In this case $-0.5 < \mathbb{S} < +0.5$ (or \$0.316 < S < \$3.16). So, the negative sign causes no problem. But, what about cell (3, 0)? In this case $-0.5 < \mathbb{P} < +0.5$ (or 0.316 < P < 3.16). But, P can never exceed unity! Therefore, these cells along the $\mathbb{S}$ axis must be half as wide in the $\mathbb{P}$ direction. Therefore, entries in the first row of Table 4.1 have been divided in half.

4.7 Problems

(1) Suppose $F(\mathcal{S}, \mathcal{P})$ is a "rect" function for some real system. Then F is defined by

$$F(\mathcal{S}, \mathcal{P}) = \begin{cases} F = 1/(\sigma_{\mathcal{S}}\sigma_{\mathcal{P}}), & \text{for } 0 \le \mathcal{S} \le \sigma_{\mathcal{S}} \text{ and } 0 \le \mathcal{P} \le \sigma_{\mathcal{P}}, \text{ where } \sigma_{\mathcal{S}} = 4, \text{ and } \sigma_{\mathcal{P}} = 3, \\ F = 0 & \text{elsewhere} \end{cases}$$

such that the integral of F overall positive $\mathcal{S}$ and $\mathcal{P}$ is unity.

(a) Set up the double integral for $\underline{R}$ in the rectangular coordinates $\mathcal{S}$ and $\mathcal{P}$.
(b) Build a table for this system like Table 4.1 for the MIS. Only consider the values of $\mathcal{S}$ and $\mathcal{P}$ from 0 to 6. For cells on the boundaries of F, divide their contribution to $\underline{R}$ (i.e., $\underline{R}_I$) by 2.
(c) What is the value of $\underline{R}$?
(d) Calculate the Norm.
(e) Is this system Gaussianally safe?
(f) Is this system a Mathematical Lemon?

(2) Suppose $F(\mathcal{S}, \mathcal{P})$ is one quarter of a "circ" function for some real system. Then, given radial coordinates $(\mathfrak{r}, \Theta)$ in the $\mathcal{S}$, $\mathcal{P}$ plane, F is defined by

$$F(\mathcal{S}, \mathcal{P}) = \begin{cases} F = 4/\pi\sigma^2 & \text{for } \mathfrak{r} \equiv \sqrt{([\mathcal{S} - 6]^2 + \mathcal{P}^2)} \le \sigma = (\sigma_S + \sigma_P)/2 = (4+3)/2 = 3.5 \\ & \text{and } 0 \le \Theta \le \pi/2 \\ F = 0 & \text{elsewhere} \end{cases}$$

such that the integral of F over all positive $\mathcal{S}$ and $\mathcal{P}$ values is unity.

(a) Set up the double integral for $\underline{R}$ in the rectangular coordinates $\mathcal{S}$ and $\mathcal{P}$. Only consider values from 0 to 6 along both axes.
(b) Build a table for this system like Table 4.1 above for the MIS. For cells on the boundary $\sqrt{\mathcal{S}^2 + \mathcal{P}^2} = \sigma$, divide their contribution to $\underline{R}$ (i.e., $\underline{R}_I$) by 2 if half the cell lies within the boundary. Ignore the contribution from cells that lie negligibly within the boundary, etc.
(c) What is the value of $\underline{R}$?
(d) Calculate the Norm.
(e) Is this system Gaussianally safe?
(f) Transform the double integral in (a) into a double integral in the polar coordinates $\mathfrak{r}$ and Θ of the $\mathcal{S}$, $\mathcal{P}$ plane.
(g) Is the double integral for $\underline{R}$ separable?
(h) Can an explicit analytic integration over $\mathfrak{r}$ be performed [8]?
(i) Can an explicit analytic integration over Θ be performed [8]?
(j) Is this system a Mathematical Lemon?

References

1. Anon. *WISE Series A*, Course 5. Retrieved December 18, 2007, from http://63.134.199.73/nw/lms/LmsCourses/w0105_0/player.html.
2. MIL-STD-882D. (2000, February 10). *Standard practice for system safety*. Washington DC: Department of Defense.
3. Meyer, P. L. (1970). *Introductory probability and statistical applications* (2nd ed., p. 199). Reading, Mass.: Addison Wesley.
4. Gradshteyn, I. S., & Ryzhik, I. M. (1965). *Table of integrals series and products* (4th ed., pp. 307 (sec. 3.32, subsec. 3.322, integrals #1 and # 2), 930 (sec 8.25, subsec. 8.250, integral # 1), 306 (sec. 3.32, subsec. 3.321, integral #1), or alternatively to pp. 306 see pp. 931 (sec. 8.25, subsec. 8.253, integral #1)). New York, NY: Academic Press.
5. Anon. (2016). Private Communications, Aircraft Maintenance, 1840 E. Valencia Ave., Tucson AZ 85706, (520) 445-6300.
6. Thomas, R. W. I., Eichelberger, M. J., & Lee, M. (2017). The theory of risk uncertainty reduction. In *35th International Systems Safety Conference Proceedings, Albuquerque*, NM, 21–25 August 2017.
7. Thomas, R. W. I., Eichelberger, M. J., & Lee, M. (2018). The theory of risk uncertainty reduction. *Journal of System Safety, 54*(2) (Summer/Fall).
8. Ref. [4] p. 310, sec. 3.351, #1 and pp. 365–522.

Part II
Electronics

Chapter 5
The Bent Pin Problem—I: Computer Search Methods

5.1 Problem Statement

Modern electrical connectors used in the aerospace industry may contain hundreds of pins protruding from the male half of the connector. Typically, during mating of such connectors, bent pins can occur one to ten percent of the time. Short circuits formed by bent pins are easily detected by continuity tests, but post-test connections can introduce dangerous new short circuits. When a bent pin event occurs, it is most likely to involve a single pin. But multiple bent pin events are also possible. If bent pins have sufficient length to reach nearby neighbor pins, it is possible for any one pin to be shorted to any other pin in a connector via a conducting chain of bent pins. Of course, the longer the path, the lower the probability of path formation. A complete solution to the bent pin problem involves: (1) identification of all possible paths between the two pins of interest, (2) computation of the single bent pin short-circuit probability, (3) computation of the probability of forming *each* short-circuit path from a chain of individual bent pins, (4) computation of the total short-circuit probability from *all* paths, (5) reduction of this total probability by ten to one hundred times to establish the real probability range of failure, and (6) assignment of a level of risk (frequent, probable, occasional, remote, or improbable).

In the past, only single bent pin probabilities have been calculated. Multiple bent pin problems have generally been addressed in a qualitative fashion. It is the object of this chapter, and the next four chapters, to deal with the multiple bent pin problem quantitatively, and to address the single bent pin problem as a special case of the multiple bent pin problem.

This first treatment will employ computer search methods to generate a complete solution to the bent pin problem. In general, a search method is the only way to generate a complete solution because, as will be seen, the bent pin problem is a "traveler's problem" for which a general analytic solution does not exist. Nevertheless, it is not always necessary to generate a complete solution to satisfy

© Springer Nature Switzerland AG 2020
R. R. Zito, *Mathematical Foundations of System Safety Engineering*,
https://doi.org/10.1007/978-3-030-26241-9_5

the demands of a particular customer application. In that case, "short cuts" may exist. One such shortcut involves matrix methods and will be treated in the next chapter.

5.2 Definitions

In all that follows, the methods of Graph Theory will provide a foundation. With that thought in mind, the definitions of Graph Theory will be adapted to our problem in the manner illustrated by the connector in Fig. 5.1. The connector on the left consists of a lattice (regular array) of pins, and three of these pins are bent. The gray pin is bent to its black nearest neighbor, and that black pin is bent to the equidistant connector shell. Another isolated black pin has also been bent to make contact with the shell. This "realistic" picture can be abstracted to the Figure on the right. The connector shell has been "cut" and shrunk to a point. Lines representing bent pins are called "edges", and each pin is rooted at a "vertex". In spite of appearances, the black and gray edges are considered to be of equal "length," in the sense that the vertices they connect are all "nearest neighbors."

5.3 Assumptions, Lemmas, Corollaries, and Theorems

In all that follows, it will always be assumed that source voltage is applied to a receptacle. Pins are on the load side of the connector. This is the usual common sense procedure to prevent electrocution of personnel by direct contact with a pin having a positive or negative potential with respect to ground. Connectors utilizing "hot" pins are outside the scope of this investigation and must be treated differently.

In this section, some basic theorems and corollaries will be derived that have far-reaching consequences for bent pin analysis. The first of these theorems is Fitzpartick's First Rule [1], *"Given a conducting path of bent pins between any two vertices, at least one pin must remain straight (not bent)."* The proof is by assuming the opposite situation. If all pins were bent, the path between the two vertices of interest would be electrically isolated, and there would be no safety problem. There

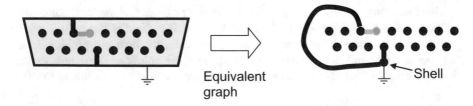

Fig. 5.1 A realistic connector with three bent pins and its Graph Theory abstract equivalent

is also a useful Lemma associated with the First Rule, "*Safety critical analysis of any two pins (vertices) can be broken down into the analysis of multiple graphs, each containing one straight pin at the end of a conducting path.*" The proof is by graph decomposition as shown in Fig. 5.2.

Next, an important corollary to the First Rule can be developed that speeds analysis of any problem where a bent pin can only contact one other pin at a time. In practice, pins are often not long enough to contact several other pins when bent, and even if they were, the probability of such an event is very low. In that case, "*No pin position (vertex) can be revisited in forming a conducting path between a straight pin and any other pin by single pin contacts only.*" The proof is by contradiction as follows. Consider a conducting path consisting of N-1 bent pins between (x_0, y_0) and (x_N, y_N). Let the pin at (x_N, y_N) be the straight pin in accordance with the previous Lemma. If the pin at the nth step, with vertex (x_n, y_n) (where n < N) is revisited, then it has already been bent once (i.e., a loop has formed). Therefore, the pin at step n + 1, with vertex (x_{n+1}, y_{n+1}), must be bent

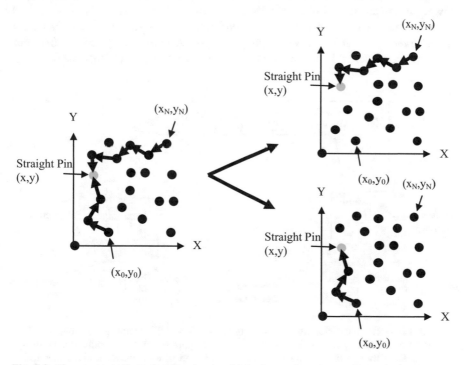

Fig. 5.2 The arrows indicate the direction in which pins are bent in a path connecting (x_0, y_0)–(x_N, y_N) in a general array of pins. The pin at (x, y) remains straight (gray). Because the original path (left) only employs contacts between pairs of pins, decomposition produces only two sub graphs. If a bent pin is allowed to contact more than one neighbor, loops can be formed in the conducting path, and decomposition results in more than two sub graphs

toward the pin at (x_n, y_n) to continue the chain, or a previous part of the path must be retraced. Note that if a previous part of the path is retraced, just continue increasing the step index. Therefore, the pin at (x_{n+2}, y_{n+2}) must be bent *toward* (x_{n+1}, y_{n+1}), or a previous part of the path must be retraced, etc. Finally, the pin at (x_N, y_N) must be bent *toward* (x_{N-1}, y_{N-1}). However, this contradicts the initial assumption that the pin at (x_N, y_N) is straight! Similarly, in the case of a multiple edge (i.e., if the pin at (x_n, y_n) is bent back to (x_{n-1}, y_{n-1})) it is clear that the pin at step n + 1 will have to be bent *toward* location (x_n, y_n) or the multiple edge will have to be retraced. Again, a contradiction will arise because, eventually, the pin at (x_N, y_N) will have to be bent.

The last theorem that will be needed is Fitzpatrick's Second Rule [1], "*A safety critical situation arises only when the straight pin is connected to a power source (i.e. not grounded).*" As with the First Rule, proof is by assuming the opposite. If the straight pin is not connected to a power source, there is no safety-critical problem. The implications here are subtle. As Fig. 5.3 shows, bending a load pin *toward* a straight pin connected to a power source is *very* dangerous, but bending a power pin toward a load pin is totally harmless!

In summary, the only problem that needs to be addressed is *a chain of bent pins in which the terminal pin is a straight power pin.* All other safety-critical bent pin problems, consistent with the assumptions, can be broken down to this last problem.

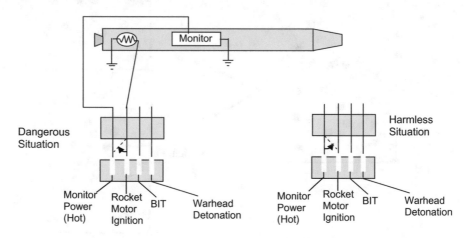

Fig. 5.3 Consider a connector with 4 pins: a monitor pin, a rocket motor ignition pin, a built-in test (BIT) pin, and a warhead detonation pin. The receptacle for the monitor pin is always energized. Even though the monitor pin (power pin) is part of a circuit for a trivial monitoring function, it can be deadly, as can be seen in this example where accidental rocket motor ignition is at risk

Fig. 5.4 The path a Knight
must follow to cover all
positions on a chessboard

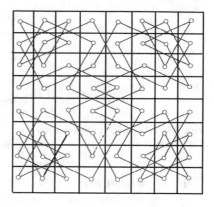

5.4 The Knight's Tour Problem

The game of chess probably had its origins in the ancient Egyptian game of *Senate* that is depicted on the wall of the tomb of Nefertari (circa 1253 B.C.) at Luxor. But, in its present form, the game of chess rightfully belongs to medieval Europe. For many years, the following curious problem circulated among gaming circles on the continent, "Starting from any location on the board, is it possible for a Knight to move in such a way as to cover every square?" The problem was solved by Euler, but the best solution was presented by Vandermonde in 1771 [2]. In his development, Vandermonde starts from a location on the chessboard labeled (x, y). Since a knight may move one step forward or backward in the x-direction, and two steps forward or backward in the y-direction, or vice versa, the following simple rule of motion emerges.

$$(x, y) \rightarrow (x \pm 1, y \pm 2) \text{ or } (x \pm 2, y \pm 1). \tag{5.1}$$

Starting from this rule, Vandermonde developed the attractive diagram of Fig. 5.4 to describe the path of the Knight. In the center of each square, a "pin" has been placed. Now, all this may seem to have nothing to do with the bent pin problem, but in fact, it is central as will be demonstrated in the next section.

5.5 Application of the Knight's Tour to the Bent Pin Problem (Identification of Paths)

The application of Vandermonde's method to bent pin problems will be demonstrated for the simple 9-pin square lattice in Fig. 5.5. The pin positions in Fig. 5.5 are identified by an ordered pair of integers that range from 1 to 3 on each axis. The problem to be solved is this, "How many conducting paths exist between (1, 1) and (3, 3)?" Even this simple problem will reveal a surprising number of paths.

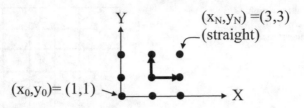

Fig. 5.5 A "simple" problem. Find the number of conducting paths between (1, 1) and (3, 3). Note that the origin of this x-y plane lies at (1, 1)

If only nearest neighbor contacts are allowed, the "rule of motion" (i.e., rule describing possible connections) for a pin at (x, y) would be

$$(x, y) \rightarrow (x \pm 1, y) \text{ or } (x, y \pm 1). \tag{5.2}$$

If the pins of our connector are long enough to allow contact with second nearest neighbors, then an additional "or" must be added; i.e., "or $(x \pm 1, y \pm 1)$", where the $\pm$ sign on both coordinates means that the pin at (x, y) may contact the pin at (x + 1, y + 1), (x + 1, y − 1), (x − 1, y + 1), or (x − 1, y − 1). In fact, rules can be devised that cover any number of possible neighbor contacts. However, in our simple example (Fig. 5.5), only *nearest neighbor* contacts will be considered.

Next, boundary conditions and end-game rules need to be taken into account. For the square lattice of Fig. 5.5, that is an easy task. Clearly, the coordinates of any pin, in a path of bent pins, can never exceed 3 or be less than 1 (i.e., $1 \leq x$, $y \leq 3$). Other lattices may require more complex boundary conditions. Furthermore, the condition (x, y) = (3, 3) marks the end of a path (straight pin), and the termination of computations for that path. Both of these conditions are trivial to code.

Figure 5.6 summarizes the method of computer search. All ordered pairs in Fig. 5.6 have been written in vertical format, with the x coordinate on top, to save space. Only ordered pairs consistent with the boundary conditions are shown. Starting from the pin at position (1, 1), a bend may be made to contact the pin at (2, 1) or (1, 2). For the second "move", or bend, the first of these last two pins may contact (1, 1), (3, 1), or (2, 2). But, wait! According to the corollary to the First Rule, *no pin may be revisited*. Therefore, (1, 1) must be eliminated from the triad. Similarly, (1, 2) may be bent toward (1, 1), (1, 3), or (2, 2). Again, (1, 1) must be eliminated. So, growing paths must continue from (3, 1), (2, 2), or (1, 3) It should be clear that this problem has a certain symmetry, because (1, 3) can be extracted from (3, 1) by interchanging the x and y coordinates. Of course, interchanging the coordinates of (2, 2) just leaves this ordered pair the same. From now on, only paths that emerge from (3, 1) and (2, 2), and successfully terminate at (3, 3), will be counted. The total number of paths will just be twice that number counted. For the third bend, connections can be made to (2, 1) and (3, 2) from (3, 1). Again, wait! It is necessary to eliminate (2, 1) because it has already been visited (i.e., bent and it cannot be rebent).

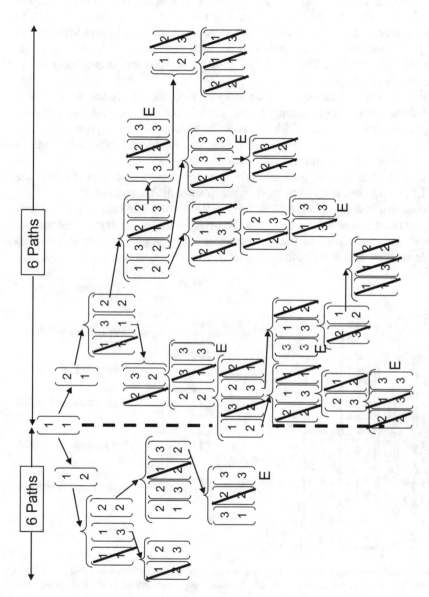

Fig. 5.6 A computer search tree for all conducting paths between (1, 1) and (3, 3). The letter "E" marks the end of a path. Paths on the left of the vertical dashed line are symmetrical to those on the right, therefore, not all paths on the left have been shown

Similarly, $(2, 2)$ can be bent toward $(1, 2)$, $(3, 2)$, $(2, 1)$, or $(2, 3)$. But, of course, $(2, 1)$ must be eliminated. Now, $(2, 3)$ can be bent toward $(1, 3)$, $(2, 2)$, or $(3, 3)$. Pin $(2, 2)$ must be eliminated, but $(3, 3)$ is our goal! One path has now been found. It consists of $(1, 1)$, $(2, 1)$, $(2, 2)$, $(2, 3)$, and $(3, 3)$. Continuing on with this type of reasoning will reveal all paths that start from pin $(1, 1)$, bent to contact $(2, 1)$, and continue on to eventually reach $(3, 3)$. There are 6 such paths, each marked by the letter "E" in Fig. 5.5. But remember, this is only half the total number by symmetry. The total number of paths from $(1, 1)$ to $(3, 3)$ is 12. All these paths are unique. They are all displayed in Figs. 5.7 and 5.8.

Notice that some paths are only four bent pins long; these paths are most likely to form. Others are six bent pins long; they are less likely to form. And, two paths are eight pins long; these paths will be least likely to form. For the square lattice, the length of the shortest path is just given by the sum of the length differences in the x and y coordinates of the end and start pins (i.e., $(x_N - x_0) + (y_N - y_0) = (3 - 1) + (3 - 1) = 4$). The longest path involves bending every pin except the last (see Figs. 5.7 and 5.8) and is given by pq-1, where p and q are the number of pins in the x and y direction, respectively (i.e., $3 \times 3 - 1 = 9 - 1 = 8$). The path lengths increase in steps of two because the smallest excursion away from a shortest path requires one step away and one step back. The next objective in the program of calculations for a bent pin analysis is to determine the probability that a single bent pin will short to a neighbor.

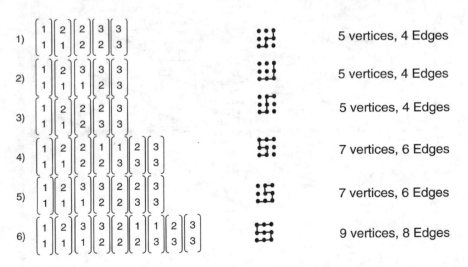

Fig. 5.7 "Dead" branches of the solution tree are discarded. "Living" branches that reach $(3, 3)$ are stored. The third sequence of pins in this figure is the path described in the text above

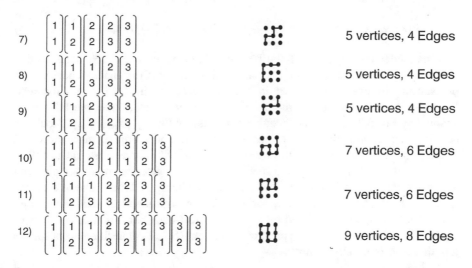

7) $\begin{bmatrix}1\\1\end{bmatrix}\begin{bmatrix}1\\2\end{bmatrix}\begin{bmatrix}2\\2\end{bmatrix}\begin{bmatrix}2\\3\end{bmatrix}\begin{bmatrix}3\\3\end{bmatrix}$ 5 vertices, 4 Edges

8) $\begin{bmatrix}1\\1\end{bmatrix}\begin{bmatrix}1\\2\end{bmatrix}\begin{bmatrix}1\\3\end{bmatrix}\begin{bmatrix}2\\3\end{bmatrix}\begin{bmatrix}3\\3\end{bmatrix}$ 5 vertices, 4 Edges

9) $\begin{bmatrix}1\\1\end{bmatrix}\begin{bmatrix}1\\2\end{bmatrix}\begin{bmatrix}2\\2\end{bmatrix}\begin{bmatrix}3\\2\end{bmatrix}\begin{bmatrix}3\\3\end{bmatrix}$ 5 vertices, 4 Edges

10) $\begin{bmatrix}1\\1\end{bmatrix}\begin{bmatrix}1\\2\end{bmatrix}\begin{bmatrix}2\\2\end{bmatrix}\begin{bmatrix}2\\1\end{bmatrix}\begin{bmatrix}3\\1\end{bmatrix}\begin{bmatrix}3\\2\end{bmatrix}\begin{bmatrix}3\\3\end{bmatrix}$ 7 vertices, 6 Edges

11) $\begin{bmatrix}1\\1\end{bmatrix}\begin{bmatrix}1\\2\end{bmatrix}\begin{bmatrix}1\\3\end{bmatrix}\begin{bmatrix}2\\3\end{bmatrix}\begin{bmatrix}2\\2\end{bmatrix}\begin{bmatrix}3\\2\end{bmatrix}\begin{bmatrix}3\\3\end{bmatrix}$ 7 vertices, 6 Edges

12) $\begin{bmatrix}1\\1\end{bmatrix}\begin{bmatrix}1\\2\end{bmatrix}\begin{bmatrix}1\\3\end{bmatrix}\begin{bmatrix}2\\3\end{bmatrix}\begin{bmatrix}2\\2\end{bmatrix}\begin{bmatrix}2\\1\end{bmatrix}\begin{bmatrix}3\\1\end{bmatrix}\begin{bmatrix}3\\2\end{bmatrix}\begin{bmatrix}3\\3\end{bmatrix}$ 9 vertices, 8 Edges

Fig. 5.8 These are the symmetrical solutions formed by interchanging the x and y coordinates of the solutions in Fig. 5.7. Note that pin sequences 7–12 are just the reflection of sequences 1–6 through a straight diagonal line defined by (1, 1) and (3, 3)

5.6   Single Bent Pin Contact Probability

What is the probability that an "ideal" pin will contact its nearest neighbor (second nearest neighbor, etc.) when bent? The geometry is summarized in Fig. 5.9. An "ideal" pin develops no curvature and maintains its length after bending only at its root.

Starting from the zeroth pin, if the nth pin in a chain at location (x_n, y_n) has diameter $d(x_n, y_n)$ and length $\mathcal{L}$, and if the contacted neighbor (nearest neighbor, or second nearest neighbor, etc.) at location (x_{n+1}, y_{n+1}) lies at a distance $\ell_{n,n+1}$ and has diameter $d(x_{n+1}, y_{n+1})$, and if $\mathcal{L} \geq \ell_{n,n+1}$, then the probability of contact, assuming that the bending process is directionally random, is given by

Fig. 5.9 Geometry for an ideal bent pin

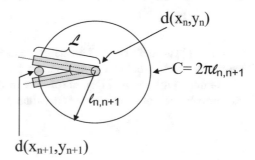

$$P(x_n, y_n, x_{n+1}, y_{n+1}) = [d(x_n, y_n) + d(x_{n+1}, y_{n+1})]/2\pi\ell_{n,n+1}. \qquad (5.3)$$

This model is reasonable if the bent pins fold nicely over each other so that contact is only made at the root of a succeeding pin, or if the chain of bent pins is propagating in one direction, or in a square array with $\mathcal{L} = \ell_{n,n+1}$. However, generally speaking, there are more opportunities for contact than are taken into account by Eq. 5.3, so that the calculated probability $P(x_n, y_n, x_{n+1}, y_{n+1})$ should be considered a minimum value in most cases [3]. The complex mechanics and mathematics of how pins bend has been solved, and will be presented in Chap. 9. It will be shown that Eq. 5.3 is just a special case of a considerably more complex and general mathematical description. With this caveat being said, the value of $P(x_n, y_n, x_{n+1}, y_{n+1})$ calculated via Eq. 5.3, for a connector with identical ideal pins 1 mm wide, 3 mm long, and separated by a 3 mm nearest-neighbor distance, is 0.106, or 10.6%. These dimensions are typical of pins in an RS 232 connector, except for the length which is usually 5 mm.

5.7   Multiple Bent Pin Short-Circuit Probability (Single Path)

The next question that must be answered is, "What is the probability $P_s(N)$ of forming a single path of length N, where $(x_N - x_0) + (y_N - y_0) \leq N \leq pq - 1$ for the square array?" Since a path is just a series chain of independent bent pins,

$$P_s(N) = \prod_{n=0}^{N-1} P(x_n, y_n, x_{n+1}, y_{n+1}). \qquad (5.4)$$

If all pins in the path have identical geometry and contact probability so that $P(x_n, y_n, x_{n+1}, y_{n+1}) = P$ for all n, then $P_s(N) = P^N$.

5.8   Subtotal Short-Circuit Probability (All Paths of Fixed Length N)

Given m_N "parallel" paths (i.e., paths connecting the same start and end pins) of length N, each labeled by an integer s such that $1 \leq s \leq m_N$, what is the short-circuit probability, $P(N)$, of forming *any* path of length N? It is just the sum of the probabilities of forming the individual paths.

$$P(N) = \sum_{s=1}^{m_N} P_s(N) \tag{5.5}$$

In the special case where all parallel paths of length N have the same single path probability, $P(N) = m_N P_s(N)$. Furthermore, if all pins have identical geometry and contact probability as well, then $P(N) = m_N P^N$.

5.9   Total Short-Circuit Probability (All Paths)

Since all paths leading from (x_0, y_0) to (x_N, y_N) are electrically parallel paths, regardless of length N, the total short-circuit probability P_{Total} is just the sum of the probabilities of forming each path. Hence, for the given problem

$$P_{Total} = \sum_{N=N_{min}}^{N_{max}} P(N)$$
$$= P(N_{min}) + P(N_{min} + \Delta_1)P(N_{min} + \Delta_2) + P(N_{min} + \Delta_3) + \cdots + P(N_{max}). \tag{5.6}$$

As in Eq. 5.5 above, P(N) is the probability of forming any path of length N, where N increases from its minimum to its maximum value in integer steps (deltas) such that $\Delta_1 < \Delta_2 < \Delta_3$...etc. For the square array when only nearest-neighbor contacts are allowed, $N_{min} = (x_N - x_0) + (y_N - y_0)$, the deltas are the even integers ($\Delta_1, \Delta_2, \Delta_3... = 2, 4, 6,...$etc.), and $N_{max} = pq - 1$. For the specific square array being discussed

$$P_{Total} = P(4) + P(6) + P(8) = 6P^4 + 4P^6 + 2P^8 = 6(0.106)^4 + 4(0.106)^6 + 2(0.106)^8$$
$$= 0.000757 + 5.67 \times 10^{-6} + 3.19 \times 10^{-8} = 0.000763. \tag{5.7}$$

It should be noticed that the contribution to P_{Total} falls off rapidly with path length. In fact, the shortest path contributes 99.2% of the total probability. The rapid convergence to truth is due to two effects. First, as paths get longer in steps of two, the value of each term in the series of Eq. 5.6 decreases by a factor of P^2. Second, as paths get longer, there is no proliferation in the number of such paths, as one might expect. Surprisingly, the opposite occurs! There are fewer acceptable paths as their length increases. This phenomenon will be discussed again in more depth in the next chapter. The two effects together result in the extraordinarily rapid convergence exhibited by this example. However, as will be demonstrated in Chap. 8, this rapid and monotonic decrease in the size of terms of the probability series (Eq. 5.6) is not necessarily guaranteed!

5.10   Probability Range of Failure and Mishap Probability Levels

Noting the usual definitions of mishap probability levels as events per lifetime (Frequent $> 10^{-1}$/life;   Probable $= 10^{-1}$–10^{-2}/life;   Occasional $= 10^{-2}$–10^{-3}/life; Remote $= 10^{-3}$–10^{-6}/life;   Improbable $< 10^{-6}$/life), it is clear that the sample problem of Fig. 5.5 just barely falls under the upper "Remote" level boundary. Even a small change in pin geometry would push this problem up into the "Occasional" level. However, it is important to remember that these calculations apply to a bent pin event that *has occurred*. In fact, as discussed in the introduction, a bent pin event will occur for only 1–10% of the connections made. However, once a bent pin event occurs, it will be assumed that any and all pins of a connector may be involved. Therefore, the real short-circuit probability is $P_{real} = f_{BP} P_{Total}$, where f_{BP} is the bent pin event frequency of occurrence such that $0.01 \leq f_{BP} \leq 0.1$. The exact value of f_{BP} depends on the design parameters of the connector (pin length and width, pin metal thickness, pin buckling strength, etc.). Therefore, the real probability of a short for the sample connector presented here is in the 7.63×10^{-5}–7.63×10^{-6} range, which is well within the "Remote" level.

5.11   Summary

The computer search methods described in this chapter provide a "complete" solution to the bent pin problem in the sense that each vertex of all possible paths, satisfying the assumptions, is known. In general, graph theory problems, involving finding paths between two points, can be divided into two types. There are "explorer's problems" in which an explorer may visit one city (vertex) more than once, just so long as he follows every path (edge), and there are "traveler's problems" in which each place of interest (vertex) is visited only once. Unfortunately, as shown by the example in this chapter, the bent pin problem is a "traveler's problem" which, in general, has no shortcut solution. Therefore, computer search is the only way to obtain a *complete* solution for a bent pin problem of any complexity. Such problems are called *hard to solve*. That is a technical term meaning that the size of the problem can always be scaled up to exceed a given computational capacity. However, many times, as will be demonstrated in the next chapter a complete solution is not necessary. It may only be required to *enumerate* paths. Under these relaxed conditions, "short cuts" to a solution are possible. One such shortcut employs matrix methods, and will be described in detail in the next chapter.

5.12 Problem

Suppose the pin array of Fig. 5.5 is refitted with longer pins so that *both* first and second nearest-neighbor contacts are allowed. Build a diagram like that of Fig. 5.6 for this new connector.

References

1. Fitzpatrick, R. G. (2010). Raytheon Missile Systems, Tucson, Arizona, 85734, Private Communications.
2. Briggs, N. L., Lloyd, E. K., & Wilson, R. J. (1976). *Graph theory 1736–1936* (pp. 22–27). Oxford: Clarendon Press.
3. Ozarin, N. (2008, September–October). *Journal of Systems Safety,* 19–28.

Chapter 6
The Bent Pin Problem—II: Matrix Methods

6.1 Problem Statement

Modern electrical connectors used in the aerospace industry may contain hundreds of pins protruding from the male half of the connector. Typically, during the attempted mating of such connectors, bent pins can occur one to ten percent of the time. This is the bent pin frequency of occurrence, f_{BP}. In this chapter, matrix methods will be employed to enumerate conducting paths between any two pin positions of interest when connectors are mismated. As shown in the previous chapter, the shortest paths are by far the most important [1]. That is fortunate, because the shortest paths between any two vertices will not contain path loops which, as previously shown, are forbidden [1]. Once a pin is bent in one direction, it can never be re-bent if it is revisited by a loop in the conducting path. If the path length (number of bent pins) of a shortest path can be easily determined by inspection, and if the number of shortest paths can be counted via matrix methods, then it is possible to compute the short-circuit probability between two pins of interest with a high degree of accuracy. It is important to note that the statement of the current problem is much more restricted than the problem defined in the pre-vious computer search treatment of the bent pin problem [1], where paths of any length are considered. Furthermore, the coordinates of the location of each pin in a conducting chain will no longer be required as in the previous chapter, and will eventually be dispensed with altogether. Therefore, in general, it will not be immediately obvious which, and how many, links in a given chain are nearest neighbor contacts, second nearest neighbor contacts, etc. Since each type of contact ("link") has its own probability of formation, the probability of formation of an entire chain, of any given length, will be affected by the mix of contact types. Hence, the techniques here are most useful for "simple" problems where, for example, all contacts are nearest neighbor, or there are only a few contacts of other types that are obvious from inspection.

© Springer Nature Switzerland AG 2020
R. R. Zito, *Mathematical Foundations of System Safety Engineering*,
https://doi.org/10.1007/978-3-030-26241-9_6

As in the previous chapter [1], certain definitions and assumptions will be employed. Bent pins will be called "edges," and the position (or "root") of each pin will be called a "vertex." It will always be assumed that the source voltage is applied to a receptacle, and that pins are on the load side of the connector. Also, because of the theorems proved in the previous chapter, the only problem that requires consideration involves an unbranched chain of bent pins that terminates in a straight pin inserted into an electrical source voltage receptacle.

Finally, it will be helpful to clarify the mathematical development with concrete examples. The first example will be a rework of the example used in the previous chapter [1]. A tin whisker variant of this problem will also be addressed. The second example, presented at the end of this chapter, will generalize the bent pin analysis to take into account foreign objects and debris (FOD). Bent pins, tin whiskers, and FOD are nothing more than conducting "edges" that can all be treated by the same general mathematical techniques [2].

6.2   Definition of the Adjacency Matrix A

Before progressing any further, a new, and essential, definition must be introduced. In particular, the notion of an Adjacency Matrix A must be defined. Figure 6.1 (left) shows the square lattice of pins used in the previous chapter. If a given pin is bent, what other pins could be contacted? The answer will depend on the length of the given pin and the distances to neighboring pins (nearest neighbors, second nearest neighbors, etc.). However, for the nearest-neighbor contact situation, the matrix to the right of the lattice in Fig. 6.1 neatly summarizes the possible contacts. If the pin

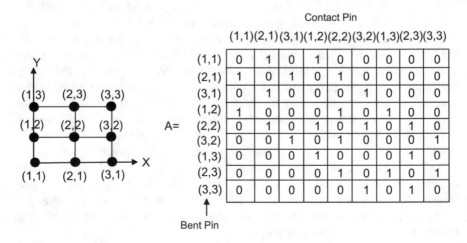

Fig. 6.1 On the left is a square lattice of nine pins with its origin at (1, 1). The corresponding Adjacency Matrix A for nearest neighbor contacts is displayed on the right. Note the high degree of symmetry in the A matrix. There are zeros all along the main diagonal, and A is symmetric with respect to this diagonal

at (1, 1) is bent away, it cannot contact itself. Therefore, the entry in the first row and first column of A is 0. However, the pin at (1, 1) can be bent toward (2, 1). Therefore, the entry in the first row and the second column of A is 1, meaning that one contact can be made. Later, when a more complex example is examined, entries other than unity will be used. Next, the pin at (1, 1) can also be bent toward (1, 2). Therefore, an entry of 1 can also be found in the first row and fourth column of A. Continuing on in this manner allows all matrix elements of A to be filled in. Consequently, *the Adjacency Matrix is the matrix of allowable edges*. With this definition in hand, Cayley's Theorem for the enumeration of paths can now be stated.

6.3   Cayley's Theorem

Circa 1889, A. Cayley proved that [3], *"The number of paths that are N edges long between any two vertices is given by A^N."* It is of central importance to note that these paths include cycles as well as multiple edges. As proved in the previous chapter [1], loops in conducting paths are generally forbidden since a pin once bent, during the mating of the male and female parts of a connector, cannot be re-bent when revisited at the crossing point of the loop. However, as proved in the previous chapter [1], it is the shortest path between two vertices that contribute the most, by far, to the short-circuit probability. Furthermore, "for any array of pins, no paths exist with cycles or multiple edges if $N = Nmin$." The proof of this lemma is by contradiction. Suppose a shortest path had a loop in it. A loop in a path necessitates a crossing point. Therefore, if the loop is removed from the path, a shorter continuous path exists. Therefore, the initial path was not the shortest. A similar argument applies to multiple edges, since these are just loops without "width."

Returning to the example of the 3 × 3 square array, Cayley's theorem can now be applied to the shortest paths between (1, 1) and (3, 3). As demonstrated in the previous chapter [1], the shortest paths for this problem involve four edges. Therefore, $N = N_{min} = 4$. The matrix A^4 is displayed in Fig. 6.2. The element in the row labeled (1, 1) and the column labeled (3, 3) gives the number of shortest paths connecting the pin at (1, 1) to the pin at (3, 3). This value is 6, which is also known from the computer search method used for the same problem in the previous chapter [1].

It is interesting to examine the element of row (1, 1) and column (3, 3) of A^6 and A^8 as well. These elements give the number of paths between (1, 1) and (3, 3) that are 6 edges (bent pins) and 8 edges long respectively. The values are 60 for the 6-pin path, and 504 for the 8-pin path. Both of these figures are much larger than the corresponding values of 4 and 2 found by computer search methods [1]. In the 6-pin case, "multiple edges" are now being counted. That is to say, a path connecting (1, 1) and (3, 3) can be envisioned that looks like a shortest path, but allows electrons to flow back one step from an intermediate pin along the path before proceeding forward to the goal at (3, 3). Or, electrons can run out to a dead-end

Fig. 6.2 Matrix elements of A^4

$$A^4 =$$

	(1,1)	(2,1)	(3,1)	(1,2)	(2,2)	(3,2)	(1,3)	(2,3)	(3,3)
(1,1)	10	0	8	0	16	0	8	0	6
(2,1)	0	18	0	16	0	16	0	14	0
(3,1)	8	0	10	0	16	0	6	0	8
(1,2)	0	16	0	18	0	14	0	16	0
(2,2)	16	0	16	0	32	0	16	0	16
(3,2)	0	16	0	14	0	18	0	16	0
(1,3)	8	0	6	0	16	0	10	0	8
(2,3)	0	14	0	16	0	16	0	18	0
(3,3)	6	0	8	0	16	0	8	0	10

Terminal Pin (column headers, top) / Start Pin (row headers, with upward arrow ↑)

neighbor pin, not part of a shortest path, and back again. Such paths would involve two extra steps for a total path length of six steps. All this, of course, is a fiction. Electrons don't move that way. And, there are no double pins rooted to vertices to provide physically parallel or dead-end paths to justify such a count. For the 8-pin paths, matters are even worse, since loops can be formed along the path connecting (1, 1)–(3, 3), and are counted together with multiple edge paths. These extra paths involving multiple edges and loops, above and beyond the "simple paths," are of no interest for the nearest neighbor bent pin problem presented here. However, their existence serves as background and motivation for the next chapter, which will deal with counting the number of simple paths longer than the minimum length necessary to complete the short circuit. Furthermore, the occurrence of non-simple paths can be of interest when a bent pin can contact more than one neighboring pin. Finally, the proliferation of non-simple paths, as path lengths increase, agrees with what would be expected on the basis of intuition. Simple paths are counterintuitive; their number can decrease as path length increases. There are two reasons for this phenomenon. The first is the constraint imposed by the prohibition on revisiting (rebending) vertices (pins), which eliminates many paths. The second reason is harder to understand. Given a pin array of finite size, and a path whose length approaches the total number of pins in the array, it becomes progressively harder to find free pins to form new paths.

6.4   Total Short-Circuit Probability

As mentioned in the introduction, and proved in the previous chapter [1], the shortest paths contribute the most (99.2%) to the total short-circuit probability of the sample problem being discussed. Since there are six shortest paths four edges long contributing a probability P(4) to the total short-circuit probability (P$_{Total}$),

$$P_{Total} \approx P(4) = A^4((1,1),(3,3))P^4 = 6P^4 = 6(0.106)^4 = 0.000757. \quad (6.1)$$

The coefficient $A^4((1, 1), (3, 3))$ is the numerical value of the element located in row (1, 1) and column (3, 3) of the A^4 matrix (i.e., 6). The probability P for single pin contact to a nearest neighbor pin has been calculated in the previous chapter for a pin 1 mm wide and 3 mm long, with a nearest-neighbor separation of 3 mm, to be 0.106 (10.6%) [1]. Since all contacts are of the same type (nearest neighbor), a single value of P will suffice. However, if the shortest paths are of mixed types (say, some paths involving only nearest neighbor contacts, and other paths of equal length involving some second nearest neighbor contacts), then two probabilities are involved. There is a single pin nearest-neighbor contact probability of P_{nn}, and a single pin second nearest neighbor contact probability of P_{snn}. In that case, in order to calculate the probability P(4), it now becomes necessary to inspect the paths and determine the number of paths of each type. In general, given a shortest path length with N_{min} edges (bent pins), some paths will contain only nearest neighbor contacts. Extending the notation used in the previous chapter [1], let there be $m_{Nmin}(nn)$ of these. Other paths will have one second nearest neighbor contact, with all other contacts being nearest neighbor. Let there be $m_{Nmin}(1\,snn)$ of these. Continue on until all contacts are of the second nearest neighbor type. Let there be $m_{Nmin}(Nmin\,snn)$ of these. For shortest paths involving N_{min} edges (bent pins), the short-circuit probability is

$$P(N_{min}) = m_{Nmin}(nn)\,P_{nn}^{Nmin} + m_{Nmin}(1\,snn)\,P_{nn}^{(Nmin-1)}P_{snn} + m_{Nmin}(2\,snn)P_{nn}^{(Nmin-2)}P_{snn}^2 + \\ \dots + m_{Nmin}(N_{min}snn)\,P_{snn}^{Nmin}, \quad (6.2)$$

where $m_{Nmin} = m_{Nmin}(nn) + m_{Nmin}(1\,snn) + m_{Nmin}(2\,snn) + \dots + m_{Nmin}(N_{min}snn).$
$$\quad (6.3)$$

For simple problems, the evaluation of this formula is not too tedious. However, for more complex problems, especially when the third nearest-neighbor contacts are allowed, the likelihood of human error in the sorting and counting process is high. In that case, the analyst may have to abandon the matrix method and return to the computer search method. This is the price that must sometimes be paid for a "short cut." Finally, it must always be remembered that the real probability of a bent pin short is $P_{real} = f_{BP}P_{Total} \approx f_{BP}P(N_{min})$.

Now, suppose a connector is properly mated, but tin solder was used to bond wires to the connector halves. Whisker growth can now begin and, after a time t, a whisker will grow to a length sufficient to contact a nearest-neighbor pin (from the back of the connector). A whisker of this kind will occur with a frequency (probability) of f_W per pin, where f_W is a function of time and temperature. The longer the time, and the higher the temperature, the greater the probability of forming a sufficiently long whisker. A chain N-whiskers long will have a frequency of occurrence equal to f_W^N. So, f_W^N plays the role of f_{BP} for whisker problems. Whiskers will grow radially from a cylindrical pin. Therefore, the single pin contact probability $P = d/2\pi l$, where d is the width of the contact pin and l is the whisker length (equal to the pin-to-pin nearest neighbor distance). P is essentially the probability that an infinitesimally thin straight whisker will grow in the proper direction. For pins with the geometry above, P = 0.053. Therefore, if the 3×3 problem presented above were a tin whisker problem, the real probability of a short between (1, 1) and (3, 3) would be $P_{real} = f_W^4$ [6 $(0.053)^4$], and f_W would be determined experimentally. So, the formalism presented above can be readily extended to tin whisker problems.

6.5   Another Instructive Example (FOD)

This section addresses a more realistic problem than the pedagogic 3×3 square lattice of pins. The example here will be an RS 232 type connector with a close packed lattice of 16 identical pins and a conducting, grounded, cylindrical shell (Fig. 6.3). What is the probability of shorting pin 1 to pin 16? As usual, pin width is 1 mm, pin length is 3 mm, and the nearest-neighbor distance is 3 mm. Therefore, the single pin-to-pin contact probability (P) is 10.6%, as before. The shell will be given an inside diameter of 17 mm so that the distance from the gray pins of Fig. 6.3 to the shell is only 2 mm. All other pins are further from the shell, and cannot reach it by bending. Let the bent pin-to-shell contact probability of the gray pins (Fig. 6.3) be P_S. It is an elementary exercise in trigonometry to show that $P_S = (1/\pi) \cos^{-1}(2/3) = 0.2677 \sim 27\%$. Furthermore, conducting Foreign Objects and Debris (FOD) will be allowed. The FOD will be modeled as thin conducting needles 2 mm long so that the two closest pins (gray in Fig. 6.3) can each be shorted to the shell with a probability P_{FOD}. It will be shown in Chap. 9 that, in this case, the FOD pin-to-shell contact probability is $P_{FOD} \approx d^3/(4l_{ps}A)$, where d is the pin diameter, l_{ps} is the pin-to-shell distance along the Sagitta, and A is the area of the connector shell (227 mm^2). Therefore, $P_{FOD} = 0.000551$ (or about 0.06%). This is a small number, but it must be remembered that the center of gravity of the FOD needle must fall in the proper location *and* with the proper orientation to make a contact.

Figure 6.3 shows the connector with its Graph Theory model. Each edge indicates a possible contact so that each interior pin has the capability of 6 nearest neighbor contacts. Figure 6.4 displays the Adjacency Matrix A for the connector in Fig. 6.3. Note that the pins are now just numbered consecutively in a left to right and bottom to top manner. Specification of the rectilinear coordinates of each pin is actually superfluous. Notice also that a 2 appears in the first row and 17th column of the A matrix. That's because there are two possible conducting paths between pins 1 and 17. One path involves a bent pin, and the other involves a conducting FOD contact. The shortest path between pins 1 and 16 is only two edges long. The edges

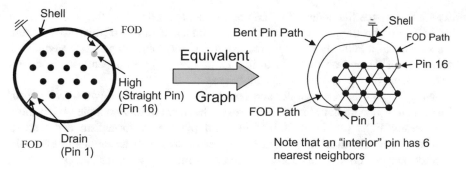

Fig. 6.3 A 16-pin RS 232 type connector with a conducting, grounded, cylindrical shell, and its equivalent Graph Theory model. Notice that the conducting shell has been collapsed to a point (vertex) in the graph (right)

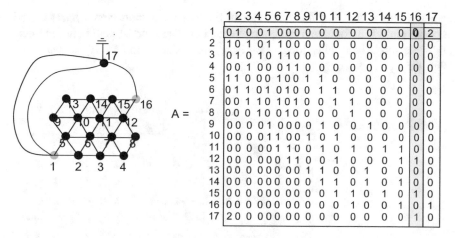

	1	2	3	4	5	6	7	8	9	10	11	12	13	14	15	16	17
1	0	1	0	0	1	0	0	0	0	0	0	0	0	0	0	0	2
2	1	0	1	0	1	1	0	0	0	0	0	0	0	0	0	0	0
3	0	1	0	1	0	1	1	0	0	0	0	0	0	0	0	0	0
4	0	0	1	0	0	0	1	1	0	0	0	0	0	0	0	0	0
5	1	1	0	0	0	1	0	0	1	1	0	0	0	0	0	0	0
6	0	1	1	0	1	0	1	0	0	1	1	0	0	0	0	0	0
7	0	0	1	1	0	1	0	1	0	0	1	1	0	0	0	0	0
8	0	0	0	1	0	0	1	0	0	0	0	1	0	0	0	0	0
9	0	0	0	0	1	0	0	0	0	1	0	0	1	0	0	0	0
10	0	0	0	0	1	1	0	0	1	0	1	0	0	0	0	0	0
11	0	0	0	0	0	1	1	0	0	1	0	1	0	1	1	0	0
12	0	0	0	0	0	0	1	1	0	0	1	0	0	0	1	1	0
13	0	0	0	0	0	0	0	0	1	1	0	0	0	1	0	0	0
14	0	0	0	0	0	0	0	0	0	0	1	0	1	0	1	0	0
15	0	0	0	0	0	0	0	0	0	0	1	1	0	1	0	1	0
16	0	0	0	0	0	0	0	0	0	0	0	1	0	0	1	0	1
17	2	0	0	0	0	0	0	0	0	0	0	0	0	0	0	1	0

Fig. 6.4 The Adjacency Matrix A for the 16-pin connector problem

can be one bent pin and a FOD contact, or two FOD contacts (note that pin 16 must remain straight—see the "Problem Statement" section).

From the matrix point of view, it is only necessary to examine the (1, 16) matrix element of A^2 matrix. The whole A^2 matrix does not need to be examined. Only one element needs to be calculated by adding the products of the corresponding entries in the shaded horizontal and vertical rectangles (left to right and top to bottom respectively). The result is 2, as expected by inspection of the graph. Since each path is of a different type, the total short-circuit probability is given by

$$P_{real} = f_{BP}f_{FOD}P_S P_{FOD} + f_{FOD}^2 P_{FOD}^2$$
$$= f_{BP}f_{FOD}(0.2677)(5.51 \times 10^{-4}) + f_{FOD}^2(5.51 \times 10^{-4})^2, \qquad (6.4)$$

where f_{FOD} is the frequency of a FOD event (i.e., the probability that a single piece of FOD will be trapped in the connector during assembly). In this problem, the importance of a FOD short circuit, relative to a purely bent pin path between pins 1 and 16, will depend on the ratio of f_{BP}/ f_{FOD}, and f_{FOD} will depend on the debris level (clean room class) in the assembly area.

Now, Eq. 6.4 is very simple, and could have been written down directly from inspection of the graph in Fig. 6.3. However, the intent of this example was to show how neatly FOD problems fit into the bent pin matrix formalism and lead to quantitative results that can be used for connector design, clean room and assembly area contamination specifications, and the determination of hazard levels.

6.6 Connector Design

Suppose pins 1 and 7 of the connector in Fig. 6.4 are a short-circuit hazard. And, suppose pin 4 is a populated but unconnected pin. The shortest path between pins 1 and 7 is three bent pins long. Therefore, the examination of A^3 is warranted (Fig. 6.5).

Terminal Pins (columns)

Start Pin (row 1) →

A³ = A^3 =

2	10	3	1	11	4	3	0	2	4	3	2	2	1	2	0	14
10	6	9	3	10	12	5	3	4	6	6	3	3	3	1	2	2
3	9	6	8	5	12	12	4	3	6	6	6	1	3	3	1	2
1	3	8	4	3	5	10	7	0	3	6	4	0	1	3	2	0
11	10	5	3	8	14	6	1	9	13	6	3	4	6	3	2	2
4	12	12	5	14	12	15	6	5	15	15	6	6	6	6	3	4
3	5	12	10	6	15	12	10	3	6	15	13	3	6	6	4	1
0	3	4	7	1	6	10	4	0	3	5	9	0	3	4	2	1
2	4	3	0	9	5	3	0	4	10	6	1	7	4	3	0	2
4	6	6	3	13	15	6	3	10	12	15	6	10	12	5	3	2
3	6	6	6	6	15	15	5	6	15	12	14	5	12	12	4	2
2	3	6	4	3	6	13	9	1	6	14	8	3	5	10	8	1
2	3	1	0	4	6	3	0	7	10	5	3	4	8	3	1	0
1	3	3	1	6	6	6	3	4	12	12	5	8	6	9	3	1
2	1	3	3	3	6	6	4	3	5	12	10	3	9	6	7	1
0	2	1	2	2	3	4	2	0	3	4	8	1	3	7	2	7
14	2	2	0	2	4	1	1	2	2	2	1	0	1	1	7	0

Fig. 6.5 A^3 for the 16-pin connector problem. Entries enumerate paths that are three pins long for the connector in Fig. 6.3. "Redundant paths" with loops and multiple edges are included in the count

Element (1, 7) of this cubed Adjacency Matrix is 3. That is to say, there are three paths, consisting of three bent pins each, that connect pin 1–pin 7. The safety situation would improve greatly if the wire connected to pin 7 were moved to pin 4. In that case, only one path three edges (bent pins) long exists (see element (1, 4) of Fig. 6.5). Therefore, examination of the A^3 matrix gives the analyst an overview of the electrical safety properties of the connector. The example presented here is very simple for pedagogic reasons, but in real connectors with hundreds of slender pins capable of making the first, second, and even third nearest-neighbor contacts, examination of powers of the adjacency matrix allow for intelligent design and modification of connectors so as to achieve maximum safety. It would be very difficult to get this same information by repeated computer searches, taking two pins at a time. In a connector with k pins, there are k choose two pairs of pins to select, or k!/[(k − 2)!2!] pairs. For a connector with 100 pins, the number of pairs to analyze is 100!/[98!2!], or (100)(99)/2 = (50)(99) = 4950 pairs. Each pair requires its own computer search. An onerous task to say the least! Therefore, matrix methods should not just be thought of as an occasional convenient short cut replacement for search methods, but as a tool, to be used with search methods, that displays global connector data in a useful and novel manner.

6.7   Summary

The result of Eq. 6.1 should be compared to the total short-circuit probability calculated by computer search methods in the previous chapter (P_{Total}= 0.000763) [1]. The difference is less than 1%, and is due to the fact that conducting paths longer than the shortest path have been ignored. As discussed in Chap. 5 [1], the rapid convergence to truth is due to two effects. First, as paths get longer in steps of two, the value of each term in the series expansion for P_{Total} decreases by a factor of P^2. Second, the number of acceptable paths also decreases as path length increases. The combined effect is *rapid convergence*. In some sense, *convergence* is a general property of bent pin solutions that will be discussed in more detail in subsequent chapters.

However, the *rate* of convergence is another matter. Thick, or long, connector pins, with high single pin contact probabilities, greatly slow convergence. For example, consider contacts between pins (2, 2) and (3, 2) in Fig. 6.1. There is only one shortest path and it is formed by bending pin (2, 2) alone. The next shortest path is three pins long, and there are two such paths. If the pin dimensions yield a single pin contact probability of 25%, which can easily happen with thick closely packed pins, then the total short-circuit probability is

$$P_{Total} = (1)(0.25) + (2)(0.25)^3 = 0.25 + 0.031 = 0.28. \qquad (6.5)$$

Now, the second term in the series contributes 12.5% to the total short-circuit probability. Furthermore, if a square array of pins is only partially populated (i.e., has missing pins), it is entirely possible that paths longer than the shortest path can make an even greater contribution to the series expansion for P_{Total}. This result is not obvious from inspection of the 3×3 problem described above. However, consideration of the 4×4 square array of pins, with pins at locations (2, 3), (3, 3), and (4, 3) removed (not populated), reveals only one shortest path (6 edges long) connecting (1, 1)–(4, 4) by nearest neighbor contacts. The next shortest path is 8 edges long, and there are seven possibilities. One simple path, and six paths containing one triple edge each. Therefore, the number of simple paths is the same for short circuits containing six or eight bent pins! Again, if the single pin contact probability is not much less than unity, the second term in the probability series for P_{Total} will make a significant contribution.

In the next chapter, a method of counting only simple paths between vertices will be presented. This procedure will eliminate the problems presented by loops and multiple edges when trying to enumerate longer paths.

6.8 Problem

For Fig. 6.1, prove that there are no possible bent pin paths five edges long connecting vertex (1, 1)–(3, 3). Do this by computing the element $A^5((1, 1), (3, 3))$ from the first row of A^4 (Fig. 6.2) and the last column of A (Fig. 6.1).

References

1. Zito, R. R. (2008). The Curious Bent Pin Problem—I: Computer Search Methods. In *29th International Systems Safety Conference*, Las Vegas, NV, August 8–12, 2008.
2. Fitzpatrick, R. G. (2010). Raytheon Missile Systems, Tucson, Arizona, 85734, Private Communications.
3. Marshall, C. W. (1971). *Applied graph theory* (pp. 178–179). New York: Wiley.

Chapter 7
The Bent Pin Problem—III: Number Theory Methods

7.1 Background

In Chap. 5, computation of the probability of a short between any two pins in a complex connector, due to a bent pin event, was accomplished by computer search methods [1]. This solution was complete, but required correctly coding potentially complex boundary conditions. Furthermore, computer search methods do not lend themselves to analytical studies on how pins in a connector must be separated and arranged in order to achieve a given safety level ("convergence problems"). To this end, matrix methods were introduced in Chap. 6 [2]. These methods enable rapid computation of the short circuit probability, and are amenable to analytic studies as well. However, matrix methods based on Cayley's Theorem suffer from one flaw. In general, it is difficult to calculate short circuit probabilities for chains of bent pins that are much longer than the shortest, and second shortest, conducting paths. The reason for this is that Cayley's Theorem does not distinguish between "simple paths" (without loops or multiple edges) and so-called "redundant paths", the latter of which are usually forbidden in bent pin analysis. This chapter will correct that deficiency by using number theoretic ideas to modify the matrix calculations so that redundant paths are subtracted off the total number of paths via Cayley, leaving only "simple paths" without redundancy.

7.2 Partitions

The origin of the pattern known as the Star of David has been the object of scholarly speculation for some years. One suggestion has been that the star's formation by upward and downward pointing triangles represents the human condition as one of decision [3]. Are we to choose good and heaven above, or evil and hell below? Or, possibly the star originated in one of the mandalas of India, where an

© Springer Nature Switzerland AG 2020
R. R. Zito, *Mathematical Foundations of System Safety Engineering*,
https://doi.org/10.1007/978-3-030-26241-9_7

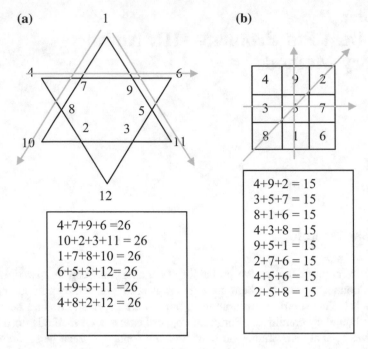

Fig. 7.1 The Star of David and Magic Square puzzles with their partitions

interlocking network of progressively larger triangles symbolizes the creation and expansion of the universe from a primeval "drop" into the world of opposites we see today (male and female, right and wrong, left and right, etc.) [4]. However, an alternate mathematical interpretation, and the one that is of interest here, suggests that the star may have had numerical significance [5]. If the first 12 integers are assigned to each vertex of the star as shown in Fig. 7.1a, then the sum of the integers along each line is 26. Each sum (order counts) is called a *partition* of 26. Another puzzle of this type is the Magic Square of Order Three [5]. When the first 9 integers are arranged as shown in Fig. 7.1b, the sum of the integers along each row, column, and diagonal is 15. Again each sum is a partition of 15. The first partition in Fig. 7.1b would be denoted (4, 9, 2), the second (3, 5, 7), etc. On the surface these simple ideas may seem irrelevant to the bent pin problem but, in fact, they are of central importance as will be shown below.

7.3   Short Paths and Partitions

The conducting path N bent pins long in Fig. 7.2 is impossible to form, using single pin contacts, without bending the final power pin. If that happens, the load circuit connected to the initial pin will be isolated from power, and there will be no safety

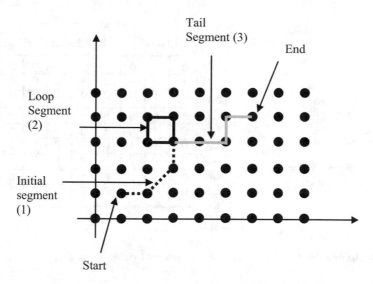

Fig. 7.2 Partitioning a redundant path. In this case $(n_1, n_2, n_3) = (3, 4, 4)$

critical problem. There will just be a damaged connector that must be repaired or cut off and replaced.

Therefore, it is desirable not to count paths of the kind displayed in Fig. 7.2. But, how can redundant paths be numerically characterized? And, once characterized, how can they be counted? The path in Fig. 7.2 can be broken down into three segments. First, there is the segment from the initial vertex (initial pin root) to the "crossing point" (revisited vertex) that is n_1 bent pins (edges) long. Next, there is the "loop" segment which is n_2 edges long. And, finally, there is the tail segment that is n_3 edges long. Now, the integers n_1, n_2, and n_3 form a partition of the path N edges long such that $N = n_1 + n_2 + n_3$, in analogy with the Star of David and Magic Square puzzles above. In fact, for any array with an Adjacency Matrix A, all paths containing loops (including paths containing multiple edges) can be characterized by their partitions, and it has been known since the 1950s and 60s how to count redundant paths of any length N that are characterized by a set of partitions [6, 7]. For large values of N the results are involved,[†] but equations for $N = 3$ and $N = 4$ can be coded in a single line using MATLAB [8]. Let R_N be the matrix enumerating all redundant paths of length N between any two pins. Then

$$R_3 = \left(A \bullet d(A^2) + d(A^2) \bullet A\right) - A \times A^T \qquad \text{and} \qquad (7.1)$$

$$R_4 = \left(A \bullet d(A^3) + d(A^3) \bullet A\right) + \left(A^2 \bullet d(A^2) + d(A^2) \bullet A^2\right) + A \bullet d(A^2) \bullet A - A \times (A^2)^T$$
$$- 2(A \times A^T) \times A^2 - \left(A \bullet (A \times A^T) + (A \times A^T) \bullet A\right).$$

$$(7.2)$$

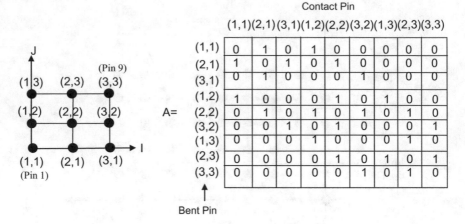

Fig. 7.3 The three by three square array and its adjacency matrix for single pin nearest-neighbor contacts

The superscript T indicates the transpose, the "•" operation is the usual matrix multiplication, and the "x" operation indicates elementwise matrix multiplication (i.e., the elements of UxV = $(u_{ij}v_{ij})$ where i and j are integers and u_{ij} and v_{ij} are elements of matrices U and V, respectively). The "d" operation on any matrix is the matrix formed by setting all off-diagonal elements to zero and retaining the *diagonal* elements unchanged. Note that for an A matrix populated with 1's and 0's, and symmetric with respect to the main diagonal, $A \times A^T = A$. Explicit expressions for R_5 and R_6 are given by Ross and Harary in the literature [6].[†]

As an example of the use of these formulas, consider the three-pin by three-pin square array used in the previous two chapters (Fig. 7.3, left). Suppose it is desired to count all redundant paths connecting the pin at (1, 2) to the pin at (2, 2); only nearest-neighbor contacts are allowed. As in the previous chapters, the adjacency matrix is displayed to the right in Fig. 7.3. Evaluation of R_3 yields the result at the bottom of Fig. 7.4. The element in row (1, 2) and column (2, 2) is 6. At the top of Fig. 7.4, the paths themselves are displayed.

None of these redundant paths are desirable to count because all require the terminal pin to be bent. Therefore, all pins in these redundant paths are bent. Therefore, there is no power source for these networks. Therefore, there is no safety-critical problem. What is important, from a safety point of view, are simple paths. That is to say, paths described by the partition $(n_1, n_2, n_3) = (N, 0, 0)$ which, unfortunately, are not directly countable. However, since the total number of paths of length N, between any two pins in an array, is given by the elements of A^N, one might think that the matrix $A^N - R_N$ is what is required. However, one thing has been forgotten. A loop can be formed that starts at any pin and loops back to itself!

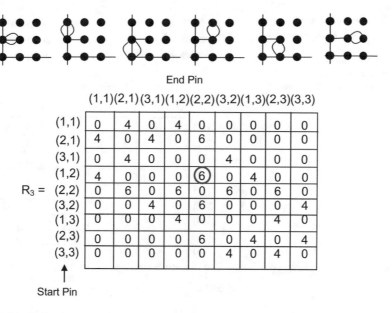

$$R_3 = \begin{array}{c} \\ (1,1) \\ (2,1) \\ (3,1) \\ (1,2) \\ (2,2) \\ (3,2) \\ (1,3) \\ (2,3) \\ (3,3) \end{array}$$

End Pin

	(1,1)	(2,1)	(3,1)	(1,2)	(2,2)	(3,2)	(1,3)	(2,3)	(3,3)
(1,1)	0	4	0	4	0	0	0	0	0
(2,1)	4	0	4	0	6	0	0	0	0
(3,1)	0	4	0	0	0	4	0	0	0
(1,2)	4	0	0	0	⑥	0	4	0	0
(2,2)	0	6	0	6	0	6	0	6	0
(3,2)	0	0	4	0	6	0	0	0	4
(1,3)	0	0	0	4	0	0	0	4	0
(2,3)	0	0	0	0	6	0	4	0	4
(3,3)	0	0	0	0	0	4	0	4	0

↑

Start Pin

Fig. 7.4 The Redundancy matrix for three edge paths of Fig. 7.3 (left). The number of redundant three edge paths connecting $(1, 2)$–$(2, 2)$ is circled

This circuit may not contain loops or multiple edges *within* itself, but it is still a loop. And, as such, it is not connected to a power source because all pins are bent. Therefore, it is necessary to ignore this type of path. The matrix enumerating all simple paths (also called "ways") of length N between any two pins is then given by Ref. [8] as

$$W_N = (A^N - R_N) - d(A^N - R_N). \tag{7.3}$$

The second term eliminates diagonal elements from $A^N - R_N$. Therefore, self-loops, paths described by the partition $(n_1, n_2, n_3) = (0, N, 0)$, are eliminated. The procedure in Chap. 6 for enumerating shortest paths can now be seen as a special case of this more general procedure in which $R_N = [0]$ (the zero matrix) and the "d" operation has been ignored [2]. As an example of the use of Eq. 7.3, calculation of W_3 for paths between $(1, 2)$ and $(2, 2)$ yields 2 for the element in the fourth row and fifth column of the matrix displayed in Fig. 7.5. The two simple paths are $(1, 2) \rightarrow (1, 3) \rightarrow (2, 3) \rightarrow (2, 2)$ and $(1, 2) \rightarrow (1, 1) \rightarrow (2, 1) \rightarrow (2, 2)$.

As a practical matter, it is *usually* not necessary to explore the effect of pins more than six edges (bent pins) away from a safety-critical pin. The reason is easy to understand. Let P be the probability of one pin contacting its neighbor, then the

Fig. 7.5 Way Matrix of simple paths that are described by the partition $(n_1, n_2, n_3) = (3, 0, 0)$. The number of ways connecting $(1, 2)$–$(2, 2)$ has been circled

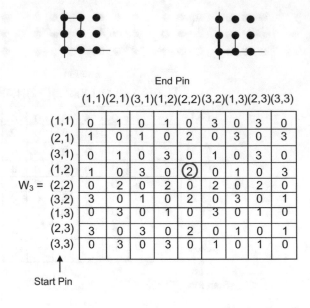

End Pin

		(1,1)	(2,1)	(3,1)	(1,2)	(2,2)	(3,2)	(1,3)	(2,3)	(3,3)
	(1,1)	0	1	0	1	0	3	0	3	0
	(2,1)	1	0	1	0	2	0	3	0	3
	(3,1)	0	1	0	3	0	1	0	3	0
	(1,2)	1	0	3	0	(2)	0	1	0	3
$W_3 =$	(2,2)	0	2	0	2	0	2	0	2	0
	(3,2)	3	0	1	0	2	0	3	0	1
	(1,3)	0	3	0	1	0	3	0	1	0
	(2,3)	3	0	3	0	2	0	1	0	1
	(3,3)	0	3	0	3	0	1	0	1	0

Start Pin

probability of building a 6-pin bridge is down by a factor of P^6. If $P = 0.1$ (10%), then $P^6 = 10^{-6}$. So, unless the number of possible six pin paths is very large, the contribution of such paths to the total short circuit probability is very small. Therefore, it will usually not be necessary to consider Redundancy or Way Matrices beyond R_6 or W_6, respectively. With that being said, it would be interesting to go further to cover any exceptions. Furthermore, higher order Way Matrices will prove useful for debugging sneak circuits, as will be discussed in Chaps. 10 and 11. Beyond W_6, it will be most practical to enumerate simple paths of length N iteratively from the enumeration of paths of length $N - 1$.

7.4   The Enumeration of "Long" Paths

The enumeration of simple paths, described by the partition $(N, 0, 0)$, of any length N can be calculated from the enumeration of paths of length $N - 1$ via the following theorem [7, 8]. *Let A[j] be the Adjacency matrix A of any connector with the jth row and jth column replaced by zeros. Then the jth column of W_N is just the jth column of $W_{N-1}(A[j])$ • A.* The expression $W_{N-1}(A[j])$ means W_{N-1} is a function of $A[j]$, as can be seen explicitly from Eqs. 7.1, 7.2 and 7.3 if A is replaced by this new matrix A[j]. Setting the jth row and jth column of the adjacency matrix equal to zero is equivalent to isolating pin j so that it cannot contact any of its neighbors. Therefore, the theorem above says that the number of simple paths of length N, in an array of pins with adjacency matrix A, can be calculated in terms of the number of simple paths of length N − 1 in the "sub-arrays" formed by isolating one pin at a time! Although this astonishing result may seem simple, it requires the computation of the $W_{N-1}(A[j])$ matrices and, if all j matrices are required, can involve about the same amount of work as the computer search methods outlined in Chap. 5 [1].

A simple example, whose answer is already known, will demonstrate the method. Suppose the number of simple paths of length 4 between pin 1 (i.e., the pin at (1, 1)) and pin 9 (i.e., the straight pin at (3, 3)) is desired. That is the same as wanting the value of the element W_4(row 1, column 9). To get this element the first row of $W_3(A[9])$ will have to be matrix multiplied by the last column of A. Here, it is important to note that row and column 9 of A will be reduced to zeros to form A[9] for the computation of $W_3(A[9])$ via Eqs. 7.1 and 7.3. The theorem above demands this because the desired element of W_4 lies in the ninth column. Following the prescription of Eq. 7.1 shows that the first row of $W_3(A[9])$ looks like 0 1 0 1 0 3 0 3 0. Only this one row of $W_3(A[9])$ is needed, since only a single element of W_4 will provide the answer to the stated problem. Next, the iterative method involves forming the matrix product $W_3(A[9]) \cdot A$. Again, since only the element $W_4(1, 9)$ is of interest, matrix multiplication of the first row of integers just calculated, with the ninth column of the matrix A displayed in Fig. 7.3, yields (0)(0) + (1)(0) + (0) (0) + (1)(0) + (0)(0) + (3)(1) + (0)(0) + (3)(1) + (0)(0) = 6. In fact, this is the correct number of paths, as is known from computer search methods [1]. Of course, the same answer could also have been found by direct computation of W_4 via Eqs. 7.2 and 7.3, followed by an examination of element $W_4(1, 9)$. Finally, it should be noted that the calculations presented above involve only simple arithmetic, and can be completed manually. However, a calculator, or an Excel spreadsheet, speeds the work.[‡]

So far, only a single element of W_4 has been determined because that is all that was necessary to solve the stated problem. But, suppose the scope of the original problem had been a little larger. Suppose it was desired to find all simple paths four bent pins long that *start at any pin and terminate in the straight pin at (3, 3)* (i.e., pin 9). In that case, the entire ninth column of W_4 would have to be calculated. This is also easy, because it only requires doing the matrix multiplication of each of the nine rows of $W_3(A[9])$ with the last column of A. If one wished to find all simple paths four bent pins long between *any two pins*, the entire matrix W_4 would have to be evaluated. In addition to the last calculation, each row of $W_3(A[8])$ would have to be multiplied by the eighth column of A to get the eighth column of W_4. Similarly, each column of $W_3(A[7])$ when matrix multiplied by the seventh column of A would yield the seventh column of W_4, etc., until the entire matrix W_4 was filled in. Therefore, the entire matrix W_4 can be constructed from the matrices $W_3(A[j])$, where $1 \leq j \leq 9$.

The elements of the matrix W_5, containing the enumeration of all paths of length 5, can now be generated as follows. Recall that the matrix A[j] was just the adjacency matrix A of a pin array with the entries in row and column j set equal to zero, thereby isolating pin j. However, each pin "sub-array" with pin j isolated can further be broken down into simpler "sub-sub-arrays" with another pin (call it "i") isolated. Now, two rows (rows i and j) and two columns (columns i and j) are zeroed out. In this way, enumeration matrices for paths of length 3 in "sub-sub-arrays" can be used to generate enumeration matrices $W_4(A[j])$ for paths of length 4 in "sub-arrays", which can finally be used to generate the enumeration matrix W_5 for all paths of length 5. That is to say:

$$W_3(A[i,j]) \cdot A[j] \xrightarrow{\forall i} W_4(A[j]) \text{ and}$$

$$W_4(A[j]) \cdot A \xrightarrow{\forall j} W_5. \tag{7.4}$$

Here, the symbol "$\rightarrow$" means "yields" and $1 \le i, j \le 9$. In practice, formulas would be used to evaluate the Redundancy and Way Matrices up to $N = 6$. Beyond that, iteration methods would be used until paths became so long that all pins are used up. It is clear that the method in this last paragraph can involve about as much computational effort as a straightforward computer search, except in certain special cases such as the evaluation of $W_4(1, 9)$ in the example above, or the evaluation of a single element of W_7 for any connector from the expression for R_6 found in the literature [6].[†]

7.5   Calculation of the Total Short Circuit Probability

As in Chap. 6 [2], the total short-circuit probability between pins (1, 1) and (3, 3) is, to a close approximation, given by

$$P_{Total} \approx 6P^4, \tag{7.5}$$

where P is the probability of forming a single bent pin connection between two adjacent pins. The exponent 4 comes from the fact that there are four bent pins in the conducting chain, and the coefficient 6 is the number of simple paths that was previously calculated. But, how does one know, a priori, that chains six or eight pins long will contribute little to the total short circuit probability? How does one know if, or where, the probability series should be truncated without doing a computer search? That will be one of the topics covered in the next chapter.

7.6 Conclusion

The numerical methods presented in this chapter for generating W_N will prove to be an important tool for understanding convergence and truncation issues associated with the calculation of the P_{Total} series. Computer search methods give an answer to practical problems, but are unsuitable for analytical studies that require a formula for the elements of W_N.

†The long and complicated formulas given by Ross and Harary [6] for R_5 and R_6 have been tested by the author and are known to be correct as stated in the literature. Their results are presented below in the high-level MATLAB user language and are displayed as code excerpts rather than as formulas in the usual mathematical notation to eliminate any possibility of transcription errors. The "d" operation was implemented by the use of the "diag(diag(arg))" instruction, where "arg" is an argument consisting of any square matrix. This instruction will extract the diagonal elements of "arg" and return a square matrix with these same diagonal elements and zeros for all off-diagonal elements. In MATLAB, the transpose operation is denoted by a period followed by a single apostrophe (. '), while element by element matrix multiplication is denoted by a period followed by an asterisk (. *). An asterisk by itself denotes ordinary matrix multiplication. The code for R_5 is displayed below with M as an Adjacency Matrix.

```
>> M= [...enter the elements of the Adjacency Matrix here...]
M2= M^2
M3= M^3
M4= M^4
M5= M^5
dM2= diag(diag(M2));
dM3= diag(diag(M3));
dM4= diag(diag(M4));
dMdM2M= diag(diag(M*dM2*M));
S= M.*M.';
PART1= M*dM4 + dM4*M + M2*dM3 + dM3*M2;
PART2= M3*dM2 + dM2*M3 +M*dM2*M2;
PART3= M2*dM2*M + 2*(M*dM2 + dM2*M);
PART4= 4*(dM2*S + S*dM2) + M*dM3*M + M2.*M.';
PART5= 3*M.*S^2 - (M*dMdM2M + dMdM2M*M);
PART6= -2*(M*(S.*M2) + (S.*M2)*M) - (S*M2 + M2*S);
PART7= -2*(M*dM2^2 + (dM2^2)*M) - (M*(M.*M2.') + (M.*M2.')*M);
PART8= -M.*(M3.') - 2*S.*M3 - dM2*M*dM2 - (M.').*M2.*M2;
PART9= -M*S*M - 2*M.*M2.*(M2.') - 4*S;
R5= PART1 + PART2 + PART3 + PART4 + PART5 + PART6 + PART7 + PART8 + PART9
```

The code for R_6 is displayed below. Again, with M as the Adjacency Matrix.

```
>>M= [...enter the elements of the Adjacency Matrix here...]
M2= M^2
M3= M^3
M4= M^4
M5= M^5
M6= M^6
dM2= diag(diag(M2));
dM3= diag(diag(M3));
dM4= diag(diag(M4));
dM5= diag(diag(M5));
dMdM2M= diag(diag(M*dM2*M));
S= M.*M.';
S2= S^2;
dMSM= diag(diag(M*S*M));
dSM2= diag(diag(S*M2));
dM2S = diag(diag(M2*S));
dM2dM2M= diag(diag(M2*dM2*M));
dMdM2M2= diag(diag(M*dM2*M2));
dMdM3M= diag(diag(M*dM3*M));
Pt1= M*dM5 + dM5*M + M2*dM4 + dM4*M2;
Pt2= M3*dM3 + dM3*M3 + M4*dM2 + dM2*M4;
Pt3= M*dM2*M3 + M3*dM2*M + M*dM3*M2;
Pt4= M2*dM3*M + M*dM4*M + M2*dM2*M2;
Pt5= 2*(M2*dM2 + dM2*M2) + 4*(M*S*dM2 + dM2*S*M);
Pt6= 4*(M*dM2*S + S*dM2*M) + M*dMSM;
Pt7= dMSM*M + 2*M*dM2*M + 4*(M*dSM2 + dM2S*M);
Pt8= 4*(M*dM2S + dSM2*M);
Pt9= M*(M.'.*M2) + (M.'.*M2)*M + 3*(M*(M.*S2) + (M.*S2)*M);
Pt10= 3*M.*(S*(M.'.*M2) + (M.'.*M2)*S);
Pt11= 3*M.*(S*(M.*(M2.')) + (M.*(M2.'))*S);
Pt12= M.*(M.'*dM2*M.') + 3*(M2.*S2) + M2.*M2.';
Pt13= 2*M.'.*(M*(M.*M2) + (M.*M2)*M) + S*M*dM2;
Pt14= dM2*M*S + 8*((S.*M2)*dM2 + dM2*(S.*M2));
Pt15= 4*((M.*M2.')*dM2 + dM2*(M.*M2.')) + 4*(S*dM3 + dM3*S);
Pt16= 2*S.*(M*dM2*M) - M3*S - S*M3;
POS= Pt1+Pt2+Pt3+Pt4+Pt5+Pt6+Pt7+Pt8+Pt9+Pt10+Pt11+Pt12+Pt13+Pt14+Pt15+Pt16;
Pt17= -M2*S*M - M*S*M2 - M2*(M.*M2.') - (M.*M2.')*M2;
Pt18= -2*(M2*(S.*M2) + (S.*M2)*M2);
Pt19= -2*(M2*((dM2)^2) + ((dM2)^2)*M2) - (M*dM2)^2 - (dM2*M)^2;
Pt20= -2*M*((dM2)^2)*M - M*(M.*M3.') - (M.*M3.')*M;
Pt21= -2*(M*(M.*M2.*M2.') + (M.*M2.*M2.')*M);
Pt22= -M*(M.'.*M2.*M2) - (M.'.*M2.*M2)*M;
Pt23= -2*(M*(S.*M3) + (S.*M3)*M) - M*(M.*M2.')*M;
Pt24= -2*M*(S.*M2)*M - 4*(M*S + S*M);
Pt25= -M*dM2dM2M - dMdM2M2*M;
Pt26= -M*dMdM2M2 - dM2dM2M*M;
Pt27= -M*dMdM3M - dMdM3M*M;
Pt28= -M*dMdM2M*M - M2*dMdM2M - dMdM2M*M2;
Pt29= -4*(M*(dM2.*dM3) + (dM2.*dM3)*M);
Pt30= -dM2*M*dM3 - dM3*M*dM2;
Pt31= -dM2*M2*dM2 - 4*(M.*(M.'*S + S*M.'));
Pt32= -M.*(M4.') - 2*M.*M2.*M3.' - 2*M.*(M2.').*M3;
Pt33= -M2.*M2.*M2.' - 2*(M.').*M2.*M3 - 8*S.*(M*S + S*M);
Pt34= -2*S.*M4 - 12*S.*M2 - 8*S.*M2.' - 4*S2;
NEG=✓
Pt17+Pt18+Pt19+Pt20+Pt21+Pt22+Pt23+Pt24+Pt25+Pt26+Pt27+Pt28+Pt29+Pt30+Pt31+Pt32+Pt33+Pt34
;
R6 = POS + NEG
```

The arrows associated with the next-to-the-last line of code indicate continuation onto the next printed line. There is a systematic procedure for generating formulas for the Redundancy Matrix R_n for still larger values of n [6]. However, the complexity of these formulas approximately doubles each time n is increased to n + 1.

After n = 6 it is easier to generate the required matrix elements of the higher order Way Matrices W_n by the iterative procedure than by direct calculation from the R_n.

‡ When using an Excel spreadsheet matrix multiplication of 9 by 9 blocks of data can be performed as follows. Fill in one block of data in cells A1 → I9, and another block of data in cells K1 → S9. You can color cells in column j to separate the data blocks visually. Next, type the following instruction into cell A19.

$$=\$K1 * A\$1 + \$L1 * A\$2 + \$M1 * A\$3 + \$N1 * A\$4 + \$O1 * A\$5 + \$P1 * A\$6$$
$$+ \$Q1 * A\$7 + \$R1 * A\$8 + \$S1 * A\$9$$

Now, drag the corner of cell A19 to cover a 9 by 9 block. The matrix product will appear in cells A19 → I27.

7.7 Problem

Given the simple pin array below, with all possible contacts indicated by lines (edges),

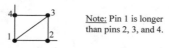

Note: Pin 1 is longer than pins 2, 3, and 4.

(a) What is the Adjacency Matrix A corresponding to this graph?
(b) Enumerate all paths three edges long by calculating A^3.
(c) How many paths exist between pin 1 and pin 2? Draw these paths.
(d) Calculate the Redundancy Matrix R_3 from Eq. 7.1.
(e) How many redundant paths exist between pin 1 and pin 2?
(f) Identify the redundant paths in the drawings of part c.
(g) Calculate W_3.
(h) Identify the ways between pin 1 and pin 2 in the drawings of part c.

References

1. Zito, R. R. (2011, August 8–12). The Curious Bent Pin problem—I: Computer search methods. In *29th International Systems Safety Conference*, Las Vegas, NV.
2. Zito, R. R. (2011, August 8–12). The Curious Bent Pin problem—II: Matrix methods. In *29th International Systems Safety Conference*, Las Vegas, NV.
3. Campbell, J. (1990). *Transformations of myth through time* (pp. 164–184). New York: Harper & Row.
4. Adkinson, R. (Ed.). (1995). *Sacred symbols* (p. 183). New York: Abrams.
5. Ouaknin, M. A. (2010). *Mysteres des Chiffres* (pp. 289–294). Paris: Assouline.

6. Ross, I. C., & Harary, F. (1952). On the determination of redundancies in sociometric chains. *Psychometrika, 17,* 195–208.
7. Parthasarathy, K. R. (1964). Enumeration of paths in digraphs. *Psychometrika, 29,* 153–1655.
8. Marshall, C. W. (1971). *Applied graph theory* (pp. 258–261). New York: Wiley.

Chapter 8
The Bent Pin Problem—IV: Limit Methods

8.1 Background

This, the fourth chapter on the bent pin problem, will deal with the limiting short-circuit behavior of large connectors during a bent pin event. In Chap. 5 [1], it was shown that for the 9-pin square array (Fig. 8.1), the probability of forming a short-circuit path from pin (1, 1) to (3, 3) was given by the finite series

$$P_{Total} = P(4) + P(6) + P(8) = 6P^4 + 4P^6 + 2P^8 = 6(0.106)^4 + 4(0.106)^6 + 2(0.106)^8$$
$$= 0.000757 + 5.67 \times 10^{-6} + 3.19 \times 10^{-8} = 0.000763.$$

$$(8.1)$$

P(4), P(6), and P(8) are the probabilities of forming paths four, six, and eight pins long, assuming random bending of all pins. The probability P is the probability of making a nearest neighbor contact for any given pin, and P is assumed to be the same for all the pins in a given chain. In the next chapter, this last assumption will be relaxed, but for now, it will be retained. The pins themselves are of the RS232 type with a length of 3 mm (or just slightly greater), a diameter of 1 mm, and a pin separation of 3 mm. The value of P was calculated to be 0.106 (or 10.6%) [1]. The probability of forming any conducting chain four bent pins long is equal to the number of paths that are four pins long (in this case there are 6) times P^4. Similarly, the probability of forming any path six pins long is $4P^6$, because there are four paths that are six pins long, etc. It is clear from the finite series of Eq. (8.1) that the first term accounts for 99.2% of the total short-circuit probability P_{Total}. Several questions naturally arise. Does the probability series for P_{Total} always terminate so rapidly? Are the shortest short-circuit paths between two pins always the most important? Can longer paths ever contribute as much, or more, to the total short-circuit probability than shorter ones (i.e., does the series for P_{Total} ever lack *monotonicity*)? Can the probability series for P_{Total} always be *truncated* after the

R. R. Zito, *Mathematical Foundations of System Safety Engineering*,
https://doi.org/10.1007/978-3-030-26241-9_8

Fig. 8.1 A 9-pin square
array. What is the probability
of forming a short-circuit
between pins at (1, 1) and
(3, 3) if pins bend randomly?

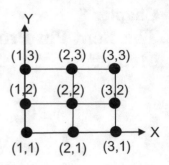

first term? Can the probability series be truncated after the Nth term with the
confidence that the finite sum is close to the true answer (*convergence*)? These are
difficult questions, but they do have answers.

8.2   The Model Infinite Connector

No connector can form more short-circuit paths between two pins then one that is
infinitely large and contains an infinite number of pins at a finite pin density ρ of n
pins per unit cell of the pin lattice. That is to say, $\rho = n/A_c$, where A_c is the area of a
unit cell. For a square array, $A_c = \ell^2$, where ℓ is the nearest neighbor pin spacing,
and n = 1, so $\rho = 1/A_c$ for a square array of pins (Fig. 8.2). Clearly, if the series for
P_{Total} is monotonically decreasing and convergent (and therefore truncatable) for
the *Model Infinite connector (MIC)*, it will be monotonic and truncatable for a large
finite connector; the two probability series being identical except for the *tail* of the
series. Suppose that a short-circuit path, N bent pins long, exists between two pins
(a safety-critical pin and a power pin) in a MIC. How many such paths (m_N), in a
MIC array of any given geometric layout, could exist? This question was answered

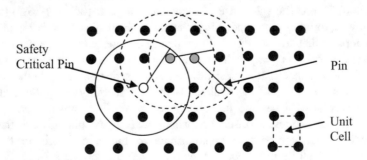

Fig. 8.2 The closed circular neighborhood above encloses **N** pins, but the unit cell contains only
four quarter pins, or one pin per cell

in the third chapter on bent pins [2]. If the safety-critical pin is labeled with index i, and the power pin is labeled with the index j, then the i, j element of the *Way Matrix* for paths of length N (i.e., $W_N(i, j)$) equals m_N; where

$$W_N = (A^N - R_N) - d(A^N - R_N).\qquad(8.2)$$

The matrix A is the *Adjacency Matrix*, R_N is the *Redundancy Matrix*, and the "d" operator retains the diagonal matrix elements of $(A^N - R_N)$ and sets all other elements to zero, as discussed in Chaps. 6 and 7 [2, 3]. For the MIC, the matrices A, R_N, and W_N have an infinite number of rows and columns. This property presents certain analytical challenges. Therefore, the question above will be relaxed. A more tractable question is, "Does an upper bound exist on the number of paths N pins long between pin i and pin j and what is its value?" The answer to the first part of this question is yes! And, the answer to the second part is as follows. Let the pin length in the array of Fig. 8.2 be $\mathcal{L}$, and assume that a pin's length is not diminished by the bending process (a model connector), then the closed circular neighborhood around each bent pin encompasses **N** (bold face type) neighboring pins, any of which can be reached to participate in chain formation during a random bent pin event. Numerically, $\mathbf{N} \approx ceil[n\,\pi\mathcal{L}^2/A_c] = ceil[\pi\mathcal{L}^2\rho]$, where the *ceil* function rounds the argument in square brackets upward to the next largest integer. It is important to note that the formula for **N** is only an *approximation* that may differ from the true value of N, especially for small values of $\mathcal{L}$, but becomes increasingly accurate as $\mathcal{L} \to \infty$.

For the square array, $\mathbf{N} \to ceil[\pi\mathcal{L}^2/\ell^2] = ceil[\pi(\mathcal{L}/\ell)^2]$. Defining the *dimensionless scaled pin length "a"* as $\mathcal{L}/\ell$, it is clear that $\mathbf{N} \to ceil[\pi\,a^2]$. So, a^2 captures the notion of pin *compactness* relative to pin length. The scaled pin length will appear many times in the subsequent discussion of connectors in the next, and last, chapter on bent pins. Now, if each bent pin can potentially make **N** contacts, and if there are N pins in a chain, then clearly, based solely on possible contacts, there can be no more than $\mathbf{N}^N$ chains; most of which will not exist because the pins may not be long enough to form a continuous path. Let P_{max} be the largest pin-to-pin contact probability of the first, second, third, etc. nearest neighbor contacts. Usually, the probability of first nearest neighbor contact will equal P_{max}. If a bent pin event occurs, it is certain that the total short-circuit probability P_{Total} for all real chains must obey the inequality

$$P_{Total} < \sum_{N=1}^{\infty} P_{max}^N \mathbf{N}^N = \sum_{N=1}^{\infty} (P_{max}\mathbf{N})^N \equiv P_{Total}(UB),\qquad(8.3)$$

where $P_{Total}(UB)$ denotes an upper bound on P_{Total}. It is important to note that Eq. (8.3) is general, and does not depend on the lattice symmetry. Equation (8.3) would apply equally as well to a hexagonal close-packed lattice of pins; in which case n = 2 (i.e., there is a central pin in the unit cell surrounded by 6 nearest neighbors contributing one-sixth of a pin each), $A_c = 3^{3/2} \ell^2 / 2$, $\rho = 4/(3^{3/2} \ell^2)$, and $N \to 4\pi \mathscr{L}^2 / (3^{3/2} \ell^2) = (4\pi/3^{3/2}) a^2$ as $\mathscr{L} \to \infty$.

8.3   Monotonicity

If the argument $P_{max} N$ of Eq. (8.3) is less than unity, each increasing term in the series will decrease in size relative to its predecessor. In that case, the series on the right side of the inequality of Eq. (8.3) is said to be *monotonically decreasing*. The *rate* of decrease can be estimated as follows. Let r be the ratio of the Nth term to the N + 1th term in Eq. (8.3). Then

$$r = (P_{max} N)^N / (P_{max} N)^{N+1} = 1/(P_{max} N). \tag{8.4}$$

Or, more generally, if path lengths increase in steps Δ, where Δ is an integer larger than unity

$$r = (P_{max} N)^N / (P_{max} N)^{N+\Delta} = 1/(P_{max} N)^{\Delta}. \tag{8.5}$$

For the square array RS232 connector described in the Background section r = 1/ $[(0.1)(4)]^2 = 6.25$. In Eq. (8.1), terms decrease by a quotient on the order of 100. The discrepancy of an order of magnitude is due to the fact that Eqs. (8.4) and (8.5) apply to a conservative *upper bound* (UB) series and not an actual series for P_{Total}.

If, on the other hand $P_{max} N > 1$, each term of Eq. (8.3) will grow relative to its predecessor. In that case, the probability series is said to be *monotonically increasing*. If the series is monotonically increasing, it will certainly diverge. And, in this case, there is no guarantee that P_{Total} will not grow as well to reach unacceptable values. That is to say, if the scaled pin length a is sufficiently large, then a bent pin event can generate so many possible conducting paths that the *upper bound short-circuit probability* $P_{Total}(UB)$ grows without bound, and P_{Total} may approach unity (i.e., a short is essentially certain). Such a connector will be said to be *unstable with respect to a bent pin event*. A plot of $P_{max} = 1/N$ is shown in Fig. 8.3, where this curve marks the boundary of the region of instability.

8.4   Convergence

In this section, proof of convergence of the series $P_{Total}(UB)$ as $N \to \infty$ will be presented. The method of attack will be the *comparison test* [4]. First, it is necessary to adjust the limits on $P_{Total}(UB)$. This task is easily accomplished by starting the probability series at N = 0 and subtracting off the first term

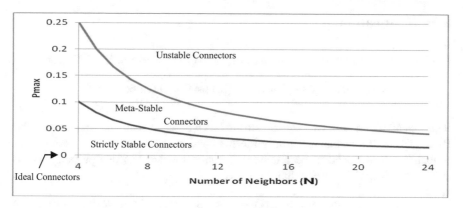

Fig. 8.3 The value of P_{max}, and the number of neighboring pins surrounding each pin, controls the stability of a connector should a bent pin event occur. Some connectors are *unstable*, in which case, there is a high probability of a short circuit between a safety-critical pin and a power pin due to a proliferation of short-circuit paths. Other connectors are stable, in the sense that short-circuit paths do not proliferate during a bent pin event. For some of these stable connectors, the probability of a short circuit is easy to calculate because it only depends on the shortest paths, and perhaps, the next longest paths. These connectors are called *strictly stable*. The short-circuit probability of the remaining stable connectors during a bent pin event is more difficult to calculate, and usually higher, than those of strictly stable connectors. These connectors are called *metastable*. *Ideal connectors* have pins that are sufficiently short to preclude contact with a neighboring pin if bent, so that $P_{max} = 0$

$$P_{Total} < \sum_{N=1}^{\infty} (P_{max}N)^N = \sum_{N=0}^{\infty} (P_{max}N)^N - (P_{max}N)^0. \qquad (8.6)$$

However, since any nonzero quantity to the zero power is unity,

$$P_{Total} < \sum_{N=0}^{\infty} (P_{max}N)^N - 1 \equiv P_{Total}(UB). \qquad (8.7)$$

The infinite series on the right is well known and, so long as $P_{max}N < 1$, it converges to yield an expression for $P_{Total}(UB)$ and an upper bound for P_{Total} [5].

$$P_{Total} < [1/(1-P_{max}N)] - 1 = P_{max}N/(1-P_{max}N) \equiv P_{Total}(UB). \qquad (8.8)$$

Since $0 < P_{Total} < P_{Total}(UB)$, the probability expansion for P_{Total} will converge. Such a connector will be called *stable*. It should be noted, however, that the upper bounding series $P_{Total}(UB)$ can converge to a number *greater* than unity. This may seem unphysical, but it must be remembered that $P_{Total}(UB)$ is an *upper bound only*, and not necessarily a physically meaningful number. In fact, as the region of instability is approached from below $P_{max} \to 1/N$ and $P_{Total}(UB) \to \infty$.

8.5   Truncation

It is instructive to compute the value of $P_{Total}(UB)$ for the example RS232 connector.

$$P_{Total}(UB) = [(0.1)(4)]/[1-(0.1)(4)] = 0.66666\ldots = 2/3. \qquad (8.9)$$

It is clear from Eq. (8.1) that strong convergence can be expected when $P_{Total}(UB) = 2/3$. In that case, only the first, and possibly the second, terms of the probability series need to be retained. A connector that has this property will be called *strictly stable*. In Fig. 8.3 the curve

$$P_{max}N/(1-P_{max}N) = 2/3 \quad \text{or} \quad P_{max} = 2/(5N) \qquad (8.10)$$

has been graphed, where P_{max} is a function of N. Ordered pairs (P_{max}, N) that lie below this curve represent strictly stable connectors. That is to say, strict stability is guaranteed if $P_{max} \leq 2/(5N)$. Above Eq. (8.10), but below the curve of instability, lies a region of *metastability*. Connectors in this region of Fig. 8.3 are stable, but computation of P_{Total} may require taking numerous terms into account. Ultimately, it will be up to the analyst to decide if he/she is *satisfied* with the answer.

For strictly stable connectors, an estimate of how close the leading term approaches the limiting probability sum can be made by examining the ratio r of the first term of Eq. (8.3) to the upper bound sum $P_{Total}(UB)$.

$$r(1^{st}\text{term/UB}) = [P_{max}N]/[P_{max}N/(1-P_{max}N)] = 1-P_{max}N. \qquad (8.11)$$

In the case of the RS232 example connector, r(1st term/UB) = 0.60 = 60%. Again, the actual rate of convergence in Eq. (8.1) is faster (i.e., the first term accounts for just over 99% of the total short-circuit probability), but it must be remembered that path lengths are increasing in steps of $\Delta = 2$ in that series. Furthermore, the value of 60% was calculated on the basis of the *upper bound* series. If the first two terms are taken into account

$$r(1st2 \text{ terms/UB}) = [P_{max}N + P_{max}^2 N^2]/[P_{max}N/(1-P_{max}N)]. \qquad (8.12)$$

In the case of the RS232 example connector, r(first two terms/UB) = 0.84 = 84%.

8.6   Conclusion and Summary

The calculations presented above indicate that so long as the values of P_{max} and N come from the strict stability region of Fig. 8.3, the analyst can truncate the short-circuit probability series with impunity after the first two terms. This is very fortunate because it greatly shortens the program of calculation. The number of shortest paths is always given by the matrix element $A^N(i, j)$, since the redundancy matrix for shortest paths is [0]. Furthermore, if the next longest path is only one bent pin longer than the shortest path, the matrix element $A^{N+1}(i, j)$ yields the proper enumeration of paths as well, because it is impossible to form a new path with a loop or a switchback if only one extra bent pin is employed [1]. It is a trivial MATLAB calculation to raise a matrix to a power. However, if tools like MATLAB are not available, it should be realized that A is a *sparse matrix*. In that case, even simple tools like EXCELL, or just a paper and pencil, suffice to make the calculation (see Appendix D).

Thus far, all calculations have assumed that a constant value of P_{max} (or more generally P), for the pin-to-pin contact probability, can be established. Actually, this is more difficult than one might suspect. And, it is the subject of the next chapter. For now, it will be stated without proof that P_{max} should be estimated from either $d/(\pi \ell)$, where d is the pin diameter, or $(1/2\pi)\cos^{-1}(\ell/\mathcal{L}) = (1/2\pi)\cos^{-1}(1/a)$, whichever is greater.[†]

In a related vein, it should be borne in mind that all the calculations in this chapter are based on a *model* connector whose pins bend *flat*. In fact, real pins when bent will possess some curvature at the root of the pin. This curvature will shorten the effective length of the pins. Therefore, a connector that might be considered metastable for model bending, may actually be a strictly stable connector. A similar caveat applies to the other connector types. A simple calculation will clarify the situation. Consider a square array of pins, each about 1 mm in diameter, 4 mm long, and separated from each other by 3 mm. As mentioned in the Background section above, $P = d/(\pi \ell) = 0.106$. But, $(1/2\pi) \cos^{-1}(\ell/\mathcal{L}) = 0.115$. As will be demonstrated in the next chapter, setting P_{max} equal to the larger of these values is tantamount to saying that "thin pin" effects are more important than "thick pin" effects for this connector. Since $2/(5N) = 2/(5(4)) = 2/20 = 0.10$, it is clear that the criterion for strict stability has not quite been met because $0.115 > 0.10$ (the inequality is the reverse. So, this connector is metastable. But, it is not unstable because $P_{max} < 1/N = 1/4 = 0.25$. However, when real connector pins bend, some of the unbent pin length is consumed by the curvature of the pin at the bend point. Suppose this effect reduces the effective value of $\mathcal{L}$ to 3.5 mm. Then, $(1/2\pi)$ $\cos^{-1}(\ell/\mathcal{L}) = 0.086$, so that "thick pin" effects have now become more important. In that case, set $P_{max} = 0.106$. The connector is now on the border between strict stability and metastability because $0.106 \sim 0.10$. Even a trivial decrease in pin width will result in strict stability. Most RS232 connectors lie on this boundary, which means that for these connectors the traditional bent pin analysis that only takes into account nearest neighbor contacts is largely correct. That is to say, the shortest conducting paths (i.e., 1 bent pin long) have the highest probability of formation by far. Of course, if the nearest power pin to a safety-critical pin is two or three bent pins away, then the probability of forming this chain will become the dominant short-circuit probability. This probability may, or may not, be small enough to ignore, depending on the particular system. But, with the addition of each link to the chain, the overall probability of short-circuit formation will decrease rapidly because each pin is not surrounded by enough neighbors **N** to support a significant amount of new path formation. However, the new high-density connectors may contain hundreds of long thin pins that are too closely packed to prevent the proliferation of paths during a bent pin event.[‡] Such connectors save space, a necessity in this high-density microelectronic age, but they lack stability. That is the unfortunate trade-off that must be made!

From a requirements and hardware assembly point of view, it may be of value to color code all connectors. Resistors are color coded with a silver or gold band, depending on their tolerance. Connectors can be color coded according to their bent pin consequences. Blue connectors could be ideal connectors, green could be reserved for strictly stable connectors, while yellow and red might be reserved for metastable and unstable connectors respectively. In safety-critical applications where life-threatening consequences are possible as a result of a bent pin event, requirements could call out the exclusive use of blue or green connectors. If only hardware damage is possible, yellow or red connectors may be permissible. Or, perhaps, the use of blue connectors might be mandated in all situations where a bent pin event can result in a Mishap Severity of "Catastrophic". Green, yellow, and red connectors would then be allowed for applications where the Mishap Severity is "Critical", "Marginal", and "Negligible", respectively. Another possibility is to associate blue, green, yellow, and red connectors with applications involving a Mishap Risk Assessment of "High", "Serious", "Medium", and "Low". Naturally, a connector of any color could be replaced by one that has a higher safety rating, but the trade-off is that the pin density will decrease. Therefore, it is most efficient to use the color code required, and not a connector with a higher rating.

The last point that needs to be discussed is a philosophical one. Most mathematicians are *Platonists*. That is to say, they believe (whether they will admit it or not) that mathematics is an idealization that can be used to describe something real (at least approximately), or can be visualized (at least approximately) in concrete terms. The MIC used in this chapter is one such idealization, an abstraction, which has no more reality than a physicist's frictionless inclined plane or weightless rope. It is a construct of the mathematician's imagination. Nevertheless, the preceding calculations tacitly assume that the MIC is a good model for a *"large"* connector. But, what does the word *"large"* actually mean? Is a connector with 9 pins large? Is a connector with 900 pins large? In the context of this chapter, *a fully populated connector is large if the greatest inscribed circle within the pin array has a diameter D that is much greater than the pin length $\mathcal{L}$ (i.e., $D \gg \mathcal{L}$).* If D is an order of magnitude larger than $\mathcal{L}$, and if $\mathcal{L}$ is approximately equal to the pin separation ℓ, a large connector would contain on the order of 100 pins. For most practical purposes a 100 pin square array is large. However, the cutoff isn't sharp because it depends on the level of descriptive accuracy demanded by the analyst. If only a rough description is satisfactory, a 30-pin connector might be considered large and properly modeled by a MIC. But, if great accuracy is required, perhaps only connectors with 300 or more pins could be modeled correctly.

Another way to define *"large"* is by sampling theory. Statistically, a sample is large (i.e., will mimic the behavior of an infinite, or very large, set of objects with "acceptable" fidelity) if the sample size is greater than about 30. For example, selecting 30 balls out of a collection of 5,000 red balls and 5,000 white balls that have been thoroughly mixed, is sufficient to establish the equality in the count of the two colors within "narrow" bounds. Mathematically speaking, the t-score values for 30 degrees of freedom (sample size—1) are close to (within 6.3%) of the t-scores for an infinite number of degrees of freedom (i.e., t-scores → z-scores as the number of degrees of freedom → ∞) [6]. Heuristically speaking, so long as a connector has at least 30 pins, the MIC is a good model for this finite connector and can be used to judge stability, metastability, or instability. Again, the cutoff isn't sharp, because a sample of 100 objects has t-scores that are even closer to z-scores. Therefore, a 100 pin connector looks even more like a MIC. As always, the analyst must decide what constitutes an "acceptable" model. The implications of the arguments, in this and the previous paragraph, concerning the definition of a large connector can be summarized by the following operational rule: *A "large" connector should contain 100 or more pins, but must contain at least 30 pins.*

† Equation 22, and the discussion that follows, of the next chapter will provide the justification for this rule.

‡ The effect can be seen in the last example by increasing $\mathcal{L}$ to $4\sqrt{2}$ mm. In that case, four second nearest neighbors can be reached in addition to the four 1st nearest neighbors, so that $N = 8$, and $1/N = 0.125$. But,

$$P_{max} = (1/2\pi) \cos^{-1}(\ell/\mathcal{L}) = 0.16.$$ And, since $P_{max} > 0.125$, this connector has become unstable!

8.7 Problem

Given a square array of pins 4 m long and having a nearest neighbor separation of 3 mm, prove that a pin diameter of 1.084 mm separates thin pin dominated bent pin behavior from thick pin dominated behavior.

References

1. Zito, R. R. (2011, August 8–12). The Curious Bent Pin problem—I: Computer search methods. In *29th International Systems Safety Conference*, Las Vegas, NV.
2. Zito, R. R. (2011, August 8–12). The Curious Bent Pin problem—III: Number theory methods. In *29th International Systems Safety Conference*, Las Vegas, NV.
3. Zito, R. R. (2011, August 8–12). The Curious Bent Pin problem–II: Matrix methods. In *29th International Systems Safety Conference*, Las Vegas, NV.
4. Rudin, W. (1964). *Principles of mathematical analysis* (p. 52). New York: McGraw Hill.
5. Gradshteyn, I. S., & Ryzhik, I. M. (1965). *Tables of integrals, series, and products* (p. 7). New York: Academic Press.
6. Fisz, M. (1963). *Probability theory and mathematical statistics* (p. 661). New York: Wiley.

Chapter 9
The Bent Pin Problem—V: Experimental Methods

9.1 Background

This, the fifth and last chapter on the bent pin problem, will deal with the frequency of bent pin events f_{BP}, and computation of the single pin contact probability P under various geometric conditions. In the previous four chapters on bent pins, it has been tacitly assumed that these fundamental numbers are either known or easily calculated. However, the problem is not as straight forward as one might think.

The frequency of bent pin events will depend on the mechanical design of the connector, the strength of materials properties of the pins, and human factors. Mechanical design factors include the number of pins in each connector, how snugly each pin fits into its female receptacle, the presence of alignment devices on the connector, the pin's width, and the thickness of the pin's metal (if hollow). Strength of materials factors associated with the buckling and bending of pins include Young's Modulus, the Shear Modulus, Poisson's ratio, and the yield points of the material. Finally, the care and patience of the assembler will affect the frequency of bent pin events, and is difficult to quantify. Clearly, with all these factors involved, an experimental, data base, approach needs to be taken to establish f_{BP}. Counting bent pin events from a database provides a *practical engineering answer*, but that answer is really only accurate for a specific type of connector and a given team of assemblers, and there is no general first principles *understanding* of the observed results. Nevertheless, the figures below justify the estimates provided for f_{BP} in the abstracts of previous work on bent pins [1]. For now, that is the best that can be done.

The calculation of the single pin contact probability P is more tractable, but very complex. Previously, it has been assumed that for a given connector design (i.e., pin spacing and pin width), the geometry of Fig. 9.1 applies [2].

In that case, P is just given by the ratio of the sum of the widths of the bent pin and the contact pin $(d_{n-1} + d_n)$ to the circumference of the circle whose radius is the pin separation distance ℓ (Eq. 9.1). Strictly speaking, this estimate

© Springer Nature Switzerland AG 2020
R. R. Zito, *Mathematical Foundations of System Safety Engineering*,
https://doi.org/10.1007/978-3-030-26241-9_9

Fig. 9.1 Geometry for an
ideal bent pin. C is the
circumference of the circle
whose radius is ℓ

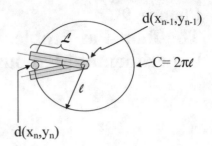

$$P \equiv P(x_{n-1}, y_{n-1}, x_n, y_n) = [d(x_{n-1}, y_{n-1}) + d(x_n, y_n)]/2\pi\ell \qquad (9.1)$$

for the single pin contact probability is correct if the bent pin can *only* contact the
"root" (bottom, base, or attachment point) of a neighboring pin *and* if bending is
nondirectional (random). For example, the last contact in a chain of randomly bent
pins is of this type, if the last pin remains straight. Another example would be the
contact probability for each pin in a straight chain of pins, provided the bending
process was truly random. In general, however, Eq. 9.1 will underestimate the
individual single pin contact probabilities contributing to the total short-circuit
probability P_s of a chain of bent pins forming a specific conducting path "s". The
real (experimentally observed) total short circuit probability of chain s is $P_{real} = f_{BP}$
P_s, where f_{BP} is the observed frequency of bent pin events. The range of values for
f_{BP} will be presented in the next section.

9.2   Observations of the Frequency of Bent Pin
Events (f_{BP})

Collected data indicates that for a hexagonal close-packed (HCP) 62-pin
D-subminiature connector, a bent pin event can be expected to occur 4.24 times
for each 1000 connections made (i.e., $f_{BP} = 0.0042$ or 0.42%). Similar statistics
apply to a 44 pin D-sub HCP connector. If a connector with the same design had
310 pins (5 times as many pins), and if the number of bent pin events is proportional to the number of pins, then 21.2 events could reasonably be expected per
1000 connections. For this large connector $f_{BP} = 0.0212$ (or 2.12%). In fact, for
large aerospace connectors having hundreds of pins, this last figure is typical, with
bent pin events being observed for 1–10% of the connections made.

A 31-pin ribbon connector, with a pin diameter half that of the D-sub connector
($\sim$0.5 mm) and 7.9 mm long, had a bent pin rate of 0.8%. Therefore, a 310-pin
connector of this same design could be expected to experience a bent pin event for

8% of the connections made. Again, this figure is a typical value observed by assemblers. So, pin diameter has a strong effect on the frequency of bent pin events.

In summary, the preliminary figures presented in this section allow *approximate* calculation of f_{BP} based on connector size and pin geometry. The next three sections will concentrate on calculation of P_s, which is simply the product of the contact probabilities P of the individual links in the chain. However, it will be found that the values of P vary from link to link, depending on the exact location of each pin in the chain s! This last effect greatly complicates calculations. Nevertheless, simplifying assumptions can be made that at least allow a reliable upper bound to be calculated for P_s.

9.3 An Instructive Children's Game

The game popularly known as "Pick-Up-Sticks" sheds some light on the true nature of a random bent pin event. The game is played as follows. First, a packet of sticks is held vertical above a horizontal surface as shown in Fig. 9.2a. Next, the packet is released so that the sticks scatter in random directions upon impact (Fig. 9.2b). In one variant of the game, each contestant tries to remove one stick each time his/her turn comes around without moving any other sticks. Moving another stick disqualifies a contestant. The last remaining contestant wins. First, the sticks on the top are removed. These sticks are not held down by the weight of other sticks, so they are like the pins described by the traditional calculation for the single pin contact probability P. What remains is a pattern like that in Fig. 9.2c. Now, if a dot (vertex) is placed at one end of these sticks, Fig. 9.2d results. If the dots are now imagined to be the roots of "bent pins," represented by sticks, then a conducting chain is formed between the first bent pin and the terminal straight pin that connects to a power source (represented by a dot contacting the side of a stick (pin 3). Notice how the sticks cross each other between their ends. Any given stick may be rotated through a considerable angle about its root and still make contact with its neighbor. Suppose an analyst wishes to calculate the probability of forming a short circuit, between the bent safety-critical pin 0 and a straight source pin 3, along the path in Fig. 9.2d. The single pin contact probability for each pin in the chain will be different. And the total short-circuit probability will be the product of these individual values of P. Like the Pick-Up Sticks in Fig. 9.2, the calculations in the next section will assume that the length of the pins is much greater than their width. That is to say, a *thin pin approximation* will be employed.

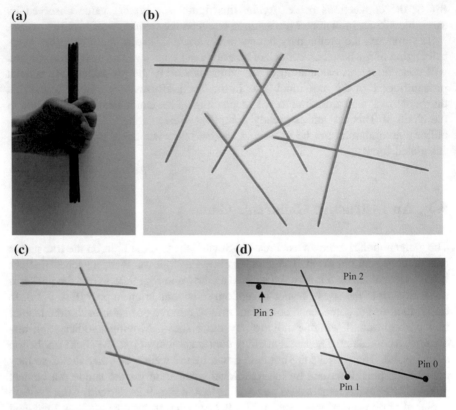

Fig. 9.2 a A packet of Pick-Up-Sticks. **b** Randomly scattered Pick-Up-Sticks. **c** A single path of Pick-Up-Sticks. **d** Pick-Up-Sticks as a conducting path of bent pins

9.4   The Thin Pin Problem (Calculating P_s)

The method of calculation to be employed for the chain of pins in Fig. 9.2d is to start at the last pin and work toward the start pin. This "backward" method allows the definition of the geometry at each contact (see Fig. 9.3a).

If pins were truly infinitesimally thin, the probability of contacting the last straight pin by the next to last bent pin would be zero. Realistically, this is incorrect. Under the assumption of random bending of identical pins, the finite contact probability is established by the width of the pins and their separation in the traditional manner. So, for this last contact the single pin contact probability, call it P_3, is given by Eq. 9.1.

$$P_3 = [d(x_3, y_3) + d(x_2, y_2)] / 2\pi\ell = d/\pi\ell, \text{ where } d(x_3, y_3) = d(x_2, y_2) = d.$$
$$(9.2)$$

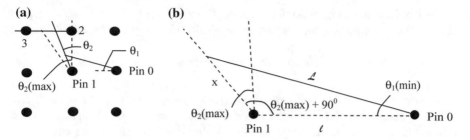

Fig. 9.3 a (left) A bridge of bent pins is formed from pin 0 (vertex 0) to pin 3 (straight). **b** (right) Geometry for the computation of $\theta_1(\min)$

Now, the calculation of the single pin contact probability for the next to the last electrical contact, call it P_2, is very different. For sufficiently thin pins, this contact probability is overwhelmingly dominated by the *sweep angle* θ_2, provided $d \ll \ell$. Therefore, random bending of pins of length $\mathcal{L}$ implies

$$P_2 = \theta_2(\max)/2\pi = (1/2\pi) \cos^{-1}(\ell/\mathcal{L}). \tag{9.3}$$

Last, P_1 must be calculated. Now, the angle θ_1 will depend on the angle θ_2. However, it is clear from Fig. 9.3b and the Law of Cosines that minimum values of θ_1 range from 0 to $\theta_1(\min)$ given by

$$\cos \theta_1(\min) = (\mathcal{L}^2 + \ell^2 - x^2)/2\mathcal{L}\ell. \tag{9.4}$$

But, what is the length of the side x? Here, it is necessary to apply the Law of Cosines again at the interior angle of vertex 1 (pin 1) when θ_2 achieves its maximum value as used in Eq. 9.3. That is to say, the upper limit of θ_1 will be a minimum when θ_2 is a maximum.

$$\mathcal{L}^2 = x^2 + \ell^2 - 2x\ell \cos(90° + \theta_2(\max)) = x^2 + \ell^2 + 2x\ell \sin(\theta_2(\max)), \tag{9.5}$$

where the identity for the sum of two angles has been used to simplify the argument of the cosine. Recollecting terms results in the second-order equation

$$0 = x^2 + (2\ell \sin(\theta_2(\max))) x + (\ell^2 - \mathcal{L}^2), \tag{9.6}$$

whose solution is given by

$$x = \tfrac{1}{2}\{(-2\ell \sin(\theta_2(\max)) \pm \sqrt{(4\ell^2 \sin^2 \theta_2(\max) - 4(\ell^2 - \mathcal{L}^2))}\}. \tag{9.7}$$

Since it is clear from Fig. 9.3(left) that $x \to \mathcal{L} - \ell$ as $\theta_2(\max) \to 90°$, only the plus sign of the last equation applies and is physically meaningful. Simplifying yields,

$$x = -\ell \sin \theta_2(\max) + \ell \sqrt{(\sin^2 \theta_2(\max) - (1 - (\mathcal{L}/\ell)^2))}. \tag{9.8}$$

It should be noted that for nearest neighbor contacts only, $\ell \le \mathcal{L} < \sqrt{2}\ell$, and since $\theta_2(\max) = \cos^{-1}((\ell/\mathcal{L})$, and since $1 = \cos^2\theta_2(\max) + \sin^2\theta_2(\max)$, this last expression for x can be transformed into an alternate form after some algebra.

$$x = -\ell(\sin \theta_2(\max) - (\tan \theta_2(\max)) \sqrt{(1 + \cos^2 \theta_2(\max))}). \tag{9.9}$$

In this form, the units of x are clearly the same as those of ℓ, as they should be. Equation 9.9 is also an occasionally useful calculation check for Eq. 9.8. Finally, after substituting Eq. 9.8 in Eq. 9.4, the minimum contact probability $P_1(\min)$ for the first pin in this chain is

$$P_1(\min) = \theta_1(\min)/2\pi = (1/(2\pi)) \cos^{-1}((1/(2\mathcal{L}\ell))(\mathcal{L}^2 + \ell^2 - (-\ell \sin \theta_2(\max)$$
$$+ \ell \sqrt{(\sin^2 \theta_2(\max) - (1 - (\mathcal{L}/\ell)^2)))^2)).$$
$$\tag{9.10}$$

For the other extreme, which occurs when $\theta_2 = 0$, the maximum values of θ_1 will range from 0 to $\theta_1(\max)$ given by

$$\theta_1(\max) = \cos^{-1}(\ell/\mathcal{L}), \text{ so that } P_1(\max) = (1/2\pi) \cos^{-1}(\ell/\mathcal{L}). \tag{9.11}$$

Under the assumption of random bending, the average value of P_1 is given by

$$P_1(\text{ave}) = [P_1(\min) + P_1(\max)]/2 = [(1/(2\pi)) \cos^{-1}((1/(2\mathcal{L}\ell))$$
$$(\mathcal{L}^2 + \ell^2 - (-\ell \sin \theta_2(\max) + \ell \sqrt{(\sin^2 \theta_2(\max) - (1 - (\mathcal{L}/\ell)^2)))^2))$$
$$+ (1/(2\pi)) \cos^{-1}(\ell/\mathcal{L})]/2,$$
$$\tag{9.12}$$

where $\theta_2(\max) = \cos^{-1}(\ell/\mathcal{L})$. For thin pins and random bending, the total probability of forming path "s" (using the naming convention of the previous chapters) in Fig. 9.3a from pin 0 to pin 3 is, to a very close approximation, given by

$$P_s = P_{0123} = P_1 P_2 P_3 = P_1(\text{ave}) P_2 P_3 = \frac{1}{2}[(1/(2\pi)) \cos^{-1}((1/(2\mathcal{L}\ell))$$
$$(\mathcal{L}^2 + \ell^2 - (-\ell \sin \theta_2(\max) + \ell \sqrt{(\sin^2 \theta_2(\max) - (1 - (\mathcal{L}/\ell)^2)))^2)) \tag{9.13}$$
$$+ (1/(2\pi)) \cos^{-1}(\ell/\mathcal{L})][(1/2\pi) \cos^{-1}(\ell/\mathcal{L})][d/(\pi\ell)]$$

when $d/\ell \ll 1$ (the thin pin approximation). At each leg of path s, the direct pin-to-pin nearest-neighbor contact probability is different. Therefore, the simplifying assumption made in Chaps. 5, 6, and 7 [3–5], viz., equal contact probability (given by Eq. 9.1) for each pin in a chain of identical bent pins with identical contact type (i.e., all nearest neighbor contacts, all second nearest-neighbor contacts, etc.), is false. Of course, a *mean* value for P (the individual link contact probability) certainly exists, although its value is complicated, and not given by Eq. 9.1. In the next paragraph, it will be shown that the simplifying assumption can be recovered, with an easy-to-calculate value of P, if the analyst is content with an upper bound on the probability P$_s$.

It has probably not escaped the readers notice that the factor $\cos^{-1}(\ell/\mathcal{L})$, or its alternate form $\theta_2(\text{max})$, appears several times in Eq. 9.13. This is no coincidence because $\mathcal{L}/\ell$ is the *dimensionless scaled pin length* (a) of the connector. As $a \to \infty$, $\ell/\mathcal{L} = 1/a \to 0$, and $\theta_2(\text{max}) = \cos^{-1}(\ell/\mathcal{L}) \to \pi/2$. Therefore, as a grows, $\theta_2(\text{max})$ also grows. That is to say, a captures the notion of *sweep angle* when pins are mashed together in a tangled mat. In fact, for a square array, and a pin length $\mathcal{L}$ such that $\ell < \mathcal{L} < (\sqrt{2})\ell$, and a conducting short-circuit path s that is N bent pins long such that all interior angles of crossed pins are obtuse ($\geq 90°$), an Upper Bound (UB) can be established for P$_s$ such that

$$P_s < P_{UB} = \{[d(x_{n-1}, y_{n-1}) + d(x_n, y_n)]/2\pi\ell\}\{(1/2\pi)\ \cos^{-1}(\ell/\mathcal{L})\}^{N-1}. \quad (9.14)$$

This formula is easy to use, generally yields a better estimate for P$_s$ than $\{[d(x_{n-1}, y_{n-1}) + d(x_n, y_n)]/2\pi\ell\}^N$, and all bent pins in the chain (except the last) now have equal direct pin-to-pin contact probability given by $(1/2\pi)\ \cos^{-1}(\ell/\mathcal{L})$. But, it is important to remember that P$_{UB}$ is only a reasonable upper bound, and not necessarily a *least* upper bound. The simplifying assumption of the first three chapters on bent pins, together with a convenient value of P, has been recovered, but at the expense of accuracy! Although derived for a square array, Eq. 9.14 also applies to hexagonal close-packed arrays. Figure 9.4 summarizes the implications of these equations.

When the pin thickness is 0.8 mm (not very thin), the ratio $d/\ell = 0.8/3 = 0.27$. This is near the upper limit of what would be acceptable as a "thin" pin. The "true" probability of forming the path 0123 via Eq. 9.13 is P$_s$ = 0.000756; where $P_{real} = f_{BP}$ P$_s$. This value is 26% larger than the traditional probability estimate based on Eq. 9.1. By the time the pin diameter drops to 0.1 mm (quite thin), the "true" probability of path formation is P$_s \approx 10^{-4}$. This value is two orders of magnitude larger than that estimated via Eq. 9.1. *The short-circuit probability due to a chain of bent pins has been greatly underestimated!* Just because a source pin and a safety-critical pin are separated by more than one pin does NOT necessarily make them safe. It is also interesting to note that the probability upper bound provided by Eq. 9.14 is 50% larger than the true thin pin probability, calculated from Eq. 9.13, over the range 0.1mm $\leq$ d $\leq$ 0.8mm (0.025 $\leq$ d/ℓ $\leq$ 0.200). If the path considered had terminated in N$_{line}$ pins in a straight line, then Eq. 9.14 could have been easily modified to yield an upper bound closer to the true value as follows:

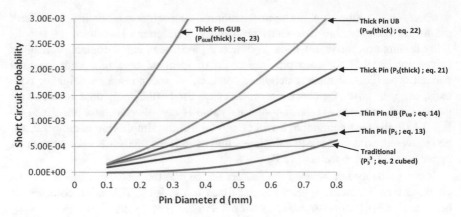

Fig. 9.4 Probability of forming a conducting path between pins 0 and 3 of Fig. 9.3a. The graph applies to a square array of pins, each 4 mm long and separated by 3 mm

$$P_s < P_{UB} = \{[d(x_{n-1}, y_{n-1}) + d(x_n, y_n)]\}/2\pi\ell\}^{Nline}\{(1/2\pi)\cos^{-1}(\ell/\mathcal{L})\}^{N-Nline}. \tag{9.15}$$

Other simple modifications of Eq. 9.14 are also possible.

The bad news with regard to bent pins isn't over. Short circuits between any two pins can follow multiple paths, and sometimes the contact probability of these multipaths, when taken together, can rival the direct 1-pin nearest-neighbor contact probability. Figure 9.5 provides an example of this effect. As usual, pin 3 is a straight power pin.

Starting from pin 3 and working backward, the probability P_3 is given by Eq. 9.1, so that $P_3 = d/\pi\ell$. And, as before $P_2 = \theta_2(max)/2\pi = (1/2\pi) \cos^{-1}(\ell/\mathcal{L})$. Now, if $\theta_2 = 0$, then θ_1 can range from 0 to $\cos^{-1}(\ell/\mathcal{L})$. But, if $\theta_2 = \theta_2(max)$, then θ_1 in Fig. 9.5 can range from 0 to $\pi/2$ radians (90°), and still make a multi-pin short between pin 0 and pin 3. Therefore, on the average,

Fig. 9.5 A 3-pin short-circuit conducting path for two nearest neighbor pins

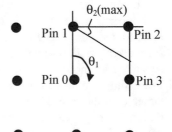

$$\theta_1(\text{ave}) = (\pi/2 + \cos^{-1}(\ell/\mathcal{L}))/2, \text{ so } P_1 = P_1(\text{ave}) = \theta_1(\text{ave})\,(1/2\pi)$$
$$= (1/2\pi)\,(\pi/2 + \cos^{-1}(\ell/\mathcal{L}))/2. \tag{9.16}$$

Therefore,

$$P_s = P_{0123} = P_1 P_2 P_3 = P_1(\text{ave})P_2 P_3$$
$$= \left[(1/2\pi)(\pi/2 + \cos^{-1}(\ell/\mathcal{L}))/2\right]\left[(1/2\pi)\cos^{-1}(\ell/\mathcal{L})\right][d/\pi\ell] \tag{9.17}$$
$$= (1/2\pi)^3(\pi/2 + \cos^{-1}(\ell/\mathcal{L}))(\cos^{-1}(\ell/\mathcal{L}))(d/\ell).$$

Note that Eq. 9.14 for P_{UB} does not apply here because of the interior acute angle $(90° - \theta_2(\text{max}))$ at pin 1. Now, there is a point to be made here. Equation 9.17 shows that for a square array of pins with a nearest neighbor separation of 3 mm, and each pin 0.8 mm in diameter and 4 mm tall, $P_s = 0.001782$;[†] about three times larger than the value calculated by Eq. 9.1. Furthermore, because there are *two* "U" shaped paths, each a reflection of the other about the line $0 \rightarrow 3$, the probability of forming either of these three pin conducting paths is about 0.0036. By comparison, the direct single pin contact probability is about 0.085. Therefore, the probability of forming a three pin short is 4.2% of the probability of forming a direct single pin contact from pin 0 to pin 3. However, a connector with pins 4 mm long that are separated by 3 mm, is a very conservative design. The problems become progressively worse for longer pins, as can be verified by slightly increasing $\mathcal{L}$ from 4 mm to, say, 4.2 mm, in which case the probability of forming a three pin short jumps to 5% of the direct contact probability. If the pin length is further increased to 5 mm, then the 3-pin short-circuit probability increases to 5.9%. Now, there is a caveat here because the rule $\ell < \mathcal{L} < \sqrt{2}\ell$ has been violated. The consequence of this violation is that only nearest neighbor pin interactions are being taken into account, even though $\mathcal{L}$ is long enough to involve second nearest neighbors. Some contributing short-circuit probability is being lost, or at least not counted. When the pin length is 7 mm, the probability of forming a 3-pin short is 7.8% of the direct nearest neighbor contact probability. But, now third nearest-neighbor interactions are being ignored as well, and it is not uncommon for high density connector pins to stretch all the way to third nearest neighbors. Given a square array and pins capable of reaching third nearest neighbors, each pin can potentially contact 12 others during random bending (four nearest neighbors, four second nearest neighbors, and four third nearest neighbors). Although there are only two possible 3-pin short-circuit chains that involve nearest neighbors, inspection, or the methods of Chaps. 5, 6, and 7 [2–4], show that there are ten 3-pin paths involving *both* first and second nearest neighbors, and 14 paths involving some sort of third nearest neighbor contact. Because of these extra 24 paths, it would not be unreasonable to multiply the 7.8% figure (which applies to 2 paths) by a factor of 12, bringing the 3-pin short-circuit probability up to about $(12 \times 7.8 + 7.8)\% = 101.4\%$ of the direct one pin nearest neighbor short-circuit probability. And, this is a conservative figure because it ignores the seven 2-pin paths which are now possible with the longer pins. Paths longer than three pins have also been ignored. When all paths are taken into account, there are so many possibilities for contact, and the sweep angles allowing contact are so great (see the Pick-Up-Sticks in Fig. 9.2), that multi-pin paths are more likely to cause a short than a single direct contact. In this case, a short between nearest neighbors is so likely that it should be considered virtually guaranteed. This is an example of an *unstable connector* [6]. Recall, from the previous chapter on limit methods, that *a connector is unstable if the maximum contact probability (P_{max}) of its pins exceeds the reciprocal of the number of neighboring pins it can reach.* Since each pin 7 mm long can reach 12 neighbors in the square array being considered, $1/12 = 0.083$. And, since $P_{max} = (1/(2\pi))$ $\cos^{-1}(\ell/\mathcal{L}) = (1/(2\pi))\cos^{-1}(3 \text{ mm}/7 \text{ mm}) = 0.18$, it is clear that $P_{max} > 0.083$. Hence, this connector is unstable. Therefore, in this case, $P_{real} \approx f_{BP}$ (1). More will be said about this topic in the summary section.

9.5 The Thick Pin Problem (Calculating P_s)

The thick pin problem involves some intricate trigonometry. The chain of pins in Fig. 9.3 will now be revisited using thick pins. First of all, it will be assumed that the tops of the pins are flat. That will simplify the calculations a little. Now, the last contact at pin 3 will have the same probability that was used for the thin pins above, so $P_3 = d/\pi\ell$. However, the contact probability between pin 1 and pin 2 is now a little more difficult to compute. The approach here will be to make small corrections to the maximum thin pin sweep angle θ_2(max; thin), where the additional arguments "thin" and "thick" will be used to distinguish thin pin and thick pin results. These corrections will be called δ_1, δ_2, and δ_3 (Fig. 9.6). If pin 1 lies to the left of the root of pin 2 (Fig. 9.6-left), and if pin 2 were thin, then the correction to θ_2(max; thin) would be $\delta_1 = \tan^{-1}(d/(2\mathcal{L}))$. If, on the other hand, pin 1 was thin and pin 2 was thick (Fig. 9.6-center), then the end of pin 1 would be depressed by an amount Δ equal to the Altitude (Fig. 9.6-center) divided by $\cos\alpha$. If pin 2 isn't *too* thick,

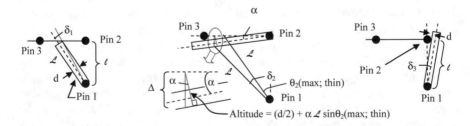

Fig. 9.6 Geometry for the computation of θ_2(max; thick) for thick pins

$\cos\alpha \approx 1$, and $\Delta \approx$ Altitude $\approx (d/2) + \alpha\mathcal{L} \sin\theta_2$(max; thin), where $\alpha = d/\ell$, so that $\Delta \approx (d/2)(1 + 2(\mathcal{L}/\ell) \sin\theta_2$(max; thin)) $= (d/2)(1 + 2a \sin\theta_2$(max; thin)). Note that for a square array with $\mathcal{L} = \ell\sqrt{2}$, $\Delta = 3d/2$, as it should. And, $\Delta = d/2$ when $\mathcal{L} = \ell$, as it should because the tip of pin 1 is depressed by the half thickness of pin 2 as θ_2(max; thin) $\to 0$. Now, Δ is also equal to $\ell - \mathcal{L} \cos(\theta_2$(max; thin) $+ \delta_2) = \mathcal{L} \cos\theta_2$(max; thin) $- \mathcal{L} \cos(\theta_2$(max; thin) $+ \delta_2) = \mathcal{L} (\cos \theta_2$(max; thin) $- \cos(\theta_2$(max; thin) $+ \delta_2)$), where again θ_2(max; thin) is the thin pin sweep angle. Expanding the cosine of the sum of two angles, and applying the approximations $\cos\delta_2 \approx 1$ and $\sin\delta_2 \approx \delta_2$, (both valid if pin 2 is not *too* thick) yields $\Delta \approx \mathcal{L} \delta_2 \sin\theta_2$(max; thin). If the two approximate expressions for Δ are equated it can be shown, after some rearrangement of factors, that $\delta_2 \approx d(1 + 2a\sin\theta_2$(max; thin))$/(2\mathcal{L} \sin\theta_2$(max; thin)), where θ_2(max; thin) $= \cos^{-1}(\ell/\mathcal{L})$. If pin 1 lies to the right of the root of pin 2, it will just make contact when the centerline through pin 1 sweeps through an angle δ_3 equal to d/ℓ. This follows from the definition of radian measure and approximating the arc length at pin 2 by the cord length d (Fig. 9.6-right). Again, this last approximation is valid if pin 2 is not *too* thick. Therefore,

$$\theta_2(\text{max; thick}) = \theta_2(\text{max, thin}) + \delta_1 + \delta_2 + \delta_3 \approx \theta_2(\text{max; thin}) + \tan^{-1}(d/(2\mathcal{L}))$$
$$+ d(1 + 2a \sin \theta_2(\text{max; thin}))/(2\mathcal{L} \sin \theta_2(\text{max; thin})) + d/\ell,$$

$$(9.18)$$

where θ_2(max; thin) $= \cos^{-1}(\ell/\mathcal{L})$. Note that as $d \to 0$, the sweep angle θ_2(max; thick) $\to \theta_2$(max; thin), as it should. Finally, P_2(thick pin) $= \theta_2$(max; thick)$/2\pi$.

Computation of P_1(thick pin) is still more difficult. First, recall the thin pin result of Eqs. 9.4 and 9.8. Since pin 0 is now thick, the expression for θ_1(ave; thin) must be modified. Since θ_1(ave; thin) $= \frac{1}{2}(\theta_1$(min; thin) $+ \theta_1$(max; thin)), θ_1(min; thin) and θ_1(max; thin) must be modified. Figure 9.7 summarizes the geometry.

The thick pin contacts are now extremely complex (Fig. 9.7-left) because θ_1(min; thick) and θ_1(max; thick) will both depend on θ_2(max; thick). But, as always, if pins are not *too* thick, so that δ_1, δ_2, and δ_3 are much smaller than

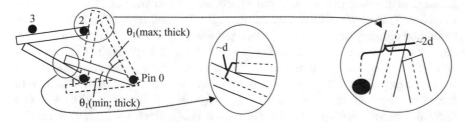

Fig. 9.7 Geometry for the computation of θ_1(min; thick) and θ_1(max; thick). On the left, dashed pin outlines indicate extreme pin positions. Dashed lines in the enlargements (center and right) mark the center of pins

θ_2(max; thin), and if $\mathcal{L}$ is only slightly larger than ℓ as in Fig. 9.7, then θ_1(min; thick) and θ_1(max; thick) can be approximated by

$$\theta_1(\text{min; thick}) \approx \theta_1(\text{min; thin}) + \delta_4 + \delta_5 \text{ and}$$
$$\theta_1(\text{max; thick}) \approx \theta_1(\text{max; thin}) + \delta_6 + \delta_7, \tag{9.19}$$

where $\delta_4 = \delta_3 = d/\ell$, $\delta_5 = d/\mathcal{L}$ (see Fig. 9.7-center), $\delta_6 = \delta_4 = \delta_3 = d/\ell$, and $\delta_7 = 2d/\mathcal{L}$ (see Fig. 9.7-right). Putting all these equations together yields

$$P_s(\text{thick}) \approx \{d/(\pi\ell)\}\{(1/(2\pi))(\theta_2(\text{max; thin}) + \delta_1 + \delta_2 + \delta_3)\}$$
$$\{(1/(2\pi))(1/2)(\theta_1(\text{min; thin}) + \delta_4 + \delta_5 + \theta_1(\text{max; thin}) + \delta_6 + \delta_7)\}. \tag{9.20}$$

After expanding the terms in { } brackets via the binomial (multinomial) theorem, ignoring terms in δ^2 or higher order, and after recollecting terms, this last equation becomes

$$P_s(\text{thick}) \approx (d/(4\pi^3\ell))[\theta_2(\text{max; thin})\theta_1(\text{ave; thin}) + \theta_2(\text{max; thin})$$
$$\{1/2(\delta_4 + \delta_5 + \delta_6 + \delta_7)\} + \theta_1(\text{ave; thin})\{\delta_1 + \delta_2 + \delta_3\}]$$
$$= P_s(\text{thin}) + (d/(4\pi^3\ell))\theta_2(\text{max; thin})\{1/2(\delta_4 + \delta_5 + \delta_6 + \delta_7)\} + (d/(4\pi^3\ell))$$
$$\theta_1(\text{ave; thin})\{\delta_1 + \delta_2 + \delta_3\} = P_s(\text{thin}) + \mathcal{O}((d/\ell)\delta_i), \tag{9.21}$$

where i = 1, 2, …, or 7, and where θ_1(ave; thin) = ½ (θ_1(min; thin) + θ_1(max; thin)). The symbol $\mathcal{O}((d/\ell)\delta_i)$ is a collection of all second-order terms (i.e., terms involving $(d/\ell)\delta_i$, where d/ℓ is small and dimensionless, and the δ_i's are all small and dimensionless). As $d \to 0$, all second-order terms are negligible so that $\mathcal{O}((d/\ell)\delta_i) \to 0$, and P_s(thick) $\to P_s$(thin), as it should. Therefore, P_s(thin) is an asymptote for P_s(thick). This phenomenon can be seen in Fig. 9.4.

The quantity d/ℓ will be called the *dimensionless scaled pin diameter b*, and is a measure of how important thick pin effects are in a connector. If d is not small relative to ℓ, thick pin effects are important. If d is small relative to ℓ, then the pin should be properly described by the thin pin problem equations above. In Fig. 9.4, the thick pin short-circuit probability is about twice as large as the thin pin probability. As in the case of thin pins, a simple upper bound exists for thick pins as well. Starting from Eq. 9.14 for P_{UB}(thin), the sweep angle must typically be increased by an amount ½($\delta_4 + \delta_5 + \delta_6 + \delta_7$) = ½(($d/\ell$) + ($d/\mathcal{L}$) + ($d/\ell$) + ($2d/\mathcal{L}$)) < ½(($d/\ell$) + ($d/\ell$) + ($d/\ell$) + ($2d/\ell$)) = $5d/(2\ell)$, since $\ell < \mathcal{L}$. If this easy to evaluate upper bound for the correction to the sweep angle θ_1 (see Fig. 9.3a) is taken as "typical", then the thick pin correction to the contact probability is $\sim(1/(2\pi))(5d/(2\ell))$. Adding this upper bound for the thick pin correction to the thin pin probability upper bound for all N − 1 bent pins yields

$$P_s < P_{UB}(\text{thick}) = \{d/(\pi\ell)\}\{(1/(2\pi))\,(5d/(2\ell)) + (1/(2\pi))\cos^{-1}(\ell/\mathcal{L})\}^{N-1}$$

$$= \{d/(\pi\ell)\}\Big\{(1/(2\pi))^{N-1}\Big[(\cos^{-1}(\ell/\mathcal{L}))^{N-1} + (N-1)(5d/(2\ell))(\cos(\ell/\mathcal{L}))^{N-2} + \cdots\Big]\Big\}$$

$$= P_{UB}(\text{thin}) + \big(5d^2(N-1)/(2^N\pi^N\ell^2)\big)(\cos^{-1}(\ell/\mathcal{L}))^{N-2} + \mathcal{O}(d^3)$$

$$= P_{UB}(\text{thin}) + \big(5b^2(N-1)/(2^N\pi^N)\big)(\cos^{-1}(1/a))^{N-2} + \mathcal{O}(b^3),$$

$$(9.22)$$

for identical pins of diameter d. In the second line of Eq. 9.22, the binomial theorem has been used to expand the first line. Ignoring terms of order d^3 (or b^3) and
higher, Eq. 9.22 can be plotted as a function of d, and is displayed in Fig. 9.4 for
comparison with the thin pin, upper bound thin pin, and approximate thick pin
calculations. Notice that P$_{UB}$(thick) is considerably larger than P$_S$(thin), P$_{UB}$(thin),
and P$_S$(thick). The term within { } brackets (first line of eq. 9.22) is an upper bound
for the contact probability of a single nonterminal link, and has a thick pin and a
thin pin part. The product $(1/(2\pi)(5d/(2\ell)))$ is just $(5/4)(d/(\pi\ell)) \sim d/(\pi\ell)$. Therefore,
the traditional pin-to-pin contact probability is approximately equal to the thick pin
effects part, as it intuitively should be. The thin pin effects part is, of course,
supplied by $(1/2\pi)\cos^{-1}(\ell/\mathcal{L})$. If either one part or the other part dominates the
probability calculations, it will form a reasonable approximation to the *actual*
contact probability of a *single link*, P. At this point, the reader may wish to review
the "Conclusion and Summary" section of Chap. 8 [6]. Equation 9.22 is a valid
overestimate so long as the pins of the chain make obtuse interior contact angles
with one another (e.g., Fig. 9.3a). In the case of acute contact angles, sweep angles
can approach 90°($\pi/2$ radians). In that case, a *Greater Upper Bound* (GUB) needs
to be established, and is given by

$$P_{GUB}(\text{thick}) = \{d(\pi\ell)\}\{(1/2\pi)(5d/(2\ell)) + \pi/2\}^{N-1}. \qquad (9.23)$$

Equation 9.23 is also plotted in Fig. 9.4 for visual and intuitive understanding of
the relationship between all these complicated formulas. In this case, since the
example path has only obtuse interior contact angles (no acute angles), P$_{GUB}$ greatly
overestimates the true value of P$_s$.

   Finally, it should be noted that the fundamental connector design constants a and
b are related through the dimensionless pin *aspect ratio* (AR), equal to the ratio of
the pin length to the pin diameter, so that A.R. = $\mathcal{L}$/d = ($\mathcal{L}/\ell$)/(d/ℓ) = a/b. A
connector designer may independently pick only two out of the three constants a, b,
and AR. The third constant will be determined from the other two.

9.6   Quasi-directional Bending

In the previous sections, random bending has been assumed. This makes sense if connectors are mated by pushing their halves together in a direction perpendicular to their faces, perhaps by means of screws. Mismated, or improperly seated, pins would then bend and buckle in random directions. The opposite of this situation is perfectly directional bending in which the pins are somehow constrained to bend in only one direction. Reality lies somewhere between these two extremes. In many cases, bent pins fall into a direction that is not well defined. For example, when coupling the two halves of a large connector, a technician will frequently inset the pins into their female apertures on one side of the connector first. Then, the technician will gently apply pressure away from this edge to seat the rest of the pins. This procedure can result in a bent pin event in which one pin falls upon another as the seating process continues. As the cylindrical surfaces contact each other, the pins undergo quasi-random bending with a preference for the "forward" direction, as defined by the locus of points at which pressure was applied, from the "back" (start) to the "front" (end). Now, if θ_i is the sweep angle for pin i, and if pins are preferentially bent into the "forward" direction (a half-disk whose radius sweeps out π radians), and if the π radians of the forward direction includes θ_i, and if bending *within* the π radians is random, then the probability of a single pin-to-pin contact is $P_i = \theta_i/\pi$. Notice that the denominator in now π, and not 2π. Therefore, the probability of a contact has increased. Suppose the bending process within the π radians is not random. What is to be done? In that case, the denominator of P_i needs to be altered, guided by experimental evidence, until an acceptable value of P_i is obtained. What if the "forward" direction is smaller or larger than π radians? Again, in that case, a value for the denominator of P_i must be selected that best matches experimental values.

9.7   The Thin FOD Problem (P_{FOD})

There are several useful short-circuit probabilities to calculate when Foreign Objects and Debris (FOD) is trapped between connector halves. First, consider Fig. 9.8(left), and suppose that the FOD is a straight conducting needle whose diameter is much smaller than the pin diameter d which, in turn, is much smaller than the pin-to-shell distance ℓ_{ps}.

Define *short FOD* to have a length ℓ_{ps} equal to the shortest pin center-to-shell distance (the sagitta) as pictured in Fig. 9.8(left). In order to short the pin to the conducting connector shell, the Center of Gravity (CG) of the FOD will have to fall into a small area between the shell and the pin that has an area equal to about half the pin cross-sectional area. The probability that the CG of randomly dispersed FOD will fall into such a small area is only $P_{c.g.} = \frac{1}{2} \pi (d/2)^2/A$, where A is the area enclosed by the connector shell. But, the orientation of the FOD must also be

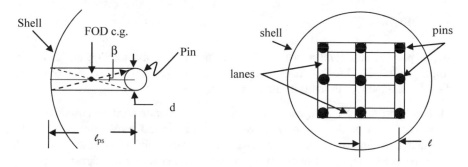

Fig. 9.8 Geometry for a pin-to-shell FOD short circuit (left). Geometry for a pin-to-pin FOD short circuit (right)

correct to form a short. Therefore, the *average* direction of the shaft of the FOD (marked by the arrowhead) must fall within the angle β, and as the FOD is rotated it passes through four angles β. Hence, the probability that the FOD will have the correct orientation is $P_o = 4 \; \beta/2\pi = (4/2\pi)\tan^{-1}((d/2)/(\ell_{ps}/2)) \approx 2d/\pi\ell_{ps}$. The approximation sign "$\approx$" is being employed because the arc tangent of a small angle is approximately equal to the angle itself in radians; a natural consequence of the thin pin approximation. The total FOD pin-to-shell shorting probability is then

$$P_{FOD} = P_{cg}P_o \approx (^1\!/_2\pi(d/2)^2/A) \, (2d/\pi\ell_{ps}) = d^3/(4\ell_{ps}A). \tag{9.24}$$

This last formula was used in Chap. 6 on bent pins [3]. The computation of the pin-to-shell shorting probability, P_{FOD}, for *long FOD* (i.e., FOD whose length $\mathcal{L}_{FOD}$ satisfies $\ell_{ps} < \mathcal{L}_{FOD} \leq 2 \, \ell_{ps}$) is considerably more difficult (see the problem at the end of this chapter), but its value is between that given by Eq. 9.24 and the approximation $P_{FOD} \approx \ell_{ps} \, d/A$. This last result might be guessed from dimensional analysis and the fact that the CG of long FOD can fall into a much larger area than that used for the computation P_{FOD} in Eq. 9.24 and still create a short. Therefore, the small area on the order of d^2 in Eq. 9.24 should be replaced by an area on the order of ℓ_{ps}^2, yielding the value of P_{FOD} for long FOD.

Finally, the probability of a pin-to-pin short by long thin FOD is calculated. Figure 9.8(right) applies. Assuming all N pins are identical, cylindrical, thin, and form a square lattice, the area of each pin lane is $[(\sqrt{N}) - 1]d \, \ell$. But, there are 6 lanes (or $2\sqrt{N}$ lanes). Therefore, $P_{cg} = (2\sqrt{N})[(\sqrt{N}) - 1]d \, \ell/A$. From the discussion of short thin FOD above, it is clear that $P_o \approx 2d/\pi\ell$. Therefore, in this case, $P_{FOD} = P_{cg}P_o \approx (4/\pi)\sqrt{N}[(\sqrt{N}) - 1]d^2/A$.

9.8 Summary

At this point it is natural to ask, "After all this detailed calculating, are there any simple 'Rules- of Thumb,' 'take-aways,' or 'lessons learned' here?" A few simple ideas for the analyst to keep in mind are as follows:

(1) It is customary for designers and fabricators to separate a safety-critical pin and a source pin by 3 or more pins. But, as demonstrated above, this may not be enough separation to ensure prevention of a potentially lethal mishap. Why? It is because long pins that can span the distance to third nearest neighbors (a common occurrence in high-density connectors) are capable of forming so many conducting multi-pin bridges, that separation by three pins offers negligible mitigation. In this case, the safety-critical and power pins might as well be next to each other! *A better criterion is separation by at least ceil(3a) pins*, where a is the connector's dimensionless scaled pin length, and the *ceil* function rounds up to the next largest integer. In a square array with 6 mm long pins and a nearest neighbor pin separation distance of 3 mm, $a = 2$. Therefore, *ceil*($3a$) = 6 pins. Clearly, a connector's scaled pin length a is a fundamental safety parameter for connector design.

(2) Separate pins in the direction of the Cartesian axis of the array so that the shortest path between pins is a straight line of N pins. The shortest path chain formation probability is then dictated by the right side of Eq. 9.1 raised to the Nth power; usually a very small number.

(3) Place dangerous pins along the edge of an array. This minimizes the number of possible multi-pin short-circuit paths between those pins.

(4) Short-circuit probabilities between pins separated by more than one pin length should be evaluated using Eq. 9.14 or 9.15, but never $(d/\pi\ell)^N$, unless N = 1 or a straight line pin path is involved.

(5) *A connector is unstable with respect to a bent pin event if the maximum contact probability (P_{max}) of its pins exceeds the reciprocal of the number of neighboring pins it can reach (this is the Connector Stability Rule)* [6]. In this case, a short between a safety-critical pin and a nearby power pin should be considered certain if a bent pin event occurs.

In Chap. 5 on bent pins, various ways of counting paths, and calculating their contribution to the short-circuit probability series, were presented. In Chap. 8 the focus was shifted to problems associated with convergence and truncation of the short-circuit probability series. Finally, this chapter was concerned with connecting calculated probabilities to shop and laboratory measurements, and with observations made in the engineering community of practice. However, in all the chapters on bent pins, there was a tacit assumption made. Namely, that once a bent pin event occurs it will involve some safety-critical pin of interest, and its local neighborhood of pins, so that a short-circuit path can form. This "worst case" assumption is pessimistic, but safe. Realistically, large connectors will usually have several safety-critical pins scattered across its pin array, so it is not unrealistic to expect that at least one of these pins and its neighbors will be affected during a bent pin event.

This chapter formally closes the solution to the bent pin problem. It is a problem of considerable interest for several reasons. First, the solution to this problem involves several branches of mathematics. And, its solution produces results that are in some measure surprising. But, most importantly, the bent pin problem is closely related to several other system safety problems of considerable interest, as will be shown in the next two chapters on sneak circuits and related system safety electrical problems. So long as systems are built up by compounding subsystems, there will always be a need for connectors. But, someday a clever engineer, perhaps using the principles discussed here, will invent a fail-safe high pin density connector. Nevertheless, the mathematical techniques presented in these last five chapters will endure and be used for other problems.

† If Eq. 9.14 is blindly applied, without regard to the restrictions imposed on interior angles, then $P_{UB} = 0.001123$. P_{UB} is now a reasonable estimate for P_s, but it is no longer an upper bound. An upper bound can be recovered if $\cos^{-1}(\ell/\mathcal{L})$ in Eq. 9.14 is replaced by $\pi/2$. This modification applies to the square array with relaxed interior angle restrictions, where sweep angles may approach 90°, and produces a *Greater Upper Bound* (GUB) that, in this case, overestimates the more exact calculation by 60%, but may considerably overestimate reality in other cases.

9.9 Problem

Prove that $P_{FOD} \approx \ell_{ps} \, d/A$ for thin FOD whose length is $2\ell_{ps}$.

Hint #1: Assume the radius of the connector is very much greater than the sagitta.

Hint #2: Assume that the allowable area within which the CG of long FOD must fall is a linear function of the angle θ between the sagitta and the FOD axis, where $0 \leq \theta \leq \pi/3$.

Hint #3: Assume the distance that is bridged by the FOD from the pin to the shell is a linear function of θ from 0 to $\pi/3$ radians.

Hint #4: Integrate over θ from 0 to $\pi/3$ radians.

References

1. Anon. (2011). *ISSC 2011 program*. ISSC, Las Vegas, NV, August 8–12, 2011, pp. 33, 38.
2. Ozarin, N. (2008, September–October). What's wrong with Bent Pin analysis, and what to do about it. *Journal of System Safety*.
3. Zito, R. R. (2011). The curious Bent Pin problem-II: Matrix methods. In *29th ISSC*, Las Vegas, NV, August 8–12, 2011.

4. Zito, R. R. (2011). The curious Bent Pin problem-I: Computer search methods. In *29th International Systems Safety Conference*, Las Vegas, NV, August 8–12, 2011.
5. Zito, R. R. (2011). The curious Bent Pin problem-III: Number theory methods. In *29th International Systems Safety Conference*, Las Vegas, NV, August 8–12, 2011.
6. Zito, R. R. (2012). The curious Bent Pin problem-IV: Limit methods. In *30th International Systems Safety Conference*, Atlanta, GA, August 6–10, 2012.

Chapter 10
"Sneak Circuits" and Related System Safety Electrical Problems—I: Matrix Methods

10.1 Background: Review of Bent Pin Analysis

Modern aerospace connectors can have hundreds of pins. When the male and female halves of such connectors are joined, a bent pin event can occur that may involve multiple pins. Many short circuit paths are possible, and the enumeration of such paths is an essential step in determining the short-circuit probability between any two pins. The enumeration of all possible "*simple*" short-circuit paths (without loops—including a self-loop—or switchbacks) between any two pins of a damaged connector, and of any length N (i.e., the number of bent pins in any conducting path), is given by (Chap. 7; [1–3]).

$$W_N = \left(A^N - R_N\right) - d\left(A^N - R_N\right). \tag{10.1}$$

The element in row i and column j (i $\neq$ j) of the matrix W_N gives the number of possible *simple* conducting paths of length N between pin i (vertex i) and pin j (vertex j) of a connector's array of pins. The two pins need not be next to each other. The "Adjacency Matrix A" is the *matrix of allowable edges* that captures the geometrical arrangement of pins in a connector. So, if a pin i can contact pin j when bent, then a 1 is entered into column i and row j of A. If pin i cannot make single pin contact, then the entry is 0. Figure 10.1 gives an example of an Adjacency Matrix for a 9-pin square array. The redundancy matrix, R_N, gives the number of paths between pin i and pin j that have loops or switchbacks. Expressions for R_N in terms of the Adjacency Matrix A can be found in the literature [3–6]. The "d" operation retains the diagonal elements of any matrix and replaces all off-diagonal elements with 0's. Notice that in Eq. 10.1, the diagonal elements have been removed from W_N, because each of these paths begins and ends at the same pin to form a *single closed loop itself*, without any other loops or switchbacks along this

© Springer Nature Switzerland AG 2020
R. R. Zito, *Mathematical Foundations of System Safety Engineering*,
https://doi.org/10.1007/978-3-030-26241-9_10

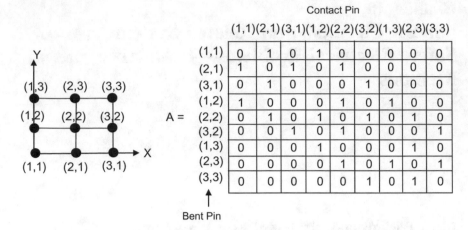

Fig. 10.1 On the left is a square lattice of nine pins with its origin at (1, 1). The corresponding Adjacency Matrix A for nearest neighbor contacts is displayed on the right. Note the high degree of symmetry in the A matrix. There are zeros all along the main diagonal, and triangular submatrices of zeros in the upper right and lower left corners

closed path. Such single loops are of no interest to the analyst for the bent pin problem because all of the involved pins are by definition removed from a power source (i.e., they are isolated) and are, therefore, not a safety threat.

10.2   Sneak Circuit Analysis

Definitions: The word "circuit," as it is generally used, means any connected network of components and wires. In what follows, such an interconnection of components and wires will be referred to as a *vague network*. A *well-defined network* (or just a *network* for short) will only include those components and wires of a vague network through which electrons can actually flow when its components are in a given *state*. The *state* of a component is a complete description of which inputs and outputs of the component are conducting, and which are nonconducting. It is important to note that one vague network may be defined by, or generate, several well-defined networks. The term *graph* will be used to mean a simplified drawing of a network schematic in which *wires or two-terminal conducting components* (e.g., resistors and other analogue components, Zener diodes in the conducting state, power supplies, etc.) are replaced by *edges*, and *electrical nodes or multiterminal solid state devices* (e.g., integrated circuits) are replaced by *vertices*. The words *edge* and *vertex* are used in the same manner in which they are used in the branch of mathematics commonly called *Graph Theory* [4]. The vertices of a network or its graph may be labeled by integers in any order convenient for the analyst. The word *circuit* will be reserved for any closed path consisting of *edges* and *vertices* that is a subset of a network. Circuits containing loops or multiple edges (switchbacks) will be called *redundant circuits*. Circuits that have no loops or switchbacks along their path will be called *simple circuits*. The number of edges and vertices in a simple circuit are equal and will be denoted by N. This is in contrast to the bent pin problem in which there is one more pin in a simple path (conducting chain without loops or switchbacks) then there are edges [1]. Note that the schematics of circuits, like networks, can also be represented by their graphs. *A "sneak circuit" is any simple circuit of a vague network that can, under certain conditions, initiate an undesired function or inhibit a desired function.* The vague network and the simple circuit can be electrical, mechanical, pneumatic, hydraulic, or a software logic flowchart. The term "certain conditions" includes, but is not limited to, the state of an electronic device (e.g., integrated circuit, transistor, diode, or switch), the unintentional use of frequencies outside operational limits, the use of discrete components with inappropriate parameters (e.g., capacitance, resistance), or an unexpected current reversal. Sneak circuits are inadvertently designed into a system because of network complexity. They are not due to hardware failures. Circuits created by parts failure will be covered in subsequent sections. In a sense, a sneak circuit is like a latent genetic defect in a living organism that can be activated to produce illness under appropriate conditions. Finally, the notion of *incidence* needs to be introduced. *The incidence of a vertex is just the number of edges that have that vertex as an endpoint.* Naturally, the higher the incidence of a vertex, the greater is the probability that it will be a part of a sneak circuit. To summarize these definitions in set notation:

$$\text{Simple circuit} \subset \text{circuit} \subset \text{network} \subset \text{vague network.} \qquad (10.2)$$

Method of Analysis: To begin, redundant circuits (circuits containing loops or switchbacks) are of no interest to the analyst. There are two basic types of redundant circuits. In Fig. 10.2a, a redundant circuit is shown in which two loops are joined at one vertex. Since a potential difference cannot exist at one conducting point, no flow of electrons can occur from one loop into the other. Therefore, this redundant circuit can be treated as two independent simple circuits. Naturally, the loop that does not involve a component of interest (such as a safety-critical device like a detonator whose terminals are marked by the two open circle vertices in this example), can be ignored, so that only a single simple circuit remains. In Fig. 10.2b a redundant circuit is shown in which two loops are joined by one edge. Now, electrons have a choice when they reach a "fork in the road" (gray vertex). They can either follow the "long road" or the "short road" back to the component of interest. That is to say, the original redundant circuit can be decomposed into two simple circuits containing the same vertices. *In fact, any redundant circuit (including a whole network) can be broken down into simple circuits.* This is the circuit analogue of the Lemma to Fitzpatrick's First Rule of the bent pin analysis [1].

When confronted with an electrical network, one of the central questions for any safety system engineer is, "How many simple conducting electrical loops contain a given safety critical part?" Ideally, that loop count should equal the number of

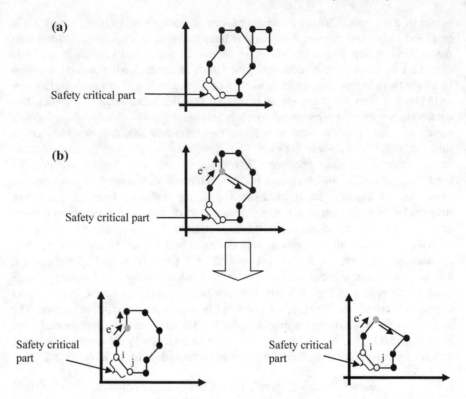

Fig. 10.2 a A redundant circuit composed of two simple circuits that share a common vertex. The "bud" can be ignored, leaving a single simple circuit. **b** A redundant circuit composed of two simple circuits that share an edge. In this case, the redundant circuit can be decomposed into two simple circuits

conditions in the analyst's mind to activate/deactivate that part. More loops mean that there may be extra circumstances, unknown to the analyst, that are capable of activating/deactivating the part in question. And, less loops might mean that some of the analyst's intended functionality can never be realized. Therefore, counting simple loops is of primary concern to the analyst. When mathematicians are confronted with a novel problem, they usually try to restate it as a problem whose solution is already known. And, that is what will be done here. It is desirable to use the Way Matrix (Eq. 10.1) above (also discussed in Chap. 7) to count loops. The Way Matrix W_N is very handy because it can be generated by a straightforward (but tedious) iterative method for loops of any "length" (number of edges) N. With the speed of modern digital computers being what it is, this method is ideal for circuit analysis. However, there is a problem. The Way Matrix is defined to have zeros on the main diagonal, so it is not possible to directly count loops that start and end at vertex i by examining the element $W_N(i, i)$. Now, the reader might think, "No problem, just ignore the d operator and examine the diagonal elements of $A^N - R_N$

to get the loop count for all vertices i." In certain cases, like N = 3, that procedure works because there is no way to produce a loop 3 edges long that contains redundancies, and the diagonal elements of R_3 are al zero[†]. Therefore, the diagonal elements of A^3 give the correct loop count, as will be seen in the second example below. However, in general, that practice will fail. It fails because the path counting procedure used to derive R_N specifically excludes paths whose partition is of the form (0, N, 0), and that is precisely the partition of a simple loop [5]. And, there is more bad news. The iterative procedure that allows the analyst to "boot-strap" up from W_N to W_{N+1} does not work for $A^N - R_N$. However, this dilemma is not as hopeless as it might at first seem. In Fig. 10.2b it is clear that counting the number of simple loops that thread through, say, vertex i, is equivalent to counting the number of nonredundant paths that start from vertex i and go to its neighboring vertex j the long way around. That is to say, the analyst need only calculate the value of $W_{N-1}(i, j)$, $i \neq j$, to determine the number of paths N edges long that contain the safety critical device between adjacent vertices i and j. The direct connection of i to j will never be counted by W_{N-1} for N − 1 >1 since all such paths are redundant by definition. Now, consider the slightly more general problem of trying to determine *all* nonredundant loops of length N passing through a vertex i that has several nearest neighbors j (not just one as in Fig. 10.2b). The allowable values (designations or labels) of j are just the column numbers (designations or labels) of the Adjacency Matrix A containing nonzero entries in row i of A. And, the total number of nonredundant loops threading through i is just $\Sigma_j W_{N-1}(i, j)$, where the summation is over all allowable j values (designations or labels).

A new matrix, the Fundamental Matrix (F_N), for paths of length N may now be defined whose diagonal and off-diagonal matrix elements are given by Eqs. 10.3 and 10.4 below.

$$F_N(i, i) = \Sigma_j W_{N-1}(i, j) \text{ for all j such that } A(i, j) \neq 0 \text{ and N} - 1 > 1, \text{ otherwise } F_N(i, i) = 0$$

$$\text{and } F_N(i, j) = W_N(i, j) = (A^N - R_N) - d(A^N - R_N) \quad \text{for } i \neq j.$$

$$(10.3 \text{ and } 10.4)$$

Since the Fundamental Matrix F_N is defined in terms of Way Matrices, it can be calculated for any value of N using the "boot-strap" iterative method. Furthermore, *given any network, the number of simple circuits N edges long that pass through vertex i of the network is given by $d(F_N)$, or $F_N(i, i)$, for all i.* Each of the simple loops counted by $F_N(i, i)$ allows current flow in one direction (from i to a given nearest neighbor j in the long direction- see Eq. 10.3). If current flow is allowed in two directions along one of these loops, this reverse circulation will be counted by the element $F_N(j, j)$ (i.e., from j to i in the long direction). A loop and its reverse are just a single *physical* simple circuit. Finally, since any directional simple circuit (directional loop) must be shared by at least one other vertex, the sum of all directional simple circuits of length N for the whole network (m_N) must satisfy the inequality

$$m_N < \Sigma_i F_N(i, i) \text{ for all } i. \tag{10.5}$$

And, the total number of directional simple circuits of any length N, and for all vertices i, is given by

$$m_{Total} = m_{Nmin} + m_{Nmin + \Delta 1} + m_{Nmin + \Delta 2} + \cdots + m_{Nmax}. \tag{10.6}$$

In this last equation, Nmin is the length (number of edges) of the shortest simple physical circuit(s), $\Delta 1$, $\Delta 2$, ...etc. increase the length of the simple circuits in integer steps (not necessarily consecutive) such that $\Delta 1 < \Delta 2 < ....$, etc., and Nmax is the length of the longest simple circuit(s). Each of the m_{Total} directional simple circuits of the network must be examined by the analyst for a possible undesired function under various conditions for a complete analysis of any given network. Of course, if the analyst is only worried about sneak circuits passing through one particular vertex because of its high incidence number, or because it is, or is directly connected to, a part that could initiate an undesired safety critical function, then it might be best to concentrate the analysis on that one vertex i. In that case, the total number of directional simple circuits, of any length N, passing through vertex i only, is given by the much smaller value of

$$m_{Total}(i) = (F_{Nmin}(i, i) + F_{Nmin + \Delta 1}(i, i) + F_{Nmin + \Delta 2}(i, i) + \cdots + F_{Nmax}(i, i)), \tag{10.7}$$

$$\text{where } m_{Total} < \Sigma_i m_{Total}(i) \text{ for all } i. \tag{10.8}$$

All of these calculations are best understood in terms of the simple examples that follow.

10.3 Analogue "Bridge Circuit" Problem

Figure 10.3 (left) shows the schematic of a fictitious "bridge circuit," which should properly be called a bridge network. In this case, since no digital components are involved, the vague network and well-defined network are the same. The schematic is composed of the symbols for wires, resistors, inductors, capacitors, and a battery and signal generator. To the right of the schematic is the equivalent graph where each node is identified by an integer from 1 to 9. Below the schematic, and its graph, is the Adjacency Matrix for the network. Suppose that the resistance connecting nodes 2 and 4 is an Electro-explosive Device (EED) for a warhead. An

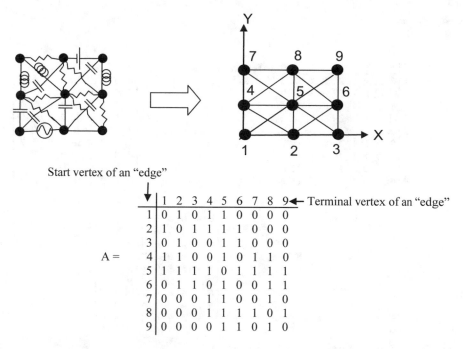

Start vertex of an "edge"

$$
A = \quad
\begin{array}{c|ccccccccc}
 & 1 & 2 & 3 & 4 & 5 & 6 & 7 & 8 & 9 \\
\hline
1 & 0 & 1 & 0 & 1 & 1 & 0 & 0 & 0 & 0 \\
2 & 1 & 0 & 1 & 1 & 1 & 1 & 0 & 0 & 0 \\
3 & 0 & 1 & 0 & 0 & 1 & 1 & 0 & 0 & 0 \\
4 & 1 & 1 & 0 & 0 & 1 & 0 & 1 & 1 & 0 \\
5 & 1 & 1 & 1 & 1 & 0 & 1 & 1 & 1 & 1 \\
6 & 0 & 1 & 1 & 0 & 1 & 0 & 0 & 1 & 1 \\
7 & 0 & 0 & 0 & 1 & 1 & 0 & 0 & 1 & 0 \\
8 & 0 & 0 & 0 & 1 & 1 & 1 & 1 & 0 & 1 \\
9 & 0 & 0 & 0 & 0 & 1 & 1 & 0 & 1 & 0 \\
\end{array}
$$

← Terminal vertex of an "edge"

Fig. 10.3 Schematic of a bridge network (top left) and its graph (top right). Wire or component crossings (edge crossings) without a black "dot" are not electrically connected (not a node). Standard electrical symbols are used for resistors, capacitors, inductors, the battery, and the signal generator in the schematic. The Adjacency Matrix (bottom) corresponds to the network graph above. Notice that the number of 1's in a row (or column) of a given vertex number is equivalent to the incidence of that vertex

analyst would like to know how many simple electrical loops contain this EED. First of all, there are no electrical *loops* one edge long that contain vertices 2 and 4. And, there are no *nonredundant* loops two edges long that contain the EED. So, the analysts next question is, "How many nonredundant simple loops 3 edges long contain the EED?" The answer to this very simple problem is given by the number of nonredundant *paths* two edges long between vertices 2 and 4; i.e., element $W_2(2, 4)$. Since there is no way to form a redundant path two edges long between two vertices i and j (in this case 2 and 4) such that $i \neq j$, Eq. 10.1 tells the analyst that $W_2(2, 4) = A^2(2, 4) = (1)(1) + (0)(1) + (1)(0) + (1)(0) + (1)(1) + (1)(0) + (0)(1) + (0)(1) + (0)(0) = 2$, where entries in the second row of A have been multiplied by corresponding elements in the fourth column of A and the result summed. That is to say, A has been matrix multiplied by itself to get the (2, 4) element of A^2. The two counted *paths* are $2 \rightarrow 5 \rightarrow 4$ and $2 \rightarrow 1 \rightarrow 4$. Therefore, the corresponding *loops* are $4 \rightarrow 2 \rightarrow 5 \rightarrow 4$ and $4 \rightarrow 2 \rightarrow 1 \rightarrow 4$. The latter loop is a sneak circuit because the impedance of a capacitor is given by $1/\omega C$, where C is capacitance, $\omega = 2\pi f$, and f is the frequency of the signal generator. Under the special conditions of high f, or a large capacitance between nodes 1 and 4, the impedance around this loop will

Start vertex
of an "edge"

$$
A^3 = \begin{array}{c|ccccccccc}
 & 1 & 2 & 3 & 4 & 5 & 6 & 7 & 8 & 9 \\\leftarrow\text{Terminal vertex} \\
\hline
1 & 6 & 11 & 6 & 11 & 16 & 9 & 6 & 9 & 6 & \text{of an "edge"}\\
2 & 11 & 12 & 11 & 16 & 20 & 16 & 9 & 12 & 9 \\
3 & 6 & 11 & 6 & 9 & 16 & 11 & 6 & 9 & 6 \\
4 & 11 & 16 & 9 & 12 & 20 & 12 & 11 & 16 & 9 \\
5 & 16 & 20 & 16 & 20 & 24 & 20 & 16 & 20 & 16 \\
6 & 9 & 16 & 11 & 12 & 20 & 12 & 9 & 16 & 11 \\
7 & 6 & 9 & 6 & 11 & 16 & 9 & 6 & 11 & 6 \\
8 & 9 & 12 & 9 & 16 & 20 & 16 & 11 & 12 & 11 \\
9 & 6 & 9 & 6 & 9 & 16 & 11 & 6 & 11 & 6 \\
\end{array}
$$

$$
R_3 = \begin{pmatrix}
0 & 7 & 0 & 7 & 10 & 0 & 0 & 0 & 0 \\
7 & 0 & 7 & 9 & 12 & 9 & 0 & 0 & 0 \\
0 & 7 & 0 & 0 & 10 & 7 & 0 & 0 & 0 \\
7 & 9 & 0 & 0 & 12 & 0 & 7 & 9 & 0 \\
10 & 12 & 10 & 12 & 0 & 12 & 10 & 12 & 10 \\
0 & 9 & 7 & 0 & 12 & 0 & 0 & 9 & 7 \\
0 & 0 & 0 & 7 & 10 & 0 & 0 & 7 & 0 \\
0 & 0 & 0 & 9 & 12 & 9 & 7 & 0 & 7 \\
0 & 0 & 0 & 0 & 10 & 7 & 0 & 7 & 0
\end{pmatrix}
$$

$$
W_3 = (A^3 - R_3) - d(A^3 - R_3) = \begin{pmatrix}
0 & 4 & 6 & 4 & 6 & 9 & 6 & 9 & 6 \\
4 & 0 & 4 & 7 & 8 & 7 & 9 & 12 & 9 \\
6 & 4 & 0 & 9 & 6 & 4 & 6 & 9 & 6 \\
4 & 7 & 9 & 0 & 8 & 12 & 4 & 7 & 9 \\
6 & 8 & 6 & 8 & 0 & 8 & 6 & 8 & 6 \\
9 & 7 & 4 & 12 & 8 & 0 & 9 & 7 & 4 \\
6 & 9 & 6 & 4 & 6 & 9 & 0 & 4 & 6 \\
9 & 12 & 9 & 7 & 8 & 7 & 4 & 0 & 4 \\
6 & 9 & 6 & 9 & 6 & 4 & 6 & 4 & 0
\end{pmatrix}
$$

Fig. 10.4 A^3, R_3, and W_3 for the network in Fig. 10.3

approach the internal resistance of the EED. The EED will then draw the current it needs from the signal generator and the warhead will detonate. This same phenomenon does not happen with path $4 \rightarrow 2 \rightarrow 5 \rightarrow 4$ because the resistance between nodes 4 and 5 limits the current. Finally, it should be noted that there are two other paths carrying current in the opposite direction. These are just counted by $W_2(4, 2) = 2$. Next, the analyst asks, "How many loops four edges long contain the EED?" The answer to this question is also very simple. It is just $W_3(2, 4)$. By Eq. 10.1, computation of W_3 requires A^3 and R_3, the latter being computed from Eq. 10.1 of Chap. 7. A^3, R_3, and W_3 are all displayed in Fig. 10.4. Notice that the element $W_3(2, 4) = W_3(4, 2) = 7$ (i.e., $2 \rightarrow 1 \rightarrow 5 \rightarrow 4$, $2 \rightarrow 3 \rightarrow 5 \rightarrow 4$, $2 \rightarrow 6 \rightarrow 5 \rightarrow 4$, $2 \rightarrow 6 \rightarrow 8 \rightarrow 4$, $2 \rightarrow 5 \rightarrow 7 \rightarrow 4$, $2 \rightarrow 5 \rightarrow 8 \rightarrow 4$, $2 \rightarrow 5 \rightarrow 1 \rightarrow 4$). Continuing the analysis for simple loops 5 through 9 edges long yields 19, 35, 37, 18, and 2 physical circuits respectively, for a total of 119 physical circuits to be inspected (see problems 1 and 2 at the end of this chapter).

Next, consider the more general problem of an analyst who wishes to detect *all* sneak circuits in the network of Fig. 10.3, starting with the shortest loops of length 3 ($N = 3$). If the network contains a sneak circuit, it is likely that a vertex with a high incidence will be part of it. Vertex (node) 5 seems particularly troublesome because its incidence is 8, the largest in the network. Hence there are 8 entries of unity in the 5th row and 5th column of the Adjacency Matrix in Fig. 10.3. The sneak circuit analysis begins by examination of all the shortest simple circuits that pass through vertex 5. Therefore, it is necessary to calculate the element $F_3(5, 5)$. Since vertex 5 makes a direct connection with vertices 1, 2, 3, 4, 6, 7, 8, and 9, it is clear that $F_3(5, 5) = W_2(5, 1) + W_2(5, 2) + W_2(5, 3) + W_2(5, 4) + W_2(5, 6) + W_2(5, 7) + W_2(5, 8) + W_2(5, 9)$. Again, since there is no way to form a redundant path two edges long between two vertices i and j such that $i \neq j$, the off-diagonal elements of the Way Matrix W_2 are just the same as those of A^2. Therefore, $F_3(5, 5) = A^2(5, 1) + A^2 (5, 2) + A^2 (5, 3) + A^2 (5, 4) + A^2 (5, 6) + A^2 (5, 7) + A^2 (5, 8) + A^2 (5, 9)$. Calculating these elements of A^2 from A follows the elementary method used to calculate $A^2(2, 4)$ above. Therefore, $F_3(5, 5) = 2 + 4 + 2 + 4 + 4 + 2 + 4 + 2 = 24$ directional loops. What do these loops look like? They are

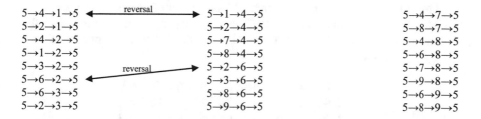

However, each one of these loops has its counterpart in the reverse direction (two reverse paths are marked above), and since the network in Fig. 10.3 does not contain any rectifying elements, there are really only 12 *physical* loops. These 12 paths are $5 \rightarrow 2 \rightarrow 1 \rightarrow 5$, $5 \rightarrow 4 \rightarrow 1 \rightarrow 5$, $5 \rightarrow 4 \rightarrow 2 \rightarrow 5$, $5 \rightarrow 4 \rightarrow 7 \rightarrow 5$, $5 \rightarrow 4 \rightarrow 8 \rightarrow 5$, $5 \rightarrow 7 \rightarrow 8 \rightarrow 5$, $5 \rightarrow 8 \rightarrow 9 \rightarrow 5$, $5 \rightarrow 8 \rightarrow 6 \rightarrow 5$, $5 \rightarrow 9 \rightarrow 6 \rightarrow 5$, $5 \rightarrow 6 \rightarrow 3 \rightarrow 5$, $5 \rightarrow 6 \rightarrow 2 \rightarrow 5$, and $5 \rightarrow 3 \rightarrow 2 \rightarrow 5$. Once all the simple circuits have been enumerated and identified. Each must be checked for possible initiation of an undesired function, or inhibition of a desired function. This is actually a rapid process because the simple circuits are just that—SIMPLE! Inspection shows that $5 \rightarrow 2 \rightarrow 1 \rightarrow 5$ is a sneak circuit because the impedance of a capacitor is given by $1/\omega C$, where C is capacitance, $\omega = 2\pi f$, and f is the frequency of the signal generator. Under the special conditions of high f, or large total capacitance around the loop, the signal generator will be shorted and fail.

To determine the number of directional loops 3 edges long that thread through the other vertices, the system safety analyst will want to calculate the other diagonal matrix elements of F_3. These are given by

$$F_3(1, 1) = W_2(1, 2) + W_2(1, 4) + W_2(1, 5) = A^2(1, 2) + A^2(1, 4) + A^2(1, 5) = 2 + 2 + 2 = 6$$
$$F_3(2, 2) = W_2(2, 1) + W_2(2, 3) + W_2(2, 4) + W_2(2, 5) + W_2(2, 6) = 2 + 2 + 2 + 4 + 2 = 12$$
$$F_3(3, 3) = W_2(3, 2) + W_2(3, 5) + W_2(3, 6) = 2 + 2 + 2 = 6$$
$$F_3(4, 4) = W_2(4, 1) + W_2(4, 2) + W_2(4, 5) + W_2(4, 7) + W_2(4, 8) = 2 + 2 + 4 + 2 + 2 = 12$$
$$F_3(6, 6) = W_2(6, 2) + W_2(6, 3) + W_2(6, 5) + W_2(6, 8) + W_2(6, 9) = 2 + 2 + 4 + 2 + 2 = 12$$
$$F_3(7, 7) = W_2(7, 4) + W_2(7, 5) + W_2(7, 8) = 2 + 2 + 2 = 6$$
$$F_3(8, 8) = W_2(8, 4) + W_2(8, 5) + W_2(8, 6) + W_2(8, 7) + W_2(8, 9) = 2 + 4 + 2 + 2 + 2 = 12$$
$$F_3(9, 9) = W_2(9, 5) + W_2(9, 6) + W_2(9, 8) = 2 + 2 + 2 = 6.$$

These matrix elements, plus the previously calculated value of $F_3(5, 5)$, give the number of 3-edge directional loops that thread through vertices 1 through 9. The total number of 3-edge *directional* loops that must be examined by the systems safety analyst for sneak circuits is $6 + 12 + 6 + 12 + 24 + 12 + 6 + 12 + 6 = 96$ loops. However, many of these loops will be identical, even with regard to direction, because any given loop is made up of several vertices, and each vertex of that loop will count that same loop again as one of its directional simple loops. Duplicate directional loops can be eliminated from sneak circuit analysis by the analyst. The sum of the diagonal elements, of F_3 is called the *trace* of F_3. And, it should be noted that this *trace* will ultimately pick up loops in both directions; a consequence of the fact that the schematic of Fig. 10.3 has no rectifying elements. Now, since the off-diagonal elements of F_3 are just given by W_3, the full Fundamental Matrix for $N = 3$ can be displayed.

$$F_3 = \begin{pmatrix} 6 & 4 & 6 & 4 & 6 & 9 & 6 & 9 & 6 \\ 4 & 12 & 4 & 7 & 8 & 7 & 9 & 12 & 9 \\ 6 & 4 & 6 & 9 & 6 & 4 & 6 & 9 & 6 \\ 4 & 7 & 9 & 12 & 8 & 12 & 4 & 7 & 9 \\ 6 & 8 & 6 & 8 & 24 & 8 & 6 & 8 & 6 \\ 9 & 7 & 4 & 12 & 8 & 12 & 9 & 7 & 4 \\ 6 & 9 & 6 & 4 & 6 & 9 & 6 & 4 & 6 \\ 9 & 12 & 9 & 7 & 8 & 7 & 4 & 12 & 4 \\ 6 & 9 & 6 & 9 & 6 & 4 & 6 & 4 & 6 \end{pmatrix}.$$

Note that for this case ($N = 3$), F_3 is just equal to $A^3 - R_3$. As mentioned above, this is not generally the case for any value of N. However, for $N = 3$, since there is no way to produce a loop 3 edges long that contains redundancies, and since the diagonal elements of R_3 are all zero[†], this simple prescription works. This point will be further clarified by problem 3 at the end of this chapter that counts directional paths of length $N = 4$. The loop analysis just presented must, of course, be repeated for simple circuits of length $N = 4, 5, 6, ..., 9$, with each step becoming more mathematically difficult and less obvious. For example, the 6-edge loop $1 \rightarrow 2 \rightarrow 5 \rightarrow 9 \rightarrow 8 \rightarrow 4 \rightarrow 1$ may also prove problematic for the same reason as loop $5 \rightarrow 2 \rightarrow 1 \rightarrow 5$. Notice that there is a battery (*ideal*-zero resistance) between vertices 8 and 9. Every *real* seat of EMF has *some* internal resistance, but if that resistance is small in the forward and/or backward direction of electron flow, the signal generator may be in peril. Some bench testing may be needed here. The point of the example in Fig. 10.3 is that there is a systematic procedure for detecting *every* sneak circuit.

10.4 "Digital Circuit" Problem

The next problem involves the vague network of Fig. 10.5 (left). In this vague network the state of the microelectronic device is such that there is no current flow through resistor r_2 from the Signal Generator (SG). Hence r_2 can be eliminated to produce the well-defined network of Fig. 10.5 (center). The simplified graph equivalent is shown in Fig. 10.5 (right). There are only two vertices and three edges in this graph. Hence, the Adjacency Matrix for this graph is just given by

$$A = \begin{pmatrix} 0 & 3 \\ 3 & 0 \end{pmatrix}. \tag{10.9}$$

Clearly, simple circuits of this well-defined network can only be two edges long. So, A^2 is of interest, and is easily calculated to be.

$$A^2 = \begin{pmatrix} 9 & 0 \\ 0 & 9 \end{pmatrix}. \tag{10.10}$$

Now, there are only three redundant "loops" (really just switchbacks) of length 2 that start from vertex 1. One loop runs from vertex 1 to vertex 2 along the edge with resistor r_1 and back. The next runs from vertex 1 to 2 via the edge with capacitor C and back. And, the last runs from 1 to 2 along the edge with the signal generator and back. Of course, analogous paths exist for vertex 2. So, the redundancy matrix must look like

$$R_2 = \begin{pmatrix} 3 & 0 \\ 0 & 3 \end{pmatrix}. \tag{10.11}$$

In this case, all matrix elements, *including diagonal elements*, are correct in spite of the fact that simple loops belong to the (0, N, 0) partition. Again, this is a special

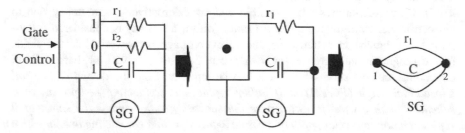

Fig. 10.5 A digital vague network (left) with one of its well-defined networks (center) and corresponding graph (right)

case in which *all* the matrix elements of the Fundamental Matrix are given by $A^N - R_N = A^2 - R_2$. Therefore,

$$F_2 = \begin{pmatrix} 6 & 0 \\ 0 & 6 \end{pmatrix}. \tag{10.12}$$

Since the $F_2(i, i)$ count current loops in both directions, dividing the diagonal elements of F_2 by 2 yields 3 simple *physical* circuits that start and end at vertex 1. What do they look like? Starting from vertex 1, move along the edge with r_1 to vertex 2 then return via capacitor C. Or, go $1 \rightarrow r_1 \rightarrow 2 \rightarrow SG \rightarrow 1$. Or, go $1 \rightarrow C \rightarrow 2 \rightarrow SG \rightarrow 1$. Three analogous loops also exist for vertex 2. The analyst must examine each loop for a sneak circuit. And, of course, the last simple physical circuit starting from vertex 1, and the analogous simple circuits for vertex 2, has the potential to destroy the signal generator under the conditions that ω or $C \rightarrow \infty$ and with the state of the microelectronic gates as shown in Fig. 10.5. It should be noted that as the analyst's focus shifts from one vertex to the next, the same simple circuit may be examined from a different starting point (as in this last example). These revisited simple circuits can be quickly eliminated once the analyst is aware of their existence.

10.5   Fail Closed/Fail High Analysis

Definitions: An analogue component is said to fail closed if its impedance becomes temporarily or permanently equivalent operationally to a conducting wire (i.e., impedance $\rightarrow 0$ relative to the component's nominal value). A digital component is said to fail high if one or more of its binary outputs incorrectly enters the high state ("1"), either temporarily or permanently. Typically, an analogue part "shorts" permanently, but intermittent failure is also possible. Failures can be due to conducting foreign objects or debris external to the device. Or, the failure can be internal, as in a transistor (amplifier) operated beyond its breakdown voltage (causing avalanche) [7, 8]. Similarly, digital components may fail temporarily, (e.g., a cosmic ray induced bit upset—a "*soft failure*"), or permanently (a "*hard failure*"). Only soft failures are reversible and nondestructive. Intermittency due to permanent hardware damage (e.g., a crack within a device) is considered a hard failure. Avalanche in a transistor due to overvoltage or powerful electric fields (electromagnetic interference, or EMI) can be either a soft or hard failure, depending on how successfully heat can be removed from the device [7]. *A Fail Closed/Fail High (FC/FH) fault either replaces a conducting component of a schematic with a wire, in which case the network graph remains the same, or it replaces a nonconducting path of a vague network with a conducting one, in which case the graph of one or more well-defined networks will change to another. Either case is capable of initiating an undesired function or inhibiting a desired function.*

Therefore, the FC/FH circuit problem is similar to the sneak circuit problem. Of course, multiple FC/FH events can occur simultaneously as well, but this is very unlikely unless the failure of one part induces the failure of another. *Let the Fundamental Matrix of the FC/FH network be called* $F_N^{(FC/FH)}$, *then the total number of directional simple circuits N edges long within this network* (m_N) *is less than the trace of* $F_N^{(FC/FH)}$, *written* $m_N < tr\,[F_N^{(FC/FH)}]$, *and the number of directional simple circuits that pass through vertex i of the FC/FH network is equal to* $F_N^{(FC/FH)}(i,\,i)$. The proof of this theorem follows directly from the analogous theorem for sneak circuits (see Eq. 10.5). As in the sneak circuit case, each simple circuit (repeats can be ignored) must be checked for initiation of an undesired function (the usual case), or inhibition of a desired function.

"Analogue Bridge Circuit" Problem: Consider the previous example in Fig. 10.3. Furthermore, suppose the capacitor between vertices 1 and 5 fails closed (fails conducting). Then the graph of the FC/FH network remains the same as that shown in Fig. 10.3. However, the analysis that follows changes because a capacitor has now been replaced by a conducting wire. Since the reciprocal of the total capacitance, of two capacitors in series, is equal to the sum of the reciprocal of the individual capacitances, it is clear that the removal of one *series* capacitance in simple circuit $5 \to 2 \to 1 \to 5$ will increase the total capacitance of that simple circuit. Therefore, the impedance of the simple circuit will decrease. Therefore, the alternating current flow through both the signal generator and the remaining capacitor in the loop will increase. Hence, the latter two parts will be at risk of an overload. As this example shows, a particularly nasty aspect of a FC/FH event is that it can cause other parts to become overloaded and fail. This cascade can spread until either a fuse is blown, or enough parts fail open/fail low (e.g., the signal generator or capacitor), to limit the damage. Effectively, an FC/FH event may initiate a "self-limiting network infection."

"Digital Circuit" Problem: Suppose the microelectronic device in Fig. 10.5 fails so that all outputs are high. In that case, the signal through resistor r_2 is a "1," and both the vague and well-defined networks are identical. The corresponding graph now has 4 edges connecting vertices 1 and 2. The matrix A^2 now has 16's along the main diagonal, and the Redundancy Matrix has 4's along the main diagonal. Therefore, the Fundamental Matrix now has 12's on the main diagonal and zeros elsewhere. Dividing 12 by 2 gives 6, the number of simple physical circuits that start and end at vertex 1. As in the sneak circuit analysis above, the simple circuit $1 \to C \to 2 \to SG \to 1$ imperils the SG under the additional conditions that ω or $C \to \infty$.

10.6 Fail Open/Fail Low Analysis

Definitions: A Fail Open/Fail Low (FO/FL) fault removes a conducting component, input/output, or a conducting wire, from one well-defined network , thereby changing it to another well-defined network , or rendering the original well-defined network inoperative. A FO/FL fault is capable of initiating an undesired function or inhibiting a desired function. Another, more compact, way to state this definition is simply to say that *an FO/FL fault removes an edge, or edges, from a network graph.* The fail-open condition in analogue components is due to a hardware failure, and constitutes permanent damage even if intermittent. The fail low condition (failure into the "0" state) of a digital integrated circuit may be either hard or soft. In practice, an FO/FL fault usually inhibits a desired function. Again, this problem is similar to the sneak circuit problem so that *the number of remaining directional simple circuits , N edges long, after failure is given by the diagonal elements of $F_N^{(FO/FL)}$, where $F_N^{(FO/FL)}(i, i)$ is the number of directional simple circuits that pass through vertex i.* The remaining simple circuits after an FO/FL fault may, or may not, include a part whose function is desired (or necessary). Therefore, the analyst must check each remaining simple circuit to see if all safety-critical network function is maintained after failure. An FO/FL event is analogous to breaking a bone and losing the functionality of, say, an arm. It should also be noted that an open parts failure in one simple circuit may cause excessive current in neighboring simple circuits with which it shares part of its path.

"Analogue Bridge Circuit" Problem: Continuing on with the analogue network of Fig. 10.3, it is instructive to examine all simple circuits that pass through vertex 9, which is a node for the battery return electron current. Element $F_3^{(FO/FL)}(9, 9)$ of the Fundamental Matrix for Fig. 10.3 is 6. Division by 2 yields 3 simple physical circuits of interest. These are $9 \rightarrow 5 \rightarrow 6 \rightarrow 9, 9 \rightarrow 8 \rightarrow 6 \rightarrow 9$, and $9 \rightarrow 5 \rightarrow 8 \rightarrow 9$. If the inductor between vertices 6 and 9 fails open, the battery will be removed from two simple circuits of length 3. Inspection of longer simple circuits reveals a similar result. Therefore, the battery function (DC source) may be lost if this inductor fails open. Therefore, a fail-safe systematic procedure exists that is capable of evaluating the consequences of a fail open.

"Digital Circuit" Problem: Continuing on with the vague network of Fig. 10.5, suppose resistor r_1 fails open. In that case, the resulting graph of the well-defined network contains only two edges that link two vertices. This is the simplest simple circuit possible. It consists of a source (SG) and a single load (C). The microelectronic device is just a gate (switch). The matrix A^2 is just a 2×2 matrix with 4's on the main diagonal (0's elsewhere), R_2 is also 2×2 and has 2's on the main diagonal (0's elsewhere), and F_2 is, again, 2×2 with 2's on the main diagonal (0's elsewhere). Dividing the diagonal matrix elements of F_2 by 2 yields 1's. That is to say, there is one simple circuit that starts from vertex 1, goes around the loop, and ends back at vertex 1. And, there is one simple circuit that starts at vertex 2, goes around the loop, and ends back at vertex 2. Clearly, this is true! There isn't much to analyze here, except to say that the safety of the SG depends only on the magnitudes of frequency f and capacitance C, and the probability of an FC fault developing on capacitor C.

10.7 Summary

Table 10.1 summarizes the solutions to the bent pin, sneak circuit, and failed parts analysis. Here, the word "solution" means a matrix (Fundamental Matrix) whose entries enumerate all possible paths between any two vertices (nodes or pins). In the special case where the start and terminal vertices (nodes) are the same, a circuit is formed, and the diagonal elements of the solution apply to simple circuits. *Notice that the solution to the bent pin problem utilizes the off-diagonal elements of the Fundamental Matrix, whereas the sneak circuit, fail closed/fail high, and the fail open/fail low problems all utilize the diagonal elements. That is to say, the bent pin problem is complementary to the other three.* The matrix elements that are discarded for the bent pin problem are precisely the elements needed to solve the other three. This curious fact is not at all obvious. It is a deep consequence of the general mathematical formalism that has been developed to solve all of these system safety electrical problems. To use the language of Category Theory, the bent pin problem is the *dual* of the sneak circuit problem. *Duality* is the observation that σ is true for some category C (of problems) if and only if σ^{op} is true for C^{op}; where "op" denotes "opposite".

In addition to the classical system safety problems above (which may collectively be referred to as network "disease"), the mathematical formalism presented in this chapter suggests and lends itself to new types of system safety analyses that have not been generally employed to date. For example, it might be useful to ask, "What is the number of simple paths between any two vertices in a network?" The greater the number of paths, the higher will be the probability of a sneak circuit. Even though this is a "circuit problem", it is phrased more like a bent pin problem topologically. Therefore, the off-diagonal elements of the Fundamental Matrix provide the number of simple paths between the start vertex (row) and the end vertex (column) (see line 5 of Table 10.1). Another useful question is, "How many simple paths exist between two vertices in a network if one part fails closed or fails high?" Again the solution is given by the off-diagonal elements of $F_N^{FC/FH}$, where $F_N^{FC/FH}$ applies to a well-defined network of a vague network with the failure in place (line 6 of Table 10.1). Finally, there is the problem of counting the number of

Table 10.1 System safety electrical problems are mathematically related

Type of analysis	Solution
Bent Pin	$F_N(i, j)\ i \neq j$
Sneak Circuit	$F_N(i, i)$
Fail Closed/Fail High	$F_N^{(FC/FH)}(i, i)$
Fail Open/Fail Low	$F_N^{(FO/FL)}(i, i)$
Number of paths between two circuit nodes	$F_N(i, j)\ i \neq j$
Number of Fail-Closed/Fail-High paths between two nodes	$F_N^{FC/FH}(i, j)\ i \neq j$
Number of Fail-Open/Fail-Low paths between two nodes	$F_N^{FO/FL}(i, j)\ i \neq j$

simple paths between two vertices in a network when a part fails open or low (line 7 of Table 10.1).

At this point, it is natural to ask a few philosophical questions. Why is it that the mathematical tools used by system safety and reliability engineers to solve electrical circuit problems seem, on the surface, so utterly different from the tools used by circuit designers? And, why do the solutions look so different? Is there any common ground? The answer to these questions lies in the motivations behind the two disciplines (i.e., safety and reliability engineering versus electrical engineering). Circuit designers are interested in *currents and voltages* in a network. Therefore, loop and node equations, which can also be arranged into matrices to solve for currents and voltages [9, 10], make sense. But, safety and reliability engineers are primarily interested in *simple paths and simple circuits*. Counting (enumerating) paths and circuits are what it's all about in safety and reliability. *As the number of simple paths and simple circuits proliferates, safety and reliability disintegrates.* So, although both approaches to circuit analysis can eventually be reduced to matrix tools, the meaning of the matrices, and their handling, is utterly different. In that sense, this chapter, its successor (Chap. 11), and three of its predecessors (Chaps. 5, 6, and 7; [1–3]), establish new methods for system safety circuit theory.

† The Redundancy Matrix R_3 for paths three edges long will always have zeros on the main diagonal. See problem 3 of Chap. 1.

10.8 Problems

(1) Compute the matrices W_4, W_5, and W_6, for the example in Fig. 10.3. Select the (2, 4) element of each and draw the corresponding paths.
(2) For the example in Fig. 10.3, build the entire matrix W_7 from the j W_6 matrices, each corresponding to a given A[j] for all j. What is the value of the (2, 4) element?
(3) In the first paragraph of Sect. 10.3 of this chapter it was stated that there are two nonredundant loops nine edges long that pass through the EED between vertices 2 and 4. Draw these loops.
(4) A given graph has Adjacency Matrix

$$A = \begin{pmatrix} 0 & 1 & 0 & 1 & 1 & 0 \\ 1 & 0 & 0 & 1 & 1 & 0 \\ 0 & 1 & 0 & 0 & 1 & 1 \\ 1 & 0 & 1 & 0 & 1 & 0 \\ 1 & 1 & 0 & 1 & 0 & 0 \\ 0 & 1 & 1 & 0 & 1 & 0 \end{pmatrix}.$$

(a) Fill in the edges of this graph on the vertices below. Use arrows to indicate the direction of "flow".

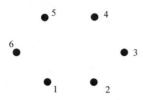

(b) Calculate A^2, A^3, and A^4.
(c) Calculate R_3 and W_3.
(d) Calculate R_4.
(e) What is the difference $A^4(1, 1) - R_4(1, 1)$?
(f) Calculate the number of directional loops 4 edges long that pass through vertex 1 [i.e., Calculate $F_4(1, 1) = W_3(1, 2) + W_3(1, 4) + W_3(1, 5)$].
(g) Draw the directional loops through vertex 1.
(h) What does the difference between parts e and f tell you?

References

1. Zito, R. R. (2011, August 8–12). The Curious Bent Pin problem—I: Computer search methods. In *29th International Systems Safety Conference*, Las Vegas, NV.
2. Zito, R. R. (2011, August 8–12). The Curious Bent Pin problem–II: Matrix methods. In *29th International Systems Safety Conference*, Las Vegas, NV.
3. Zito, R. R. (2011, August 8–12). The Curious Bent Pin problem—III: Number theory methods. In *29th International Systems Safety Conference*, Las Vegas, NV.
4. Marshall, C. W. (1971). *Applied graph theory* (pp. 258–261). New York: Wiley.
5. Ross, I. C., & Harary, F. (1952). On the determination of redundancies in sociometric chains. *Psychometrika, 17,* 195–208.
6. Parthasarathy, K. R. (1964). Enumeration of paths in digraphs. *Psychometrika, 29,* 153–1655.
7. Cowles, L. G. (1966). *Analysis and design of transistor circuits* (pp. 259–262). Princeton, NJ: Van Nostrand.
8. Lo, A. W., et al. (1955). *Transistor electronics* (p. 299). Englewood Cliffs, NJ: Prentice-Hall.
9. Desoer, C. A., & Kuh, E. S. (1969). *Basic circuit theory* (pp. 381–392). New York: McGraw Hill.
10. Slepian, P. (1968). *Mathematical foundations of network analysis*. New York: Springer.

Chapter 11
"Sneak Circuits" and Related System Safety Electrical Problems-II: Computer Search Methods

11.1 Background

Traditionally, sneak circuit analysis was accomplished by reducing a circuit to a topological "tree" format, and then identifying patterns within this tree that lead to common sneak circuit failures [1]. Many modern automated sneak circuit software packages still work on this principle. Although the automation greatly speeds up the analysis process and eliminates human errors, available tools still focus primarily on analog circuits. Digital effects such as timing problems and incorrect logic are not considered part of the sneak circuit problem. And, there is no provision for handling the related failed parts problems (e.g., memory failure). The search methods presented here depend on game theory decomposition of any type of electrical network (analog or digital) into the simple circuits (loops) that have been counted by matrix methods described in the previous chapter on sneak circuits and related system safety electrical problems [2]. Each easy-to-understand simple circuit is then presented to the analyst for inspection, or evaluated via artificial intelligence, for its sneak circuit and/or parts failure risk potential. No reference is made to historical sneak circuit patterns or information, and the "tree" concept has been completely discarded. In these ways, the sneak circuit analysis methods presented in this chapter and its predecessor, employ new methods.

11.2 Definitions

In this section, the basic definitions used in the previous chapter on sneak circuits will be reviewed. In general, a "circuit", and its schematic diagram, will include solid-state digital devices with multiple inputs and outputs. Such a "circuit" (the word is being used here in its colloquial sense) will be referred to as a *vague network* because a sneak circuit analysis cannot be conducted on such an ill-defined

© Springer Nature Switzerland AG 2020
R. R. Zito, *Mathematical Foundations of System Safety Engineering*,
https://doi.org/10.1007/978-3-030-26241-9_11

"circuit". However, once the *state* of each digital device is specified, so that the electrically active inputs and outputs are all known, then a *well-defined network* (or just a *network* for short) results. Many possible networks can correspond to any given vague network, and each network must be examined independently. Once all the well-defined networks are known, a simplified representation, called a *graph*, is constructed from the schematic for each network. Graphs are made up of lines (called *edges*) and dots (called *vertices*). The solid-state devices are replaced by vertices, as are all electrical nodes. And, electrically active input and output lines of solid-state devices are represented by edges. Two-terminal analog conducting devices (resistors, capacitors, batteries, etc.) are also replaced by edges. However, the resulting graph is still too complex for analysis. So, each network is further broken down into *simple circuits* (or *loops*) which are equivalent to electrical loops in traditional circuit analysis involving loop and node equations. Simple circuits have no loops of their own. It is simply a chain of edges and vertices that closes on itself so that the start and end vertices are identical. It is now possible to define a sneak circuit. *A "sneak circuit" is any simple circuit of a vague network that can, under certain conditions, initiate an undesired function or inhibit the desired function.* Finally, the Graph Theory concept of the *incidence* (or *degree*) of a vertex will prove useful. *The incidence of a vertex is the number of edges that it contacts.* And, a new definition, *the normalized incidence (or normalized degree) is the incidence of a vertex divided by the total number of vertices in the network graph G, and will be denoted by $r_i(v)$ for any vertex v.*

11.3 Elements of Game Theory with Three Examples

A "Game of No Chance" is a game in which probabilistic events are not involved. "Victory" depends only on strategy. Locating all sneak circuits in a well-defined network is an example of a Game of No Chance. One either wins or loses. In discussing the Bent Pin Problem [3], an analogy was made between the strategy of Vandermonde, used to find a continuous simple circuit covering all squares on a chessboard by a Knight, and the strategy used to find all possible short circuit simple paths between any two pins during a connector bent pin event. In this section, Vandermonde's strategy will be revisited for three games, but only the results, and the mathematical properties of the solutions, will be presented. Later sections will exploit the Vandermonde strategy in detail for finding simple circuits in a network.

Figure 11.1a summarizes the situation for Chess. Allowable moves are defined by

$$(x, y) \rightarrow \quad (x \pm 1, y \pm 2) \quad \text{or} \ (x \pm 2, y \pm 1). \tag{11.1}$$

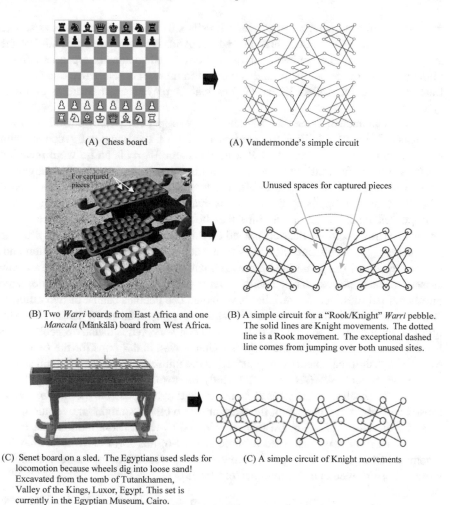

(A) Chess board

(A) Vandermonde's simple circuit

(B) Two *Warri* boards from East Africa and one *Mancala* (Mănkălă) board from West Africa.

(B) A simple circuit for a "Rook/Knight" *Warri* pebble. The solid lines are Knight movements. The dotted line is a Rook movement. The exceptional dashed line comes from jumping over both unused sites.

(C) Senet board on a sled. The Egyptians used sleds for locomotion because wheels dig into loose sand! Excavated from the tomb of Tutankhamen, Valley of the Kings, Luxor, Egypt. This set is currently in the Egyptian Museum, Cairo.

(C) A simple circuit of Knight movements

Fig. 11.1 a Chessboard in a starting configuration (left; from Pixabay at http://pixabay.com) with Vandermonde's simple circuit (right) for a Knight. **b** East African *Warri* boards (left) with a simple circuit for a pebble piece (right). **c** Ancient Egyptian Senet board (left) with a simple circuit for the dome or drum pieces (right). In this last case, the circuit can be completed with Knight moves only. For all these games (**a**, **b**, and **c**) each position (square or cup) is marked by a vertex. All vertices are covered once by a piece before returning to the start, so the incidence of each vertex is 2

The Knight may take three steps in each move; one in the x-direction and two in the y-direction, or vice versa. Of course, the moves cannot extend beyond the edges of the board, so there are boundary conditions on the Knight's movements as well. Figure 11.1a-right shows the *unique* simple circuit a Knight can follow to cover all squares. So, there is no place where an opponent can hide. A Knight can eventually track down, and destroy, an opponent who sits still. Now, a Chessboard is square with 8 by 8 (or 64) squares (Fig. 11.1a-left). And, it might be suspected that the ability of a Knight to complete a circuit that covers all squares is dependent on the high symmetry of the board. With this thought in mind, it is instructive to discuss an East African game called *Warri* (Wărē) that is played on a board with lower symmetry.

Warri is an ancient game that originated in the upper Nile valley of Sudan during the Kush (Nubian) Civilization; contemporaries of the ancient Egyptians, with whom they traded and were often at war. The name *Warri* is an Ijo word meaning "houses" (as in "factions" of a feud) that has become part of the Swahili language. The board is shown in Fig. 11.1b-left. The two heads represent the two opponents, who also represent opposing forces in this dualistic world in which we live—pairs of opposites. And, it is natural to think that the pieces, which are just pebbles, might occupy starting positions near each head on this rectangular board. Actually, this is *not* the case. The opponents sit across from each other in the narrow direction and, in one version of the game (and there are many), each pebble cup in the long row closest to each player is filled with several pebbles. Each player has the same number of pebbles, but that number is variable. The pebbles can be of two different colors, one for each player, but that is an unnecessary refinement that is generally ignored. Furthermore, the length of the board can vary, but is usually 8 pebble cups for the long rows closest to the players (outer rows), and 7 cups for the two inner rows, with "captured" pieces being placed in the unused curved quadrilateral spaces of the two inner rows (see Fig. 11.1b-left) for a total of 30 pebble cups. Now, moves in *Warri* allow three steps to be taken, so that two possible motions resemble those of a Knight in Chess. But, there are also two other "straight" moves along the x- and y-directions of the board that are allowed as well. These motions are like a "Rook" in Chess, only restricted to three steps. So, this form of *Warri* somewhat resembles Chess, with each piece being a kind of Rook/Knight—a *very* complex game. These movements are summarized by Eq. 11.2 below.

$$(x, y) \rightarrow \underbrace{(x \pm 3, y) \text{ or } (x, y \pm 3)}_{\text{Rook—like moves}} \text{ or } \underbrace{(x \pm 2, y \pm 1) \text{ or } (x \pm 1, y \pm 2)}_{\text{Knight's moves}} \qquad (11.2)$$

Whatever move a player makes, he captures an opponent's pebble in the terminal pebble cup. Now, it is natural to ask, "Can a pebble be moved in a simple circuit so as to reach every cup?" The answer is *yes*. The pattern of movements is depicted in Fig. 11.1b-right. Furthermore, this simple circuit is *not* unique, another can be formed by rotating the pattern of Fig. 11.1b-right by 180° (i.e., applying the rotation operator C_2). This phenomenon will be seen again when electrical circuits are discussed below. Clearly, there is a relationship between the lower half of the "Knight's Tour" pattern in chess (Fig. 11.1a-right) and the Warri pattern (Fig. 11.1b-right). However, the symmetry has been partially upset due to the removal of one cup from each of the inner rows. The fact that the removal has been staggered is even more symmetry breaking. When a player gets to a curved quadrilateral used for captured piece storage, he simply jumps over it and proceeds to the next nearest cup in the x- or y-direction, violating the rule of Eq. 11.2 (top-center of Fig. 11.1b-left). Just as in Chess, there is no place for an opponent to hide if he sits still. One last example of a game with a rectangular (non-square) board will be useful; the ancient Egyptian game of *Senet*.

No one is really certain how the game of *Senet* was played, although there are two leading theories [4–6]. These suggestions are based on the notion that *Senet* was played much like Backgammon, a game that probably originated in Iran about 3000 BC as evidenced from excavations in the Middle East. However, an examination of a *Senet* board, and many (~40) have been found in tombs, reveals the truth (Fig. 11.1c-left). First of all, the board is rectangular, always being three rows wide, and usually 10 rows long, with 30 squares total. The squares themselves were all white, inlaid with ivory imported from Nubia. The similarity of the *Senet* board to the modern *Warri* board is striking; even more so because ancient Egypt had so much trade contact with the kingdoms of East Africa. However, it is difficult to know if *Warri* was the predecessor of *Senet*, or vice versa. There are two types of pieces, hemispherical or bullet-shaped pieces, and drum-shaped pieces. The color of all pieces was normally white; unless all pieces had the same shape, in which case they were black and white as in the Senet set in the Nubian Museum, Aswan, Egypt. And, although there was the same number of pieces of each type, that number was variable, ranging from as few as 5 to as many as 10. *Warri* suggests that the pieces were initially lined up along the opposite long rows with opponents sitting across from each other in the narrow direction. The *Senet* board depicted on the wall of the Tomb of Nefertari (1253 BC) in the Valley of the Queens, Luxor, Egypt implies this arrangement, but Nefertari is sitting in profile at the short end of the board so as not to obscure the pieces [7]. It must always be remembered that ancient Egyptian art portrayed objects in their most recognizable orientation, regardless of reality [8]. Hence, there are images of men with two left arches and two right hands, etc. In cases when the board is made of alternating ebony and ivory squares, one type of piece may have been placed on the black squares and the other type on the white squares. Sometimes, a few squares have hieroglyphs inscribed on them. For example, one square might have three *ankh* signs, where the *ankh* symbolizes "life". Another square might have an eye of Horus (God of time), or a royal falcon. It is as if the Egyptians provided a kind of "safe" square for pieces because they realized that a stationary piece had no place to hide from an opponent. Perhaps you could use the *ankh* (life) zone three times, perhaps you could be under the protection of Horus, or pharaoh, once. There is one other aspect of *Senet* that deserves mention. *Senet* was not *purely* a game of strategy. An element of chance was introduced by the use of sticks that had one curved side and one flat side. Perhaps these represented the "heads and tails" of a "coin toss" (the Egyptians had no coins until they were introduced by invading foreigners late in their history). Such an arrangement may have been used to decide the number of moves for each player, or who would make the first move. *Under the likely assumption that Warri and Senet are related*, with differences being confined primarily to minor modifications of board shape, the use of "safe" squares, and the use of "head/tail" sticks, *one might reasonably guess that the rules of motion for Senet are also given by* Eq. 11.2. As usual, the all-important question is, "Do these rules of motion permit reaching every square with a simple circuit?" The question is important for electrical circuit theory, and game theory as well. Again, a simple circuit can be found. The pattern is shown in Fig. 11.1c-right. There is no place for a stationary piece to hide on this board unless, of course, the piece is resting on a "safe" square! Also, notice that the beautiful symmetry of the circuit pattern has returned due to the symmetric nature of the playing board. Clearly, a second simple circuit covering all board positions can be formed by rotating the first pattern by 180° (applying C_2), or the reflection operator σ across a horizontal line defined by the central row of vertices. Finally, it should be noted that some archaeologists would disagree with this analysis of *Senet* because it is not based on Backgammon. The author's position is that Backgammon is not an indigenous game of East Africa, where the Egyptian Empire was located, but *Warri* is. In any case, this academic debate makes little difference for this chapter, since these games are intended to exhibit certain features of game theory that will be useful for the network analysis of subsequent sections.

To summarize, the pieces of each of the three games above, in spite of different board shapes and rules, could be moved along a simple circuit such that each vertex is visited, but no vertex is *re*visited. Why is that so? Unfortunately, there is no simple answer to this question but, in part, it is because the imaginary vertices at the center of every playing board square, or pebble cup, form a *lattice*. More precisely, *when the vertex set N(G) of a graph G consists of k-tuples of integers from a Euclidian k-space, G is called a lattice graph* [9]. The following theorems explain a lot, and will prove useful in the next section. First, *"Given a lattice whose vertices are connected by nearest neighbor edges, a simple circuit containing all vertices exists if, and only if, the number of vertices in G is even"*. Now, the squares of Chess and *Senet* form a lattice and are linked by edges longer than the nearest neighbor distance, but the theorem above [10] suggests that it is no accident that these games, each of which has an even number of squares, has a simple circuit that covers the board. The situation for *Warri* is more complex, the board doesn't quite form a regular lattice owing to the two missing pebble cups of the inner rows. However, a simple circuit can still be formed because the rules have exceptions and, again, the even number of cups helps. Furthermore, *"Let G be a graph with m vertices, and let u and v be any pair of non-edge connected vertices. Then a simple circuit containing all m vertices exists if the incidences (or degrees d) of all u and v satisfy $d(u) + d(v) \geq m$."* That is to say, if the vertices of a graph are sufficiently connected together by enough edges ($r_i(u) + r_i(v) \geq 1$), then a simple circuit that contains all vertices is guaranteed to exist. It is important to realize that this last theorem [11] provides a *sufficient* condition and not a *necessary* one. Therefore, a graph may violate this last theorem and still have a simple circuit containing all m vertices. This is the case with the three board games above. A corollary of the last theorem that was first proved by Dirac [9] is often useful when studying networks,

"Given a graph G with m vertices, a simple circuit involving all m exists if the incidence of each vertex is not less than m/2." That is to say, if $r_i(v) \geq \frac{1}{2}$, for all vertices v, then a simple circuit exists that contains all m vertices.

11.4   How the Computer Search Works for a Well-Defined Network

Suppose it is desired to find all simple circuits that contain vertex (1, 1) of the well-defined network of Fig. 11.2 (left). Vertex (1, 1) is important because it is connected to a signal generator that must be protected. This network can be modeled by the graph in Fig. 11.2 (right). Redundant paths with multiple edges will not be allowed since any double edge simply means that electrons in the redundant circuit have returned to the same vertex; stationary electrons are unphysical. And, a triple edge just means that electrons have progressed one step to the next vertex; this case is automatically covered by any simple circuit with one edge between each vertex. Comparable arguments can be made for still higher order multiple edges. From the discussion of board games above, it is natural to label each vertex with coordinates (x, y) and demand that electrons progress according to the rule:

$$(x, y) \rightarrow \underbrace{(x \pm 1, y)}_{x \text{ motion}} \quad \text{or} \quad \underbrace{(x, y \pm 1)}_{y \text{ motion}} \quad \text{or} \quad \underbrace{(x \pm 1, y \pm 1)}_{\text{diagonal motion}} \qquad (11.3)$$

Furthermore, for simple circuits, no vertex can be revisited except, of course, the starting vertex at which time one search is finished. Finally, there are boundary conditions to model. These are very simple for the symmetrical circuit of this example. Namely, $x \leq 3$ and $y \leq 3$. In general, however, specification of boundary conditions can be very complex. Now, there is a caveat here. The game cannot end on the first move, because then electrons are stationary—not allowed! Nor can the starting vertex at (x_0, y_0) be revisited on the second move, because that

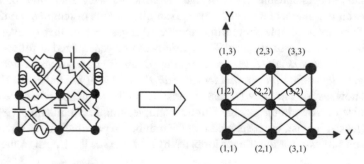

Fig. 11.2 A Bridge Network schematic (left) and its graph (right)

would be a double edge. Again, electrons would remain at the start—not allowed! At least three moves must be taken before terminating this "Game of No Chance".

As shown in Fig. 11.2 (right), starting from the vertex at (1, 1), electrons can progress to (1, 2), (2, 2), or (2, 1). Next, from (1, 2), there is a conducting path to (1, 1), (2, 1), (2, 2), (2, 3), and (1, 3). Wait! The vertex at (1, 1) cannot be revisited. So, there are now only four possibilities. From (2, 1) electrons can move to (1, 1). But, that is the end of one search, because a circuit has been formed with the starting point that is three edges, or more, long. This first simple circuit has vertices at (1, 1), (1, 2), (2, 1), and finally a return to (1, 1). Continuing on in this manner, all loops can be identified. There are a total of 3 simple circuits that start and end at (1, 1) and are 3 edges long. No account is taken of the same three simple circuits in the reverse direction. The other two simple circuits are (1, 1) → (1, 2) → (2, 2) → (1, 1), and (1, 1) → (2, 1) → (2, 2) → (1, 1). Notice that for these last two simple circuits, one can be generated from the other by interchanging the x and y coordinates. A similar phenomenon was observed in the Bent Pin Problem due to the high symmetry of the example's graphical representation. These results can be checked by matrix methods. As shown in the previous chapter [2], $F_3(1, 1)/2 = A^3(1, 1)/2 = 6/2 = 3$, as expected.

There are also 7 simple circuits that start and end at (1, 1) and are 4 edges long. They are (1, 1) → (1, 2) → (2, 2) → (2, 1) → (1, 1), (1, 1) → (2, 1) → (1, 2) → (2, 2) → (1, 1), (1, 1) → (1, 2) → (2, 1) → (2, 2) → (1, 1), (1, 1) → (2, 2) → (2, 3) → (1, 2) → (1, 1), (1, 1) → (2, 1) → (3, 2) → (2, 2) → (1, 1), (1, 1) → (1, 2) → (1, 3) → (2, 2) → (1, 1) and (1, 1) → (2, 1) → (3, 1) → (2, 2) → (1, 1). And, there are 17, 35, 53, and 50 unique physical simple circuits 5, 6, 7, and 8 edges long, respectively, that start and end at (1, 1). Although this network (unlike the game boards previously discussed) contains an odd number of vertices (9) there exists a simple circuit 9 edges long that involves all 9 vertices. In fact, there are 16 such simple circuits. Why? It's because the vertices of the network's graph are so highly connected with edges. In fact, all the edges, except the four on the corners of the network, satisfy the Dirac criterion. And, even for these corner vertices $r_i(1, 3, 7, or\ 9) = {}^1/_3$, which is close to the value of ½ needed to *guarantee* a simple circuit containing all 9 vertices; which, of course, must also contain 9 edges. Furthermore, as in *Warri* and *Senet*, all of these 16 patterns are related to one another by rotation and reflection operators. Hence, the result for a simple circuit containing 9 edges is at least heuristically understandable. If the Bridge Network had been less connected by, say, removing all the diagonal elements, then no simple circuits of length 9 would exist. Finally, there are no simple circuits that involve 10 edges because then one vertex will have to be revisited—*not allowed*! These results can be generalized as follows, *given a well-defined network with N vertices, the longest simple circuit contains either N − 1 or N edges, depending on the incidence of all the nodes in the network.* Thus, the analyst knows when and why (at least heuristically) the sequence of simple circuits will terminate. Now, in the Bridge Circuit example, simple circuits involving 10 or more edges can't exist, but it seemed that the number of possible simple circuits was tapering off anyway for the longer loops. Why? At first, as the length of the simple circuit increases, the number of possible options for circuit shape grows. However, at some point, for a finite graph of vertices, it becomes progressively harder to find free vertices to form new simple circuits, and their number begins to drop. Recall that a similar phenomenon occurred in the treatment of the Bent Pin Problem [4].

11.5 A More General Procedure

The problem with the procedure above is that it requires the vertices of a network's graph to form a lattice. Although this is usually the case for the pins of a connector in a bent pin problem [3] it will, in general, not be true for a network's vertices. In that case, the rules of electron motion (Eq. 11.3), and the search boundary conditions, will quickly become very complicated. To remedy this difficulty, a hybrid search method will now be employed that uses the matrix methods of the previous chapter on sneak circuits [2], together with the search strategy above. First, it will be important to simplify notation by relabeling the vertices of the graph of Fig. 11.2 (right). Notice that in Fig. 11.3 (left) the connection between vertices and coordinates has been completely broken.

Each vertex has simply been assigned a sequential integer, but the order is irrelevant and letters could have been used instead of numbers. Next, it will be necessary to build an Adjacency Matrix A (Fig. 11.3-right). Recall, from the previous chapter on sneak circuits, that the adjacency matrix captures which vertices are directly connected to each other in a network. So, for example, vertex 1 (row 1) is connected to vertices 2, 4, and 5 by edges. These connections are denoted by placing a 1 in columns 2, 4, and 5 of the first row. The remaining elements of the first row are zeros. The rest of the matrix can be filled in accordingly. Suppose, as in the previous section, all simple paths 3 edges long are required. Starting at node 1 (vertex 1) it is legal to progress to vertices 2, 4, or 5. Now what? From vertex 2 it is, in principle, legal to progress to any vertex labeled 1, 3, 4, 5, or 6 because row 2 has

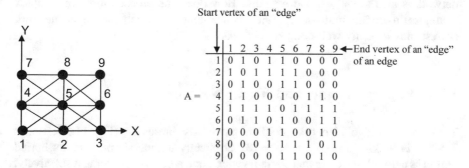

Fig. 11.3 The graph of Fig. 11.2 with simplified labeling of vertices (left), and its Adjacency Matrix A (right)

Fig. 11.4 Summary of possible loops that are 3 edges long. Only 3 of these 6 loops are physically unique

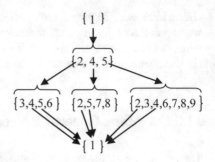

a 1 in columns 1, 3, 4, 5, and 6. But, wait! Vertex 1 must be eliminated because returning to the start on the second step is not allowed. So, the set of possible vertices for the second step after leaving vertex 2 is {3, 4, 5, 6}. If vertex 4 had been the departure point, then an inspection of the Adjacency Matrix indicates that the second step could have been to any vertex in the set {2, 5, 7, 8}. Again vertex 1 is not in this last set for the reason previously stated. Finally, if the first step had been to vertex 5, then the second step would have been to the vertices in the set {2, 3, 4, 6, 7, 8, 9}. From these last 3 sets, how will the circuit be closed for the third and last step? From the first set (i.e., {3, 4, 5, 6}), vertices 4 and 5 can both lead back to 1. From the second set, vertices 2 and 5 can lead back to 1. From the third set, vertices 2 and 4 can lead back to 1. Therefore, the six possible loops are $1 \rightarrow 2 \rightarrow 4 \rightarrow 1$, $1 \rightarrow 2 \rightarrow 5 \rightarrow 1$, $1 \rightarrow 4 \rightarrow 2 \rightarrow 1$, $1 \rightarrow 4 \rightarrow 5 \rightarrow 1$, $1 \rightarrow 5 \rightarrow 2 \rightarrow 1$, and $1 \rightarrow 5 \rightarrow 4 \rightarrow 1$. Figure 11.4 summarizes all this information. However, an inspection of these simple circuits shows that three of them are just the reverse of the other. Hence, as in the previous section, there are only 3 independent simple circuits.

In this generalized method, the Adjacency matrix, and the information it contains, has replaced the "rules of motion" and the "boundary conditions" of the previous method. All the simple circuits that pass through vertex 1, as revealed by the previous method, can also be discovered by this adjacency matrix method. However, the Bridge Network of this example has the property that its vague network is also a well-defined network. How does one proceed to more complex examples involving multi-state digital components in which the vague network corresponds to many well-defined networks?

11.6   Expert Systems and Vague Networks

The analog Bridge Network problem previously presented (Fig. 11.2-left) has $3 + 7 + 17 + 35 + 53 + 50 + 16 = 181$ physically unique simple circuits buried within its network structure that contain vertex 1 and are 3–9 edges long, respectively. Each of these simple physical circuits, displayed in Fig. 11.5 up to 6 edges, is a potential sneak circuit. Each must be evaluated for unexpected consequences.

Simple circuits of length 3:

Simple circuits of length 4:

Simple circuits of length 5:

Simple circuits of length 6:

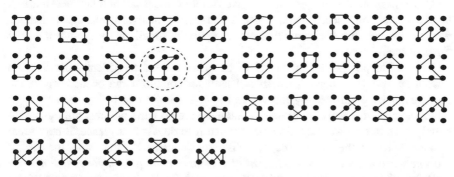

Fig. 11.5 The physically unique simple circuits that involve vertex 1 (i.e., lower left corner) of the Bridge Network, sorted according to length up to 6 edges long and topology. Edge crossings without a vertex are conductive "jumpers", and not nodes. Of all these 62 simple physical circuits, $1 \to 2 \to 5 \to 1$ is certainly a sneak circuit. However, $1 \to 2 \to 5 \to 9 \to 8 \to 4 \to 1$ may be a sneak circuit if the battery between vertices 8 and 9 is ideal (i.e., zero resistance). Both of these simple circuits are circumscribed by a dashed circle. Even this "elementary" problem can be amazingly complex

If survival of the signal generator on edge $1 \to 2$ is a safety-critical issue, a simple automated impedance check around each simple circuit reveals that $1 \to 2 \to 5 \to 1$ is a sneak circuit in the limit as $\omega \to \infty$. This problem is easily cured by the introduction of a fuse, or resistor, in series with the signal generator, should it be driven inadvertently into high-frequency operation. If this process of analysis and correction is continued for longer paths up to and including length 9, the Bridge Network would be free of sneak circuits. However, if the Bridge Network example also contained digital solid-state devices, then all the simple circuits of *each* well-defined network would have to be examined for sneak circuits. Here is where expert systems can help. One simplifying factor is that *only those simple circuits that contain a part, or parts, that directly initiate, or retard, a safety-critical function may need to be examined by the system safety engineer.*

Upon visual presentation of a candidate circuit to the analyst, the system needs to ask the question, "Are you aware that this same simple circuit occurs $n^{(1)}$, $n^{(2)}$, $n^{(3)}$, … times within the well-defined networks described by the sets $S^{(1)}$, $S^{(2)}$, $S^{(3)}$, … of digital device states?" If, for example, the Bridge Network schematic contained 10 gates, then the Bridge could be in 1024 possible states, each being represented by a well-defined network described by a set $S^{(\#)} = \{i_1, i_2, …, i_{10}\}$, where each of $i_1, i_2, …i_{10}$ are either 1 or 0 depending on the state of the gates 1, 2, …10 respectively in the vague network, and the superscript in brackets (#) is the well-defined network numerical label. Once the analyst is presented with *parallel functionality* to activate/deactivate a safety-critical device/function, the analyst must decide if this parallel functionality is wanted. If not, then one or more sneak circuits have been found. These will have to be tagged by color coding and eliminated from further consideration. Of course, if no other parallel functionality has been detected by the expert system, then no circuits need to be presented to the analyst for a given safety-critical part or function at this first level of inspection. Clearly, expert systems can save the analyst a great deal of tedious work. However, sneak circuits that produce parallel functionality are not the only types of sneak circuits.

Next, a deeper inspection needs to be made for safety-critical parts that are *always energized or never energized.* This latter condition can occur because the part cannot be controlled by the vague network's logic due to a possible design flaw. The circuit is working properly, and there are no parts failures, but the circuit simply does not do what the designer wanted it to do. At first glance, it may seem impossible for mathematics to capture a designer's intent, but that is not the case. *If the safety-critical part p is an element of at least one unique simple circuit of each well-defined network, then it is always energized. Similarly, if p is not part of any simple circuit for any well-defined network, then it can never be energized (i.e., some of the circuitry of the vague network is never used)!* Unwanted circuits that keep p perpetually on, or unused circuitry containing p, need to be tagged by color coding and removed from further consideration.

At the third level of inspection *additional functionality* (as opposed to parallel functionality) will be identified. Additional functionality consists of other (non-identical) circuits that can initiate/retard a safety-critical device and its function. Assuming the first level of examination has been done correctly, each circuit that provides additional functionality is unique. Furthermore, if the second level of examination has been done correctly, the number of unique simple circuits that contain part p (call it $\aleph_p > 0$) should equal the number of unique conditions that the designer wants for activation (or deactivation) of p. If less, then some desired functionality is not present. If more, one or more of the simple circuits maybe a sneak circuit. Let n_p be the total number of well-defined networks that contain p. Ideally, $n_p - \aleph_p = 0$. If $n_p - \aleph_p$ is a negative integer, then two or more of the designer's conditions that can energize/inhibit p are satisfied by a single vague network logic state (the "Pigeonhole Principle"). This is probably an error in logic. Therefore, these circuits will also have to be tagged.

Finally, the fourth, and last, level of inspection reveals the deepest sneak circuit problems. Every component in a vague network has nominal specifications *and* a tolerance. This tolerance can be problematic. For example, a safety-critical device, that controls a safety-critical function, may require voltage within a certain range. But, due to production tolerances for resistors, inductors, etc., the impedance of one or more parts within a simple circuit (electrical loop) may be too high. In that case, the voltage drop across the safety-critical component may be too small for it to operate properly. The circuit itself is operating properly, there are no parts failures, the circuit logic is correct, and the designer's intentions have been satisfied, but the natural manufacturing tolerances of components combine to occasionally produce an unexpected functionality (*intermittent functionality*). So, this too must be considered a kind of sneak circuit according to the definition in Sect. 11.2. At this point, it must be assumed that the circuit in question was bench tested (breadboarded) at least once so that the optimal parameters for each component are known. It is also assumed that the component tolerances can be supplied by the manufacturer. Now, the analyst must examine a suspect loop on his/her monitor and calculate worst-case impedances and source voltages to determine if the safety-critical device will ever be out of range for logical control. If all four of these levels of inspection are past, the vague network is free of sneak circuits. Next, related electrical problems will be considered.

Once the sneak circuit analysis has cleaned the vague network of parallel functionality, always/never energized safety-critical parts, additional functionality, and intermittent functionality, parts failure problems can be addressed. First, consider the Fail Open/Fail Low (FO/FL) scenario. If a conducting line (e.g., one IC output, a two-terminal analog part, or a circuit board printed conductor) fails open, then a new vague network will result that can be decomposed into a new set of well-defined networks. No new simple circuits will be produced by this action, although some may be removed. Therefore, it will be necessary to rerun the level II and IV sneak circuit analyses. In a way, the situation for the Fail Closed/Fail High scenario is less complex. The worst-case failure occurs when all digital channels fail closed (i.e., conducting). In that case, the vague network and the well-defined network are the same. Furthermore, the well-defined network will have maximum complexity, and many new unexpected paths will be created. These new paths can all be revealed by repeating steps I, II, and III of the sneak circuit analysis. The new paths are not really sneak circuits, but a sneak circuit analysis will display them. This is why there is so much confusion about what constitutes a sneak circuit.

11.7 Summary

The advent of the digital age has made the problem of sneak circuit analysis considerably more involved. However, the mathematician's trick of reducing a new problem to an old problem, whose solution is already known, still holds. Digital circuitry creates a new level of complexity, the vague network, that can be reduced to a set of well-defined networks, each of which can be characterized (or defined) by a set of switch states characterizing the condition of each digital line (i.e., a given line is either on [1], or off [0]). So, a vague network can be represented as a superset of sets, each containing an ordered $(\log_2 n_p)$-tuple of 1's and 0's, where n_p is the number of all possible well-defined networks containing part p. Each well-defined network can be treated as a kind of pseudo "analog circuit" that can be attacked by traditional methods already available; such as searching for characteristic network patterns that have a history of leading to sneak circuits. However, here the method of choice for solution is to break down each well-defined network even further into simple circuits involving the safety-critical part p because, strangely, and as demonstrated in the section above, the count of these simple circuits, plus the side information afforded by their $(\log_2 n_p)$-tuples, is all the information that is needed to find every sneak circuit of a vague network with 100% fidelity—a mere book-keeping procedure! Some computer scientists would consider the search for simple circuits to be artificial intelligence (AI). This author takes the view that such a search is nothing more than "number crunching". The real magic is in the packaging and presentation of the simple circuits to the human analyst. Only the top-level information is presented in a seamless (i.e., doesn't require any additional look-ups) and error-free format that includes the simple circuit with each of its components and their parameters, the "next highest assembly" (i.e., the simple circuit's well-defined network), and the numerical set expression of the switch settings in the well-defined network. And, at each level of analysis, the software asks the right questions, and there is less and less to discuss, so that the problem (sneak circuit identification) is quickly resolved. In this sense, communication with the software is equivalent to talking to an expert human about sneak circuits containing p. Figure 11.6 summarizes the human/machine interaction as a pyramid. There is a huge amount of number crunching at the bottom. Next, there is an intermediate AI-expert decision maker to package and display information that is

Fig. 11.6 The task pyramid for the sneak circuit analyzer

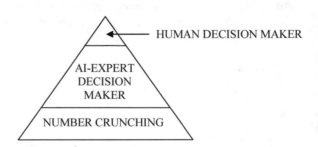

HUMAN DECISION MAKER

AI-EXPERT DECISION MAKER

NUMBER CRUNCHING

under examination at each level, and to store information that is no longer under consideration. Finally, at the top of the pyramid is the human decision maker. The analyst has the smallest information load to deal with but makes the most difficult decisions.

Artificial Intelligence is traditionally divided up into 8 subdisciplines. These are problem-solving, logical reasoning, languages, programming, learning, expertise, robotics and vision, and systems and languages [12]. So far, only "expertise" has been discussed. This manifestation of AI occupies an intermediate position in the sneak circuit detection process; just before interaction with the human analyst. However, robotics and vision systems can also play a key role in sneak circuit analysis. In this case, the robot is just an optical scanner for schematics. Each conducting path may be followed and converted to an "edge" of a graph, and each microelectronic device is converted to a vertex, etc. In its final state of evolution, the author envisions a sophisticated double AI system for sneak circuit analysis.

11.8 Problems

(1) Section 11.6 of this chapter opens with a claim that there are 16 physically unique loops 9 edges long that involve vertex 1 (see Fig. 11.2). Prove that this is so.

(a) Calculate $W_8(1, 2)$ from $W_7(1,i) = W_7(1, 1)$, $W_7(1, 2)$, $W_7(1, 3)$, $W_7(1, 4)$, $W_7(1, 5)$, $W_7(1, 6)$, $W_7(1, 7)$, $W_7(1, 8)$, and $W_7(1, 9)$. You will have to use the iterative method of Chap. 7 to "bootstrap" yourself up from the formula calculated values of $W_6(A[i, 2])$.

(b) Do the same for $W_8(1, 4)$ and $W_8(1, 5)$.

(c) Sum $W_8(1, 2) + W_8(1, 4) + W_8(1, 5)$ to get 32.

(d) Since each edge conducts in *both* directions the number of *physical loops* is $32/2 = 16$.

(e) Draw all 16 physical loops that are 9 edges long for the network in Figs. 11.2 and 11.3.

(2) Referring to Fig. 11.5, there are 7 loops that are 4 edges long. However, three of these loops have the property that if one resistor fails closed (conducting) the signal generator may overload. A piece of FOD could accomplish this. Can you identify these loops? Which resistors would have to fail closed to cause a problem?

References

1. Anon. (2006). SCAT, *Sneak circuit analysis tool user's guide—Ver. 5.0*. Culver City, CA: SoHar Inc.
2. Zito, R. R. (2012). 'Sneak circuits' and related system safety electrical problems—I: Matrix methods. In *30th International Systems Safety Conference*, Atlanta GA.
3. Zito, R. R. (2011, August 8–12). The curious Bent Pin Problem—I: Computer search methods. In *29th International Systems Safety Conference*, Las Vegas.
4. Zito, R. R. (2011, August 8–12). The curious Bent Pin Problem—II: Matrix methods. In *29th International Systems Safety Conference*, Las Vegas.
5. Kendall, T. (1978). *Passing through the netherworld: The meaning and play of Senet, an ancient Egyptian funerary game*. Belmont: The Kirk Game Co.
6. Bell, R. C. (1979). *The board game book*. London: Marshall Cavendish Ltd..
7. Weeks, K. R. (2005). *Luxor* (p. 376). Vercelli, Italy: White Star.
8. Fazzini, R. A., Romano, J. F., & Cody, M. E. (1999). *Art for eternity—Master works from ancient Egypt* (p. 26). Brooklyn, NY: Brooklyn Museum of Art.
9. Marshall, C. W. (1971). *Applied graph theory* (pp. 30, 115). New York: Wiley.
10. Kotzig, A. (1964). Hamilton graphs and hamilton circuits. In M. Fiedler (Ed.), *Theory of Graphs and Its Applications—Smolenice Symposium* (pp. 63–82). New York: Academic Press.
11. Ore, O. (1962). *Theory of graphs*. American Mathematical Society.
12. Barr, A., & Feigenbaum, E. A. (1981). *The handbook of artificial intelligence-V. 1* (pp. 7–11). Los Altos, CA: Kaufmann, Inc.

Part III
Software

Chapter 12
Predicting Software Performance

12.1 Problem Statement

Complex software systems are so prone to unexpected or undesirable behavior that it is customary in systems safety science to assign a "failure" probability of unity to such software systems when evaluating the overall failure probability of a software/firmware/hardware chimera.[†] Here, the word "failure" is used in a manner different from the way it might be used for hardware, or even firmware, because software doesn't "break", or even experience bit "upsets". But, software does do unexpected things in the sense that an electric circuit might do unexpected things because of a "sneak circuit" inadvertently embedded into the design. In this sense, a software "failure" (or "defect") might be considered a kind of "logical sneak circuit." This theme will be developed further in the next two chapters.

For now, it is sufficient to realize that assigning a probability of unity to software failure is a "worst-case" approach to estimating the failure probability of any complex chimera. However, there are occasions when an analyst may wish to know "what can be expected" of software. So, a more accurate estimate of failure probability may be desired than the assignment of unit probability. The following argument is due to Muniak [1].

> Suppose a motorist goes into a gas station to fill up his tank. He reaches into his wallet, pulls out his credit card, and swipes it at the pump. Now, why would he do that if he really believed that the probability of a software failure was unity? His personal information could have been lost! Or, perhaps a great deal of money could have been removed from his account and not recorded on his receipt! In fact, any number of dire scenarios could be imagined that are *possible*. But, after perhaps a decade of inconsequential experience with this particular system, our hypothetical motorist has come to *expect* a failure probability that is much lower. In fact, so low, that he is willing to accept the risks, and repair any damage that might occur, after the fact. Certainly, there are back up systems in place should a catastrophic failure occur. Credit card companies have fraud limits on their cards, and arrests can be made if necessary. Also, transactions can be tracked so that corrections can be made. All of these remediation methods are slow, time consuming, and inconvenient, but

© Springer Nature Switzerland AG 2020
R. R. Zito, *Mathematical Foundations of System Safety Engineering*,
https://doi.org/10.1007/978-3-030-26241-9_12

they do correct *possible* 'Black Swan' failures, even if these failures are not what a motorist might *expect*. And, it is *expectation* that often shapes our behavior! So, there is a place for non-unit defect prediction.

The Muniak Argument is an example of a general situation in which a software designer wishes to know if the failure (defect) rate has fallen below some threshold level that allows a given software system to be released. A closely related scenario is to demand some numerical guarantee that the probability of unexpected software behavior over a system's foreseeable useful life (cumulative defect probability) lies below some threshold; say one part in a million. These are very practical questions, but their answer is complicated by the fact that the failure rate for any given software system *varies with time*.

Newly released software is prone to failure (exhibits defects), whereas mature, extensively field tested, software is stable and displays few failures (defects) per unit time. That is to say, there exists a generic failure rate law by which the failure rate grows rapidly at first, and then after reaching a maximum value, recedes as time progresses. *Once the failure rate as a function of time is known for any particular piece of software, a prediction can be made about its future reliability.* In the next section the generic failure rate law, and its special cases will be reviewed.

12.2   Failure Rate Laws

In Chap. 2 [2], it was shown that complex systems, including complex software systems, can fail according to a variety of laws that seem unrelated at first. However, it was shown that any complex system can be carved up into a set of many "simple" subsystems. Each subsystem is "simple" in the sense that its *normalized* failure rate law can be described by a *Rayleigh distribution*. So, the normalized failure rate law for a complex system is really just a composite of the normalized failure rate laws for the individual simple subsystems. Each subsystem's Rayleigh curve is characterized by a unique value of its "mode," or peak location in time. The failure rate law (normalized or un-normalized) for some simple subsystems will "peak-out" early. Others will not peak-out for a very long time. So, the modes of subsystems are themselves distributed in time. And, the overall complex system normalized failure rate law $\rho_{global}(t)$ may be thought of as a *distribution of distributions* [3]. Mathematically, this overall (generic) normalized failure rate law may be written

$$\rho_{global}(t) = \int_0^\infty (t/\sigma^2) \exp\left(-\tfrac{1}{2}[t/\sigma^2]\right) f(\sigma)\, d\sigma, \qquad (12.1)$$

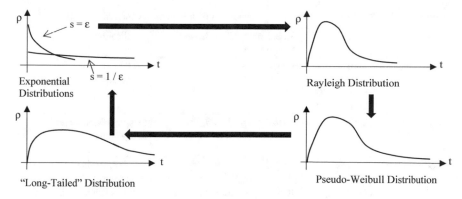

Fig. 12.1 The shapes of normalized failure rate histograms, where ε is a small number

where f(σ) is the normalized distribution of subsystem modes σ, and t is time. Depending on the selection of f, various specific normalized failure rate laws emerge as special cases. These are reviewed in the table below.

The four models above form a more or less continuous spectrum of curves for failure probability density laws (Fig. 12.1). The exponential model starts at some finite value at t = 0 and then can decay rapidly if the decay constant λ = 1/s is large. This is followed by the Rayleigh model, which starts at zero, pops-up to a maximum value, and then decays away more slowly than the exponential. Next comes the pseudo-Weibull model with a still more slowly decaying tail. The unnamed "Long-Tailed" model has the slowest decaying tail of all. Notice that all of the equations above express $\rho_{global}(t)$ as a function of overall system parameters (s, σ_m, or M). For example, the exponential distribution is characterized by the standard deviation s, which is twice the spread of subsystem modal values (σ) above 0, the Rayleigh distribution is characterized by a single value σ_m, whereas the pseudo-Weibull distribution is characterized by the standard deviation s of subsystem modes about a mean value σ_m. Finally, the unnamed "Long-Tail Distribution" is characterized by the location of the peak M of subsystem modal values that are themselves Rayleigh distributed in time. These system values make the selection of a model for hardware systems easy, but they are usually not known a priori for software. Nor is it possible to judge a priori which (normalized) failure rate model is applicable based on the "applications column" in Table 12.1 with any *certainty*. Instead, *change control data* (i.e., the history of failures per unit time, say per month) must suggest a failure model for any given piece of software as discussed in the next section.

Table 12.1 Failure laws (s is the standard deviation of a Gaussian centered on σ_m)

Model	$f(\sigma)$	Global probability density function for $\rho_{global}(t)$	Application
Exponential	Half Gaussian with s $\neq$ 0 and σ_m = 0	$= (1/s) \exp(-t/s)$	Well tested post-development systems
Rayleigh	Delta Func. @ location σ_m	$= (t/\sigma_m^2)\exp(-\frac{1}{2}[t/\sigma_m]^2)$	Simple systems and subsystems
Pseudo-Weibull	Narrow Gaussian (s $\ll \sigma_m$)	$\cong (t/\sigma_m^2)\exp(-\frac{1}{2}[t/\sigma_m]^2)$ $\{1 + \frac{1}{2}(s/\sigma_m)^2 [6-7(t/\sigma_m)^2 + (t/\sigma_m)^4]\}$ where $s/\sigma_m \ll 1$	Efficient designs with peak subsystem failure rates tightly clustered about a mean value
Unnamed "Long-Tailed Distribution"	Rayleigh with a peak @ M	$= [(2\sqrt{\beta}/M)] \sum_{j=0}^{\infty} \{[\beta^j/(j!)^2]$ $[\psi(j + 1) - \ln\sqrt{\beta}]\}$, where $\beta = (t/2M)^2$, and $\psi(j + 1) = -c + \sum_{n=1}^{j} (1/n)$, and c = 0.577..., and $j \leq 9$	New poorly tested designs

12.3 Picking a Coarse Model for Future Reliability Prediction (an Example)

Figure 12.2 shows the raw data (not normalized to a probability density function) for the first 22 months of the failure rate (defect rate) history recorded for a particular software project [4]. Let t_{max} be the time when the maximum failure rate occurs during system testing. By the time t = 2 t_{max} = 22 months, it is clear that the exponential model is an unacceptable model for this particular project, at least in the short term, because $t_{max} \neq 0$. The next decision is, "Should the normalized software failure rate model be Rayleigh or 'Long-Tailed'?" It might seem natural to simply fit each of these two models to the data and see which works best. However, this is not possible because the total amount of data (total number of defects) is unknown. Therefore, this *piece* of defect history cannot be normalized. And, normalization affects the fit. What a problem! However, there are several ways out of this dilemma. The most robust method will be described in detail here.

If the data in Fig. 12.2 was *truly* Rayleigh, it would have to have a peak at t = 10.5 = σ_m; where σ_m is the standard symbol used for the peak (or mode) of a Rayleigh distribution. Furthermore, since $\rho_{global}(t)$ is such a simple function for the Rayleigh distribution, it should be easy to calculate the fraction of the total unit area trapped under the curve for $0 < t < 2t_{max} = 2\sigma_m$. This fraction $\mathcal{F}$ is just

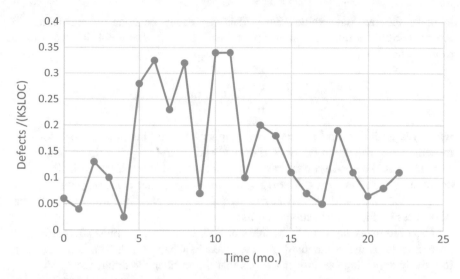

Fig. 12.2 A 22-month software defect rate history, where the defect rate is measured in Kilo Software Lines of Code (KSLOC) per month. The raw data is from Module C of the study by Peterson and Arellano [4] and involved a total of 74.4 KSLOC

$$
\mathcal{F} = \left[\int_0^{2\sigma_m} (t/\sigma_m^2) \exp\left\{-\tfrac{1}{2}(t/\sigma_m)^2\right\} dt \right] \Big/ \left[\int_0^{\infty} (t/\sigma_m^2) \exp\left\{-\tfrac{1}{2}(t/\sigma_m)^2\right\} dt \right]
$$
$$
- \int_0^{2\sigma_m} (t/\sigma_m^2) \exp\left\{-\tfrac{1}{2}(t/\sigma_m)^2\right\} dt,
$$
(12.2)

where the denominator in the ratio above has unit value. Noting that $d(t/\sigma_m)^2 = 2(t/\sigma_m^2)dt$, and defining $u \equiv t/\sigma_m$ yields

$$
\mathcal{F} = \tfrac{1}{2} \int_0^{t=2\sigma_m} \exp\{-\tfrac{1}{2}u^2\}du^2 = \tfrac{1}{2} \int_0^{t=2\sigma_m} \exp\{-\tfrac{1}{2}x\}dx = -\exp\{-\tfrac{1}{2}x\} \Big|_0^{t=2\sigma_m}
$$
(12.3)
$$
= 1 - \exp\left\{-\tfrac{1}{2}(2\sigma_m/\sigma_m)^2\right\} = 0.864,
$$

where an additional substitution $x \equiv u^2$ has been made to evaluate the first integral in Eq. 12.3. Now, this result is quite nice because it does not depend on the value of σ_m, and it implies that at time $t = 2t_{max} = 2\sigma_m$, the normalized experimental data should have an area of about 0.864 under it. Provided, of course, that the data is Rayleigh. Normalization now proceeds as follows. First, calculate the total area

A under the raw data values $\mathcal{D}(t_i)$. This can easily be done by evaluating the Riemann–Stieltjes integral via an approximating sum, where each interval $\Delta t = 1$ month. Therefore,

$$A = \sum_{i=1}^{n} [\mathcal{D}(t_i)(1 \text{ mo.})], \tag{12.4}$$

where i is an index, and n is the total number of time bins (i.e., 22, one for each month). A normalization factor $\mathfrak{N}$ is desired such that $A/\mathfrak{N} = \mathcal{F} = 0.864$. Therefore, $\mathfrak{N} = A/\mathcal{F} = A/0.864$. After each raw data point is divided by $\mathfrak{N}$, the normalized data curve in Fig. 12.3 emerges. *Clearly, if the normalized data is Rayleigh, a Rayleigh analytical model should fit this normalized data best.* Now, the question is, "What 'Goodness of Fit' criteria should be used?"

There are many tests for "Goodness of Fit". However, the most straightforward procedure is to use a standard χ^2 test given in Chap. 1 [5]. This simple test is intuitively appealing because it implies that the smaller the difference

$$\chi^2 \approx \sum_{i=1}^{n} \left\{ \left[\mathcal{D}(t_i) - \mathcal{N}\rho_{\text{global}}(t_i)\right]^2 / \mathcal{N}\rho_{\text{global}}(t_i) \right\}, \tag{12.5}$$

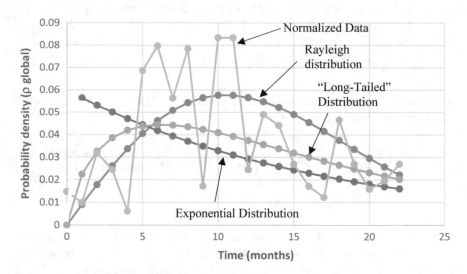

Fig. 12.3 A data set with a hypothetical Rayleigh normalization, a Rayleigh curve with $\sigma_m = 10.5$, a "Long-Tailed" distribution with $M = 10.5$, and a best fit exponential distribution are all plotted together for comparison. As expected, the exponential model can be readily eliminated as a global model for the data. It is harder to decide whether the Rayleigh distribution or the "Long Tailed" distribution is a better fit to the data by visual inspection. However, the "Long-Tailed" distribution is clearly the best model for the "tail" of the data (large t). The χ^2 Goodness-of-Fit criterion selects the "Long Tailed" distribution as the best overall fit

between the data and the model, the smaller the value of χ^2. There are, however, problems associated with curve fitting via the χ^2 statistic. First, the defect count during the early stages of any software project tends to be very noisy, so that minimizing χ^2 in Eq. 12.5 emphasizes minimizing the numerator for the early data since these terms are usually the biggest contributors to the sum. Furthermore, by the time $t = 2\, t_{max}$, the earliest data is too obsolete to be useful for future prediction of software performance. Both of these difficulties can be overcome by only fitting the data after $t = t_{max}$.

If a Rayleigh model is generated with a peak that matches the normalized data peak (i.e. $\sigma_m = 10.5$) then the curve in Fig. 12.3 results in a $\chi^2 = 67.573$. The contribution to this value from the tail of this Rayleigh model ($\sigma_m < t < 2\sigma_m$) is 35.456. By comparison, setting $M = 10.5$ for the "Long-Tailed" distribution yields a slightly lower overall $\chi^2 = 65.745$, with a much lower contribution of 19.591 from the tail. Note that the true *optimal* "Long-Tailed" fit actually lies at $M = 10.9$ with an almost identical $\chi^2 = 65.511$. Clearly, the assumption that the data is Rayleigh is false, because the Rayleigh model does not give the best possible fit. The "Long-Tailed" distribution is doing a better job of fitting the data (especially in the tail) and is, therefore, the model of choice.

One last point deserves discussion with respect to Fig. 12.3. The best fit exponential curve has also been plotted with the other curves. Aside from the obvious fact that there is no match for small values of time, the exponential's slope in the tail is about the same as that for the "Long-Tailed" distribution. However, the curve is too low. The standard deviation of the half-Gaussian distribution of sub-system modes associated with this exponential complex system model is a rather large 16.7. This one parameter must control both the curvature and the probability density level in the tail. The compromise between these conflicting properties results in an overall poor fit in this case ($\chi^2 = 67.834$). The contribution to the overall χ^2 from the tail is 22.275, intermediate between the Rayleigh and "Long Tailed" models. Therefore, although the Model C system is still too immature at $t = 22$ months for an exponential failure (defect) rate model to apply, such a model may be useful in the far tail ($t \rightarrow \infty$) for this and other complex systems. Peterson [6] has suggested that many very mature complex systems display a more or less steady-state defect rate, because as one defect is corrected new defects are introduced into the code. In other words, new Rayleigh failure rate curves are introduced into the code, with their modes spread over a wide interval in time. The large value of the standard deviation of subsystem modes in this example, and the slowly decaying tail of the exponential fit of Fig. 12.3, mimics the strange behavior of the Peterson Conjecture.

12.4 Fine Structure Splitting Between Model Tails (an Example)

The last curve-fitting problem to consider is detailed modeling of the far tail region of the data distribution. As previously mentioned, modeling of the tail is important because this is where the "Black Swan" probability is trapped. However, modeling of the tail of a distribution is a subtle problem complicated by the fact that the "Long-Tailed" distribution and the pseudo-Weibull distribution can have a very similar shape in the tail, as can the pseudo-Weibull distribution and the Rayleigh distribution (Fig. 12.1). The only solution is to wait for more data to come in. At $t = 3t_{max}$ there is just barely enough data to make these fine distinctions. Continuing with the present example, it is desirable to know if the "Long-Tailed" or pseudo-Weibull distributions are best for modeling the far tail region of the data. If σ_m is set equal to 10.5 as before, the best fit pseudo-Weibull curve has a standard deviation of subsystem modes $s \approx 3.0$. A much larger value of s yields a curve with a distorted shape that is not a good data model. In that case, the value of $\rho_{global}(t)$ is 0.005 for the pseudo-Weibull curve at $t = 33$ months. By contrast, the "Long-Tailed" distribution yields $\rho_{global}(t) = 0.008$ while the data curve has a value of 0.010 at $t = 33$ months. Clearly, the pseudo-Weibull distribution lies below most of the experimental data (Fig. 12.4). It is decaying *too* fast to be a viable model for

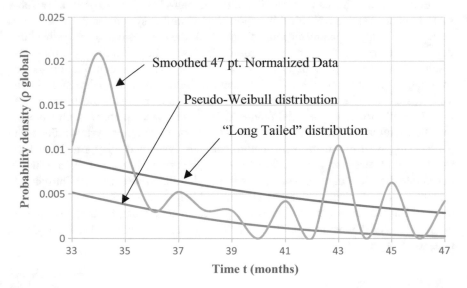

Fig. 12.4 Projected behavior in the tail from $T = 33$ months ($\approx 3\sigma_m$) to the end of the data set at $t = 47$ months. The data has been normalized using the normalization factor for the entire 47-month dataset. The "Long-Tailed" distribution (M = 10.5) is the best data fit, while the pseudo-Weibull distribution is generally under-estimating the probability of failure

the far tail. At least, that is the conclusion for a pseudo-Weibull model that is truncated after fourth-order terms in t/σ_m. *Therefore, the "long-Tailed" distribution remains the model of choice in this example.*

At this point, one might naturally ask, "Why not use the simple method of comparing endpoints to perform the course splitting in the same way that the fine splitting was performed?" The answer is that there is an unfortunate curve crossing at $t \approx 2\sigma_m$ for the "Long-Tailed" and Rayleigh distributions (see Fig. 12.3). Therefore, it is necessary to make the first splitting using the available data from the interval $t_{max} < t < 2t_{max}$, not just the end point at $t = 2t_{max}$.

12.5   Calculated Software Failure Probabilities—Theory Versus Experiment

For the defect data presented in this chapter, a good model was found early in the development of the software that accurately fit future data. The first 22 months of system behavior (~ 2 years of data) as recorded by change control was sufficient to predict mean behavior of the system over the next 25 months (~ 2 years into the future). But what does this model say about the behavior of a piece of software in any particular month, or over any particular interval of time? Can a definite probability of failure (or more precisely the probability of defect detection) be calculated? The graph in Fig. 12.4 shows that it is reasonable to expect maximum failure probabilities about twice as large as predicted by the model for any given month. But, failure probabilities in any given month can also have a minimum as low as zero! However, as will be demonstrated in the next paragraph, software behavior over time is more stable.

Once the selection of a model has been made by a "Goodness-of-Fit" test and a fine structure splitting test, the probability of software failure ($P_{failure}$) between any two times T_1 and T_2 is just given by

$$P_{failure} = \int_{T_1}^{T_2} \rho_{global}(t)dt \approx \sum_{i=m=T_1/\Delta t}^{n=T_2/\Delta t} \left[\rho_{global}(t_i)\Delta t_i \right], \qquad (12.6)$$

where T_2 can (in theory) include ∞, i is an index, m and n are integers, and all the $\Delta t_i = \Delta t = a$ unit data collection interval (say, one month). For example, the actual probability of failure calculated from the data between $T_1 = 33$ months and $T_2 = 34$ months is 0.0313 (i.e., 3.13%). The probability trapped between the same two time points under the a priori model curve is 1.70%. The difference between these probabilities is 1.43%. But, more importantly, the *relative* error (relative to the model probability) is 84.1% (see Fig. 12.4). The probability trapped under the data curve between $T_1 = 22$ months and $T_2 = 47$ months is 28.6%. And, the probability trapped under the same two points of the a priori model curve is 23.8%. The probability difference is now 4.8%, and the relative error has been reduced to only 20.2%. So, the longer the integration time, the smaller the relative error between the a priori model and the data. This phenomenon is an example of "noise reduction" by integration. Table 12.2 gives the cumulative probability ($P_{failure}$) of both the data and model curves as a function of time by integrating from a starting point of $T_1 = 22$ months to terminal points of $T_2 = 23$–47 months. Initially, the relative error is very unstable. But, as time T_2 increases, the cumulative noise between successively longer integrations averages out and decreases, as does the value of the relative error itself. As $T_2 \rightarrow \infty$ the relative error should ideally approach zero, meaning that the model is a perfect representation of the total failure probability of the actual system for large t. However, this goal will never actually be realized because a "Long-Tailed" model with $M = 10.5$ is not *exactly* optimal for this system example, as discussed above.

Table 12.2 Future probability of defect detection

T_2 (mo.)	$P_{failure}$ (data) in % $T_1 = 22$ months	$P_{failure}$ (model; $M = 10.5$) in % $T_1 = 22$ months	Probability difference (%)	Relative error (%)
23	4.60	3.91	0.69	17.7
24	8.78	5.66	3.12	55.07
25	10.87	7.29	3.58	49.08
26	14.00	8.80	5.20	59.07
27	15.05	10.21	4.84	47.43
28	16.51	11.51	5.00	43.47
29	17.14	12.71	4.42	34.80
30	17.97	13.83	4.14	29.96
31	20.48	14.86	5.62	37.80
32	20.48	15.82	4.66	29.48
33	21.53	16.70	4.83	28.89
34	23.62	17.52	6.10	34.82
35	24.66	18.27	6.39	34.98
36	24.97	18.96	6.01	31.69
37	25.50	19.61	5.89	30.04
38	25.81	20.20	5.61	27.78
39	26.12	20.74	5.38	25.93
40	26.12	21.25	4.88	22.95
41	26.54	21.71	4.83	22.25
42	26.54	22.14	4.40	19.89
43	27.59	22.53	5.05	22.43
44	27.59	22.90	4.69	20.48
45	28.21	23.23	4.98	21.45
46	28.21	23.54	4.67	19.86
47	28.63	23.82	4.81	20.18

Even though there is very little probability in the far tail region. These small probabilities are, nevertheless, extremely important because they take into account those rare and unexpected events that are so destructive. The draconian consequences of these "Black Swan" events occur because of a lack of contingency planning due to user complacency and overconfidence in software that may seem to be perfect—*for a while*! Both *probability* and *possibility* ("Black Swan" events) need to be considered for a balanced assessment of future software performance. The ability to calculate small "Black Swan" probabilities is one of the things that makes the model probability distribution so useful!

12.6   Conclusion

Software is, in many ways, fundamentally different from hardware. It is the lack of knowledge about $f(\sigma)$ that requires a different plan for the evaluation of Eq. 12.1. For hardware, $f(\sigma)$ can often be determined from manufacturer's data, or the performance of legacy subsystems. Information of this kind is usually not available when building software. Instead, the problem of predicting future software performance must rely on experience acquired during development (change control data). As time progresses, the (normalized) failure rate (defect rate) model becomes more and more refined (Fig. 12.5). The software example used in this chapter was the most problematic of four software projects that have been carefully analyzed [4]. It is the most problematic example because its change control data has proved to be

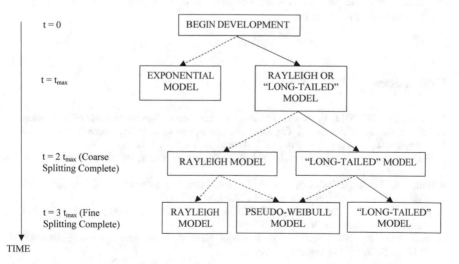

Fig. 12.5 Normalized defect rate model splitting diagram for software development. The solid arrows indicate the chain of decisions for the example in this chapter. Dashed lines indicate other possible decisions

the most difficult to fit. Thirty or more software projects need to be examined before a final conclusion can be reached regarding the robustness of the methods presented in this chapter. Finally, as pointed out by Fitzpatrick [3], it must always be remembered that performance predictions of the kind presented here are only valid, strictly speaking, for a particular software environment. If the environment (application) is changed, a discontinuity will likely develop in the normalized failure rate curve.

In any case, once a defect has been detected, how does the analyst go about removing it? Or, more importantly, how does the analyst remove defects even before they manifest themselves? Is it possible to build *perfect* software? Of course, that will depend on what one means by *perfect*. These difficult questions will be addressed in the following two chapters on software.

† The word "chimera" refers to a "monster made of body parts from a variety of animals." In biochemistry, it refers to a macromolecule made up of smaller macromolecules of diverse origin. Here, the term will be used to refer to a system composed of some combination of software, firmware, and hardware.

12.7   Problems

(1) From month 47 to $t \to \infty$, what is the probability of failure for the system in Fig. 12.4 if the "Long-Tailed" distribution is an appropriate model for system behavior in this time interval?
(2) From month 47 to $t \to \infty$, what is the probability of failure for the system in Fig. 12.4 if the pseudo-Weibull distribution is an appropriate model for system behavior in this time interval?

References

1. Muniak, C. G. (2014, February). Private communications. Stevens Institute of Technology.
2. Zito, R. R. (2013, August 12–16). How complex systems fail-1: Decomposition of the failure histogram. In *Proceedings of the ISSC Annual Meeting*-2013, Boston, MA.
3. Fitzpatrick, R. (2012, January). In *29th ISSC Conference Chair*, Private communications, no. 520, pp. 647–1608.
4. Peterson, J., & Arellano, R. (2002, March). Modeling software reliability for a widely distributed, safety-critical system. *Reliability Review, 22*(1).
5. Meyer, P. L. (1970). *Introductory probability and statistical applications* (2nd ed., pp. 328–332). Reading, MA: Addison-Wesley.
6. Peterson, J. R. (2013, February 7). Raytheon Network Concentric Systems, Fullerton, CA, 714-446-2927, Private Communications.

Chapter 13
Alternative Flowcharts
for a Mathematical Analysis of Logic

13.1 Introduction

Traditionally, Flowcharts have been used either as a general guide for the designer/programmer during software construction, or as a documentation tool after software construction. The intent of this work is to exploit the topological properties of flowcharts as a mathematical tool for debugging software logic. However, flowcharts are typically presented as an in-line sequence of steps, or as a "tree". This format is not useful for mathematical analysis. Therefore, the *Flowchart (FC)* as commonly conceived will be redesigned. First, it will be made to look more "circuit-like" so that the methods of circuit theory can be exploited. However, here the "current" will be a flow of logic rather than a flow of electrons (charge). And, a software logical flaw will correspond to a hardware *sneak circuit*. Next, the *"Circuitized" Flowchart (CFC)* will be transformed into a *Flowchart Graph (FCG)* so that the methods of graph theory can be exploited to transform the graph into a *Numerical Flowchart (NFC)*, or *Adjacency Matrix*. The methods of abstract algebra can then be used to analyze the NFC and, therefore, the original software logic. Each transformation adds a new level of abstraction and insight into the problem of debugging software logic. The theoretical discussion presented in this chapter will be concretized by the use of two examples. The first is a simple generalized program in the form of Fig. 13.1. The second example will be a rather specific program containing nested IF Statements.

© Springer Nature Switzerland AG 2020
R. R. Zito, *Mathematical Foundations of System Safety Engineering*,
https://doi.org/10.1007/978-3-030-26241-9_13

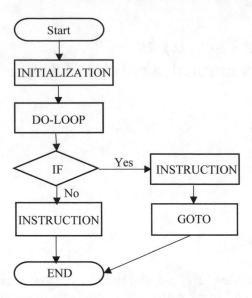

Fig. 13.1 A simple generalized FC that will be used as an example and transformed into a CFC, an FCG, and an NFC

13.2   The Circuitized Flowchart (CFC)

The Start/End Block: The first step in *circuitization* is simply to connect the *root* (or "End") of an FC tree to its *top* (or "Start"). That is to say, the operator, who initiates the software, acts as a *battery* to start the flow of logic. This may seem like a trivial step but, in fact, it is of central importance for the *loop counting* procedure that will be employed in this and the next chapter. Now, some software is naturally circuitized. Control software is typical of this type since it depends on repeated measurements followed by tests and corrections. Termination of this cycle is usually not wanted. Replacing the Start/End Blocks of Fig. 13.1 by a battery results in Fig. 13.2a.

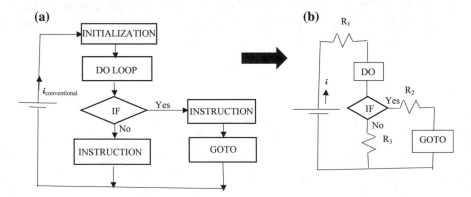

Fig. 13.2   a Step 1 of circuitization—removal of Start/End Blocks (Terminator Blocks). **b** Step 2 of circuitization—replacement of equations by resistors

Initialization Statements and Equations: Setting the initial values of constants (integers, real numbers, etc.) used in calculations and counters constitutes a set of simple equations. More complex instructions are also equations. All of these equations can be modeled as resistors in the CFC. Figure 13.2b shows this replacement.

The IF Statement: The IF Statement is a decision that can cause the logic flow to move along one path or another, depending on its input value(s). Similarly (analogously), an electron reaching a junction will decide on the path to follow by the downstream resistances it will encounter. A large downstream resistance in one branch of a circuit will create a pileup of electrons in that branch that will force newcomers into an alternative direction having lower resistance, and ultimately a higher current. Therefore, an IF Statement can be replaced by a simple junction, or *vertex*.

The GOTO Statement: A GOTO Statement is just a *feedback*, or *feedforward*, for logic flow, and can be replaced by a simple *wire* in the circuit of Fig. 13.2b.

The Delay Statement: A *delay* instruction can also be thought of as a wire since it makes data that *will* be computed or collected in one branch of a network, available to another branch that has not, and cannot, progress any further without the required future data. The notion of delay has not been used in the example of Fig. 13.2.

The DO-LOOP: Any DO-LOOP can be built up from primitive instructions involving an integer counter C, an instruction(s) whose output may be sent to memory, and an IF Statement involving C that decides if the sequence should be repeated, or not. The net result is that a series of instructions are executed, and the details of the internal structure of the loop are not important. A DO-LOOP can be packaged together with R_1 in Fig. 13.2b to form an equivalent "resistance" R_{eq}.

Logic Flow Direction: Software is fundamentally different from hardware in the sense that electrons can potentially flow in two directions in a sneak circuit embedded in a complex electronic circuit. Only the presence of a diode, or some other rectifying device, in such a circuit can prevent reverse flow. However, for software, the flow direction in each logic loop is always prescribed. That is to say, diodes must be introduced into Fig. 13.2b.

The Final CFC: When the notions in the subsections above are introduced into Fig. 13.2b, the CFC of Fig. 13.3 results. Figure 13.3 looks quite different from Fig. 13.1. However, in this form, the methods of circuit analysis can be applied to the problem of debugging logic [1]. For example, certain circuit patterns are notorious for producing sneak circuits (in this case sneak *logic* circuits). And, commercially available off-the-shelf software packages are capable of detecting these patterns in electronic hardware [2]. The CFC makes these tools practical for the analysis of software as well.

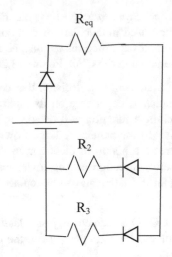

Fig. 13.3 The final CFC for the FC of Fig. 13.1

13.3 The Flowchart Graph (FCG)

The next type of Flowchart useful for mathematical debugging of software is the *flowchart graph (FCG)*, also called a Control Flow Graph (CFG) as introduced by Allan [3]. A graph is nothing more than a collection of dots (vertices) connected by lines (edges). The edges of the flow chart do not contain any circuit elements (i.e., battery symbol, resistors, etc.) embedded in them. The reason for this is that symbols like R_1, R_2, and even R_{eq}, represent an/are instruction(s) that can be debugged at the first (*syntactic*) level of correction. Such individual statement errors (punctuation errors, spelling errors, undefined variables, etc.) are usually easily identified by computer compilers, and are quickly corrected by the analyst/ programmer. This chapter concerns the next higher level of correction; i.e., *logic errors*. Logic errors can be completely passed over by a compiler because a code may be operating completely correctly. The problem is that a program may not be doing what a programmer intended. The next chapter will capture the notion of *intent* more completely, but for now, it is sufficient to note that *logic errors* are errors in *logic flow*. Capturing logic flow is independent of R_1, R_2, and R_{eq}, but it is not independent of the three diodes that restrict logic flow to one direction. Therefore, the FCG for Fig. 13.3 is just Fig. 13.4. Note that *logic junctions*, that is

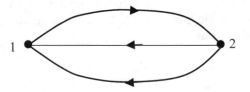

Fig. 13.4 The FCG of the FC in Fig. 13.1. Logic junctions are represented by vertices (dots) 1 and 2, whereas logic flow paths are represented by arrowed edges (lines)

to say locations where logic branches meet because of the presence of a battery or IF Statement in Fig. 13.3, are marked by the integers 1 and 2. In this form, the FC can now be reduced to a matrix representation.

13.4 The Numerical Flowchart (NFC or Adjacency Matrix A)

The *Adjacency Matrix* is a purely *numerical flowchart (NFC)* that captures the *topology*, or interconnections, between various parts of an FC. In fact, the Adjacency matrix is also called a Connectivity Matrix [4]. The Adjacency Matrix is square, and has as many rows and columns as there are vertices in the FCG. The elements of the Adjacency Matrix are filled in as follows. First, the analyst asks, "How many paths connect vertex 1 to itself?" Well, there are no self-loops attached to vertex 1, so the answer is none. Therefore, a 0 is filled in at location (1, 1) of the Adjacency Matrix A. Next, the analyst asks, "How many paths start from vertex 1 and end at vertex 2?" At first, one might say 3, but this is quite incorrect because direction counts in software. So, the proper answer is just 1, which comprises the value of element (1, 2) of A. Now, moving to vertex 2, "How many paths connect vertex 2 to vertex 1?" This time there are 2. So the value of element (2, 1) of A is just 2. Finally, "How many self-paths connect vertex 2 to itself?" None! Therefore, the value of element (4, 4) is 0. The Adjacency Matrix A, or NFC, for the FC in Fig. 13.1 should look like that in Eq. 13.1 below,

$$A = \begin{pmatrix} 0 & 1 \\ 2 & 0 \end{pmatrix} \tag{13.1}$$

Unlike the Adjacency Matrix for many electronic circuits, A will always be *asymmetrical* with respect to the main diagonal for software. The reason for this asymmetry is that software flow is unidirectional. It may seem that this numerical representation of the FC in Fig. 13.1 is too abstract to be useful, but, in fact, it is the most powerful representation possible. And, the NFC will allow the identification and solution of a variety of complex logic problems.

13.5 A More Complex Example

As a second more specific and complex example of how to generate a CFC, FCG, and an NFC from an FC, consider the nested sequence of IF Statements in the FC of Fig. 13.5. This program controls a hypothetical rocket motor throttle via the value of the integer variable IU. If IU is zero, the motor is lit, but produces negligible

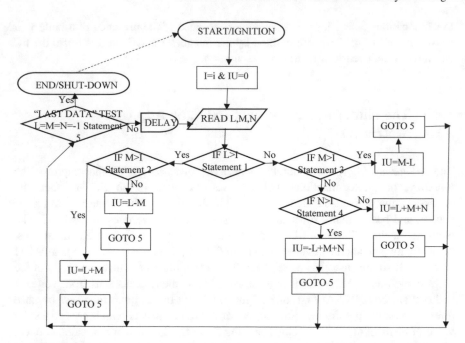

Fig. 13.5 Code with a nested sequence of IF Statements. The value of IU controls the throttle of a rocket motor. When IU is negative, motor shut down occurs

thrust. If IU is greater than zero, the motor burns at a proportional rate. If IU is negative, the motor shuts down. Clearly, this program is much more complex than that described by Figs. 13.1 through 13.4.

The code works as follows. First, at the start of the program (ignition), a test integer variable (I) is set equal to some *positive* value i, and the integer variable of interest (IU) is set equal to zero (motor lit). Next, the values of three other integer variables (L, M, and N) are read from a file; and could, for example, be related to target speed, range, and maneuverability. These integer values are normally all positive, except for the last data triplet, and will be used to shape the thrust profile (thrust vs. time). After these integers are read, the first IF Statement is encountered (labeled "Statement 1" in Fig. 13.5). If the value of the integer variable L exceeds that of I, then the logic moves to another IF Statement labeled "Statement 2". But, if the test in Statement 1 fails, then the logic proceeds to the IF Statement labeled "3". Now, if the test in Statement 2 is satisfied, then a computation is completed (i.e., IU = L + M, which is necessarily a positive integer—motor on). After this calculation, a GOTO Statement sends the logic to a "LAST DATA" test and a delay before reading the next triplet of L, M, N values. On the other hand, if the test in statement 2 is *not* satisfied, then IU = L − M (a positive integer—motor on, but at a different level than in the previous test). And, again, logic flow returns to the "LAST

DATA" test and a delay. Now, if the test of statement 3 is satisfied, then $IU = M - L$ (positive) and logic passes to a GOTO Statement that sends the flow back to the "LAST DATA" test. If not, then another IF Statement (4) is encountered. If the test on N is satisfied in 4, then $IU = -L + M + N$ (positive) and a GOTO returns logic to the "LAST DATA" test (5). If not, then $IU = L + M + N$ (positive) and a GOTO returns logic to 5. Of course, if $(L, M, N) = (-1, -1, -1)$, then the tests at statements 1, 3, and 4 all fail and $IU = L + M + N = -3$. In that case, the rocket motor shuts down because IU is negative.

Examination of Fig. 13.5 shows that the statements between "read" and GOTO form a natural sub-circuit, *provided* the "LAST DATA" test and the delay can be passed. Suppose an analyst wishes to build an Adjacency Matrix for this sub-circuit. Replacing the "Last Data" IF Statement test and the delay by a wire, substituting a battery symbol for the "read" block, resistor symbols for computational statements, junctions for IF Statements, and inserting diodes to insure proper flow direction, results in the CFC of Fig. 13.6a. Only the black dots mark true junctions (vertices). Other apparent vertices merely indicate the return of logic back to the read statement (battery), and ultimately to vertex 1. This distinction is important when the CFC is translated into an FCG (Fig. 13.6b). The FCG clearly identifies vertices that are nearest neighbors—a necessity for building the NFC.

Finally, the Adjacency Matrix A can be constructed as follows. Starting from vertex 1 of the FCG, there are no self-loops. That is to say, starting from vertex 1, the logic does not loop back on itself and end at vertex 1 in a single step. However, logic does flow to vertex 2 and vertex 3. But, there is no direct connection from vertex 1 to vertex 4. Therefore, the first row of the Adjacency Matrix is "0 1 1 0" (remember, direction counts). Similarly, the other rows of A can be filled in to yield Eq. 13.2. Each row refers to a different vertex, and each column refers to the nearest neighbor to which it is connected. So, the entry at (row = 2, column = 1), which has a value of 2, means that two paths exist that allow logical flow from vertex 2 to

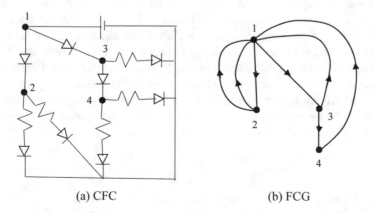

(a) CFC (b) FCG

Fig. 13.6 The CFC (**a**) and FCG (**b**) for the FC of Fig. 13.5. Notice how the returns from vertex 2 have been placed on the other side of the $1 \rightarrow 2$ edge in the FCG for clarity

its nearest-neighbor vertex 1. Notice that this Adjacency Matrix is asymmetrical with respect to the main diagonal, just as in Eq. 13.1.

$$A = \begin{pmatrix} 0 & 1 & 1 & 0 \\ 2 & 0 & 0 & 0 \\ 1 & 0 & 0 & 1 \\ 1 & 0 & 0 & 0 \end{pmatrix}. \tag{13.2}$$

13.6   How to Use the Adjacency Matrix

At this point one may ask, "All right, so it is possible to capture software architecture in numerical form, but WHAT GOOD IS IT?" The answer is that *the Adjacency Matrix can be easily manipulated algebraically to count logic loops in a given code, or part of a code.* To understand why counting logic loops is important, consider the following problem. An analyst is confronted with a complex program that contains a particular safety-critical instruction (say, for example, a missile self-destruct instruction). He knows that there is only one condition for which this instruction should be activated (say a missile "off-course" condition). However, if there exists two logic loops in the code that contain this instruction, then there must exist one other condition that is capable of activating this dangerous instruction. Either there is a deeply buried logical flaw in the software (a sneak logic circuit), or there is a flaw in the designer's thinking. Nothing can be done about the latter problem, but the former is tractable because each logic loop can be traced and unwanted logic removed. There are other ways of using the count of logic loops as well. The algebra involved in loop counting is as follows.

Consider the program of Figs. 13.1, 13.3, and 13.4 (the FC, CFC, and FCG of our first example). The FCG shows that this program has no *redundancy*. That is to say, there is no way to go from either vertex 1 to vertex 2 and return by traveling over a given edge more than once. The same is, of course, true for vertex 2. In that case, the so-called *Redundancy Matrix* (called R_2 in this case) is the zero matrix. More will be said about the Redundancy Matrix in the next chapter. But, for now, the number of nonredundant circuits is simply given by the diagonal entries of A^2. Here, the square of the Adjacency Matrix, refers to the *matrix multiplication* of A by itself, where the elements $A^2(i, j)$ of the square matrix are just the sum of the products of the corresponding entries in row i of the first A matrix by column j of the second [5]. The result is

$$A^2 = \begin{pmatrix} 0 & 1 \\ 2 & 0 \end{pmatrix} \begin{pmatrix} 0 & 1 \\ 2 & 0 \end{pmatrix} = \begin{pmatrix} 2 & 0 \\ 0 & 2 \end{pmatrix}. \tag{13.3}$$

The diagonal entries are just 2's, meaning that starting from either vertex 1, or vertex 2, there are only two nonredundant logic loops to follow that will bring you back to your starting point. And, the examination of Fig. 13.4, as well as Figs. 13.3 and 13.1, shows that this result is clearly true! Calculations for the second, more complex, example (Fig. 13.5) are more interesting.

The FCG (Fig. 13.6b) for the program of Fig. 13.5 shows that there are three logic loops that contain vertex 1 (the first decision after the read statement) and are two edges long. There is also one logic loop that contains vertex 1 and is three edges long. Each of these loops contains a safety-critical statement (i.e., the calculation of IU). And, if there were a typographical error in one of these statements (say, IU = −L − M instead of IU = L − M; an error that no compiler could detect), then IU would become negative for positive L and M, and the hypothetical rocket motor of this example would shut down unexpectedly during mid-flight. This, in turn, would cause departure from the intended course, as the rocket fell back to Earth along an unexpected parabolic trajectory.[†] Figure 13.6b is easy to understand. But, what if the program in Fig. 13.5 were 1000 times more complex, or even just 10 times more complex. Would an analyst feel confident that he could count and identify all logic loops by inspection of the FCG without missing any? Once identified, each logic loop can be *proved* by *Mathematical Induction* (called *Inductive Assertion* in the software context [6]). This procedure is particularly suitable for the rocket motor example because only integer variables are used in its safety-critical algorithms, and the analyst knows IU must always be positive during flight. Therefore, a numerical computation of the number of logic loops in a piece of code is central to the debugging process. Again, in this as in the previous example, loops that start from vertex 1 and return to that same vertex, and that are two or three edges long, cannot be redundant. That is to say, no edge is traversed more than once in completing the loop. Furthermore, there are no logic loops whose length is four edges long, or longer. Therefore, the redundancy matrices for loops 2 or 3 edges long are, again, the zero matrices (i.e., $R_2 = [0]$ and $R_3 = [0]$). In that case, the number of logic loops that are two edges long, and start and end at vertex 1, are supplied by the $(1, 1)$ matrix element of A^2, and those that are three edges long, and start and end at vertex 1, are supplied by the $(1, 1)$ element of A^3. That is to say,

$$A^2 = \begin{pmatrix} 0 & 1 & 1 & 0 \\ 2 & 0 & 0 & 0 \\ 1 & 0 & 0 & 1 \\ 1 & 0 & 0 & 0 \end{pmatrix} \begin{pmatrix} 0 & 1 & 1 & 0 \\ 2 & 0 & 0 & 0 \\ 1 & 0 & 0 & 1 \\ 1 & 0 & 0 & 0 \end{pmatrix} = \begin{pmatrix} 3 & 0 & 0 & 1 \\ 0 & 2 & 2 & 0 \\ 1 & 1 & 1 & 0 \\ 0 & 1 & 1 & 0 \end{pmatrix} \quad (13.4)$$

$$A^3 = A^2 A = \begin{pmatrix} 3 & 0 & 0 & 1 \\ 0 & 2 & 2 & 0 \\ 1 & 1 & 1 & 0 \\ 0 & 1 & 1 & 0 \end{pmatrix} \begin{pmatrix} 0 & 1 & 1 & 0 \\ 2 & 0 & 0 & 0 \\ 1 & 0 & 0 & 1 \\ 1 & 0 & 0 & 0 \end{pmatrix} = \begin{pmatrix} 1 & 3 & 1 & 0 \\ 6 & 0 & 0 & 2 \\ 3 & 1 & 1 & 1 \\ 3 & 0 & 0 & 1 \end{pmatrix}. \quad (13.5)$$

The boxed number 3 in the rightmost matrix of Eq. 13.4 (i.e., the (1, 1) element of A^2) means that there are three logic loops that start at vertex 1 and return to vertex 1 after two steps. These are $1 \rightarrow 2 \rightarrow 1$, another $1 \rightarrow 2 \rightarrow 1$ returning through a different path, and $1 \rightarrow 3 \rightarrow 1$. Similarly, the boxed number 1 in the right most matrix of Eq. 13.5 (i.e., the (1, 1) element of A^3) means that there is only one path that is three edges long and passes through vertex 1. This path is $1 \rightarrow 3 \rightarrow 4 \rightarrow 1$. All total, there are four safety-critical logic loops (3 + 1) in the FC of Fig. 13.5, as there should be. If there were five safety-critical loops that would mean that an extra unwanted calculation of IU could be performed—a very serious safety critical programing error!

Now, in the case of logic loops that pass through vertex 1, only the (1, 1) element is needed from the matrices on the right of Eqs. 13.4 and 13.5. Therefore, the analyst can take a short cut by computing the sum of the multiplications of corresponding values enclosed within the rectangles of matrices on the left of Eqs. 13.4 and 13.5; i.e., $(0)(0) + (1)(2) + (1)(1) + (0)(1) = 3$ for the (1, 1) element of the rightmost matrix of Eq. 13.4, and $(3)(0) + (0)(2) + (0)(1) + (1)(1) = 1$ for the (1, 1) element of the rightmost matrix in Eq. 13.5. But, of course, the other elements in the first row of A^2 were necessary for computation of the (1, 1) element in A^3. Furthermore, calculating other elements of A^2 and A^3 can be very helpful because of the additional information they contain. For example, the 1 in location (3, 3) on the rightmost matrix in Eq. 13.5 means that there exists only one logic loop that passes through vertex 3 and is three edges long. Inspection of Fig. 13.6b reveals that this is in fact so. Off-diagonal elements do not refer to logic loops, but to the number of logic paths between any two vertices within a code. This can be useful, but it can also produce illogical or trivial results. For example, the entry (3, 4) of A^3 is 1, meaning that there is a path with flow in the proper direction, that will lead from vertex 3 to vertex 4. That path is $3 \rightarrow 1 \rightarrow 3 \rightarrow 4$ (see Fig. 13.6b). Notice that vertex 3 is visited twice! In that case, a new set of L, M, and N values will be read, and there is no guarantee that these new values will ultimately send the logic back to the IF Statement 3, although it could. Furthermore, the path fragment $3 \rightarrow 1 \rightarrow 3$ is a loop that would be counted by the (3, 3) element of A^2 (see Eq. 13.4). Ultimately, the edge $3 \rightarrow 4$ must be traversed, therefore the off-diagonal element (3, 4) of A^3 tells the analyst nothing new about going from statement 3 to statement 4, other than to send the analyst around a pointless loop, in this case. Finally, it should be noted that the Adjacency Matrix used in this example (Eq. 13.2) is still quite small. Real NFCs for real code can be enormous. In that case, computation of powers of A must be accomplished by commercial off-the-shelf software [7] since manual computation is impractical.

13.7 Conclusion and Summary

Each new type of Flowchart discussed in this chapter (the CFC, FCG, and NFC) provides the analyst with new insights about logic flow within a program. However, by far, the most abstract Flowchart, the NFC, is the most powerful of all. The Adjacency Matrix (NFC) captures the topology of a software system at an intermediate structural level. It tells the designer which statements, or blocks of statements, are connected to which others. While the diagonal elements (i, i) of the matrices A^2, A^3, ... A^n, or more properly the Fundamental Matrices (F_2, F_3, ... F_n) of Chaps. 7 and 10, tell the designer how many logic loops exist that are 2, 3, ... n edges long and contain vertex i. In the examples presented in this chapter, the Redundancy Matrices were all zero matrices. However, in general, that will not be the case. And, the Redundancy Matrix, which can be calculated from the Adjacency Matrix, becomes harder and harder to calculate as n increases (see Chaps. 7 and 10; [8–10]). Additionally, there is more to say about the Adjacency Matrix that is of a philosophical nature.

The asymmetry of the Adjacency Matrix says something about the flow direction of logic in a complex piece of software. The Adjacency Matrix is often symmetrical for real electronic hardware because current can often flow in both the forward and reverse directions in many circuits; such as when an alternating current is applied to an L-C-R circuit. But, software is different. The Adjacency Matrix for software is asymmetrical. This *broken symmetry* reflects the fact that software progresses in a given direction. As in particle physics, broken symmetry produces an *arrow of time*. Software starts from an initial instruction and as time progresses, more and more instructions are involved. Software logic spreads with time, like gas molecules spreading into an evacuated box. So, software has an *entropy* change associated with it. The idea of entropy for a network is not new. It is part of Complexity Theory [8], and it may be defined in several ways depending on the problem at hand. Here the most useful definition of entropy, or more properly, the change in entropy by physical analogy, ΔS, seems to be

$$\Delta S = \frac{1}{2} \sum_{\substack{\text{All } i, j \\ i \neq j}} |A(i,j) - A(j,i)|, \qquad (13.6)$$

where the pre-factor of ½ prevents double counting when the values of i and j are interchanged. The larger and the more complex a code is, the larger will be ΔS. Entropy may sound like an abstract idea that is too theoretical to be useful. In fact, it is down to earth and practical. The higher the entropy of a piece of code, the more prone it will be to unexpected or undesirable behavior (such as timing errors, for example). Given the choice between several pieces of code, or several software designs, the one with the lowest entropy should be preferred, since it will be the most robust.

The final questions are, (1) where does the loop counting procedure, described in this chapter, fit in with other debugging procedures and (2) can truly perfect software be written? These are deep waters, but the next, and last, chapter of Part 3 of this work will attempt to answer these very questions.

† There is currently a debate going on within the system safety community as to whether a mishap of this kind, beyond the safe separation distance, should be considered. Some analysts feel that anything that happens beyond safe separation is a "reliability problem". Others feel that the impact of a rocket or missile on an unintended location is a significant system safety issue that should be addressed. Undoubtedly, this argument will continue for years to come!

13.8 Problem

What do the elements $A^3(2, 2)$, $A^3(3, 3)$, and $A^3(4, 4)$, tell the analyst? What are the corresponding paths?

References

1. Zito, R.R. (2012). "Sneak Circuits" and related system safety electrical problems-I: Matrix methods. In *Proceedings ISSC—2012*, Atlanta, GA, August 6–10, 2012.
2. Anon. (2006). *SCAT, sneak circuit analysis tool user's guide—Ver. 5.0*. Culver City, CA: SoHar Inc.
3. Allen, F. E. (1970). Control flow analysis. *SIGPLAN Notices, 5*(7), 1–19.
4. Prosser, R. T. (1959). Applications of Boolean matrices to the analysis if flow diagrams. In *Proceedings of the Eastern Joint IRE-AIEE-ACM Computer Conference* (pp. 133–138), December 1–3.
5. Selby, S. M. (1968). *Standard mathematical tables* (16th ed., p. 108). CRC: Cleveland, OH.
6. Nance, D. W., & Naps, T. L. (1995). *Computer science* (3rd ed., pp. 820–825). St. Paul, MN: West Publishing Company.
7. Hanselman, D., & Littlefield, B. (1998). *Mastering MATLAB 5*. Upper Saddle River, NJ: Prentice Hall.
8. Marshall, C. W. (1971). *Applied graph theory* (pp. 258–261, 235–240). New York: Wiley.
9. Ross, I. C., & Harary, F. (1952). On the determination of redundancies in sociometric chains. *Psychometrika, 17*, 195–208.
10. Parthasarathy, K. R. (1964). Enumeration of paths in digraphs. *Psychometrika, 29*, 153–165.

Chapter 14
Fail-Safe Control Software

14.1 Introduction

The central question of software engineering is simply this, "*Is it possible to write, or correct, software so as to produce 'perfect' code?*" It is a controversial question with a history that spans many decades. And so, only a brief survey can be presented here of material that is directly relevant to the task at hand. If by "perfect" code one means a code that can be *proved* to yield truth under *all* circumstances, perfection may not be achievable. In a seminal paper by Fetzer [1], the author points out that a computer program as a collection of algorithms to be run on a machine may, or may not, correctly model the real-world system which it is intended to mimic. Furthermore, our confidence in a complex software system rests on an "empirical, inductive investigation to support inconclusive relative verification." After all, no analyst can test an infinite number of cases. But, is that really so? There is certainly a class of algorithms that can be conveniently tested by mathematical induction. The basic idea (which is really deduction) is this [2]: "If an algorithm involving the positive integer n can be proved to have the following properties:

(A) The algorithm is correct for n = 1,
(B) If k is any value of n for which the algorithm is true, and the algorithm is also true for the next value n = k+1,

then the algorithm is true for all positive integral values of n."

Examples of the use of The Axiom of Induction in programming can be found in the literature [3]. In the context of computer work, the axiom is usually referred to as the Method of Inductive Assertion, and can successfully verify "perfect" code (*prove* code) provided [4].

(1) the input and output assertions are, in fact, true,
(2) no exceptional conditions are encountered that deviate from the input assertions (e.g., n becomes negative!),

© Springer Nature Switzerland AG 2020
R. R. Zito, *Mathematical Foundations of System Safety Engineering*,
https://doi.org/10.1007/978-3-030-26241-9_14

(3) the logic within an individual module is being tested, and not how different modules interact,

(4) the semantics (language) of the algorithm is correct and has not been misinterpreted,

(5) there are no practical machine limitations such as roundoff and overflow errors, or "hard"/"soft" failures,

(6) each "proven" individual algorithm produces no side effects that result in overall system errors,

(7) non-numeric algorithms like searching and sorting are not involved as these are more difficult to "prove",

(8) the algorithm does what the user actually needs, and

(9) the human beings that use the code behave correctly (i.e., inputs correct data, no transcription errors, etc.).

Because of these limitations, the Method of Inductive Assertion has been limited to fairly simple codes (numerical algorithms). For more complex codes, the analyst has been traditionally forced into the trap of "relative verification" as outlined by Fetzer [1]. Therefore, "formal correctness proofs must currently be considered a technique that is mostly of theoretical importance. Such proofs are simply not sophisticated enough to find practical application in programs that solve realistically complex problems" [3]. *The object of this chapter is to bridge the "complexity gap" and present a method of verification that examines a code, not just as a collection of individual statements, or even modules, but as a logical network.* This approach, when combined with Inductive Assertion, should allow the production of "perfect code" for a specific type of software; subject to only a few of Myer's 9 caveats listed above (whose negatives are errors which will be denoted by M1 through M9). In particular, this chapter will focus on control code used in a hardware/firmware/ software *chimera.*[†]

14.2   Sneak Logic Circuits

There is a saying among programmers who are testing code, "First, you have to eliminate the problems, then you have to eliminate the real problems, and finally there are the 'real-real' problems." What this humorous bit of advice refers to are the "stages" of debugging.

At the first stages, problems are usually syntactic. These are usually caught by the computer compiler (e.g., unmatched parenthesis, two adjacent arithmetic operation signs, etc.). However, some of these errors can be more subtle; like a typographical error in an equation (M8). These can be eliminated by Inductive Assertion. A "point test" with one or more sets of input data can indicate to the analyst how a piece of code might behave if inputs are unusual or unexpected (e.g., a wrong sign, a non-integer, an out of range input, or an input with peculiar numerical properties like 0, 1, and 2 [i.e. if r is real then $0! = 1! = 1$, $0^r = 0$ for $r > 0$, $1^r = 1$ and $r^0 = 1$ for $r \neq 0$, $2 + 2 = 2 \times 2 = 4$, and $1/0$, 0^0 and $\log_r(0)$ for

$r > 0$ are all undefined] [M2, M9]). Sensitivity to roundoff errors (M5) can also usually be detected by a simple computational test. If an input dataset is increased/decreased by some small amount ε, the output should change by a small amount δ. Sensitivity to initial conditions is a sign of instability that can make roundoff errors serious. Using integer variables and values can mitigate unwanted behavior from roundoff errors. Overflow errors (M5) are also easily defeated by integer arithmetic "maxint" testing procedures [3].

Next, come the "real errors". These are *much* more subtle, and usually, involve some flaw in the logic of a program. That is to say, there is some logical path in the code that was unintended by the programmer/analyst (M3, M5, M6, and M8). The analogous situation occurs in circuit design where an engineer may inadvertently design a path into a complex circuit that causes it to behave in an unexpected way. With this thought in mind, the matrix techniques developed to uncover "sneak circuits" in hardware [5] will be adopted to uncover "sneak logic circuits" in modified logic flowcharts. Control software flowcharts are particularly attractive for this approach because, topologically, they are naturally "circuits", as opposed to "trees". *The purpose of this chapter is to address sneak logic circuit problems.*

Finally, there are "real–real" problems. These are the most difficult to understand. And, *misunderstanding* is the cause of most of them (M4 and M8). That is to say, the model constructed by the engineer might not correctly mimic the physical system of interest, or might not have high enough fidelity. These types of errors are very deep and difficult to remove. They are outside the scope of this chapter.

In addition to the problems associated with these three stages of debugging, there are also user problems (M9) in which incorrect data is input into a software system. For example, in a missile guidance/autodestruct system, a hardware/firmware failure in the navigation system (sensors, processors, memory, etc.) can result in input errors. However, these types of errors will not be addressed here as they are not software in nature.

14.3 An Instructive Example: A Missile Guidance and Autodestruct System

Consider the flowchart (FC) (Fig. 14.1) for a hypothetical missile guidance and autodestruct system.

A designer has *one, and only one*, criterion in his mind for activation of the autodestruct system, viz. *that the missile should destroy itself if it is off course*. With that thought in mind, the designer constructs a code that works as follows. First, ignition takes place at time $t = 0$ (start), and three-coordinate arrays and four error arrays of dimension $N \times 1$ are initialized; where the integer $N = \text{RoundUp}\,(T/\Delta t)$, and T is the maximum time of flight of the missile. After a delay of Δt, a counter k is set equal to 1. Next, a navigation check is made for current values of the x, y, and z coordinates of the missile (call them $x_k(\text{true})$, $y_k(\text{true})$, and $z_k(\text{true})$), where $k = 1$

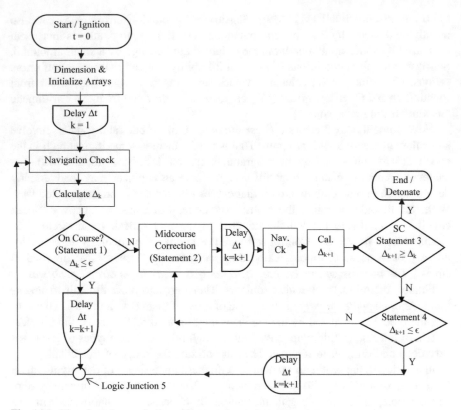

Fig. 14.1 Flowchart for a missile guidance and autodestruct system

after the first delay. Errors are then computed between the true and planned missile coordinates.

$$\Delta_k(x) = x_k(\text{plan}) - x_k(\text{true})$$
$$\Delta_k(y) = y_k(\text{plan}) - y_k(\text{true})$$ (14.1)
$$\Delta_k(z) = z_k(\text{plan}) - z_k(\text{true}).$$

The total error Δ_k is given by

$$\Delta_k = \sqrt{(\Delta_k^2(x) + \Delta_k^2(y) + \Delta_k^2(z))}.$$ (14.2)

This information is passed on to an IF Statement (Statement 1) which makes a test. If the kth delta, $\Delta_k \leq \epsilon$, where ϵ is some small tolerance, then do nothing but pass logic to a delay block, after which there will be another navigation check and test at Statement 1. But, if the test fails due to some flight perturbation, then a midcourse correction will be calculated (Statement 2) and implemented based on the values of $x_k(\text{plan})$, $x_k(\text{true})$, $y_k(\text{plan})$, $y_k(\text{true})$, $z_k(\text{plan})$, and $z_k(\text{true})$. This last action will then be followed by a delay. Some time will be needed for a high-speed missile with a great deal of momentum to correct its course, and then another navigation check and calculation, Δ_{k+1}, of the midcourse error will be made. Now, a safety-critical (SC) IF Statement is encountered (Statement 3). If $\Delta_{k+1} \geq \Delta_k$, then the midcourse correction hasn't worked and the missile must be destroyed, lest it should "lock-on" and attack an unintended target.[‡] On the other hand, if this last test fails, then midcourse corrections are working and the only question is whether further corrections need to be made. Therefore, the next IF Statement (4) tests to see if $\Delta_{k+1} \leq \epsilon$. If not, then return to Statement 2 and make another midcourse correction. If affirmative, do nothing but wait (delay), update the counter, and return for another navigation check and test at Statement 1. All variables and calculations should be considered to be of the integer type to avoid roundoff errors. The logic is so simple, what could go wrong?

A lot! This simple system contains a subtle flaw that is best explained by an examination of the "Circuitized" Flowchart (CFC) described in the previous chapter [6]. In Fig. 14.2 below, the counters k = k + 1 (associated with each delay Δt) drive the logic forward and act like batteries, calculations for Δ_k and midcourse corrections are represented by resistors, logic flow directions are denoted by diodes, logical decisions are marked by junctions, navigation checks have been replaced by wires, and Start/End instructions have been represented by a switch. The black dots represent logic junctions.

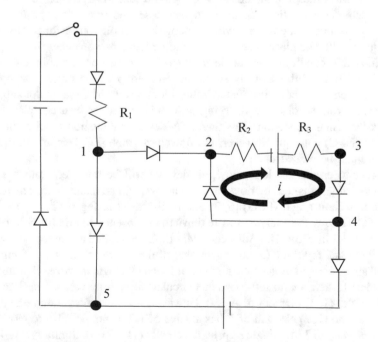

Fig. 14.2 The CFC for the FC of Fig. 14.1

The inner loop $2 \to 3 \to 4 \to 2$ should immediately attract an analyst's attention because if R_2, $R_3 \to 0$ a short circuit will develop (a sneak short circuit) that will drive the current $i \to \infty$. What is the meaning of such a statement in terms of code logic? By way of analogy, $R_2 \to 0$ means that the midcourse correction (Statement 2) is far too small. And, $R_3 \to 0$ means that Δ_{k+1} has hardly changed from Δ_k. That is to say, R_3 did almost nothing to the value it received (i.e., it acted almost like an ideal straight wire with zero resistance). In fact, as long as Δ_{k+1} remains in the interval $\Delta_k > \Delta_{k+1} > \epsilon$, logic will continue to circulate within this inner loop. Possibly for a very long time before escaping. Or, to state this argument differently, each time around the loop an abnormally small correction is made, perhaps due to a poor midcourse correction algorithm design. This kind of error might not be caught by Inductive Assertion, but it has subtle system side effects (M2, M6, M8). And, after each delay Δt, the updated value of Δ_k (i.e. Δ_{k+1}) is only slightly diminished from its previous value, so that logic must go around the inner loop again and again.

In time, after many delays Δt, the missile could have reached the target area, in which case two things can happen. Either the missile will miss its target (a reliability issue), or it will lock on to something unexpected (a system safety issue).[‡] Basically, this problem arises because there are two logic loops that contain safety-critical Statement 3. And, one of these loops allows the safety-critical action (autodestruct) to be bypassed. The problem could, of course, be easily fixed simply by not allowing logic to circulate in the $2 \to 3 \to 4 \to 2$ loop more than a fixed number of times (i.e., the Statement 3 test should read "$\Delta_{k+1} \geq \Delta_k$, or n consecutive midcourse corrections made"). In that case, the inner "escape loop" will eventually disappear, the missile will be recognized as off course for too long, and autodestruct will take place. The number of logic loops now matches the designer's intent (i.e., one condition for autodestruct—*off course*). For this simple example, visual inspection of the CFC is sufficient. But, for more complex examples, a numerical method of counting (enumerating) loops is desired that is not subject to the human error associated with trying to identify all loops in a complex logical network by visual inspection. This then provides the motivation for the numerical flowchart (NFC) concept. However, the flowchart graph (FCG) is an intermediate step that must be passed first.

In order to emphasize logic flow, all devices will be removed from the CFC. Only lines (edges) marked by flow direction arrows will be maintained. The result is called a flowchart graph (FCG) [6, 7] and is displayed in Fig. 14.3.

The FCG allows the analyst to write down the Adjacency Matrix (or NFC) for the FC of Fig. 14.1 directly. The Adjacency Matrix A, also called a Connectivity Matrix [8], captures the *topology* (interconnections) of flow in the flow chart of Fig. 14.1 and will allow computation of the number of loops in a logic network. Starting from vertex (dot) 1, there are no self-loops that circulate logic from vertex 1 back to itself. Therefore, the (1, 1) element of A is 0. However, logic can flow along one path to vertex 2, so the (1, 2) element of A has a value of 1. It is not possible to move from vertex 1 *directly* to 3 in a single step, so the element (1, 3) = 0. Similarly, (1, 4) = 0. But, (1, 5) = 1. Note that even though 2 edges (lines) connect vertices 1 and 5, the

Fig. 14.3 The FCG for the
FC in Fig. 14.1

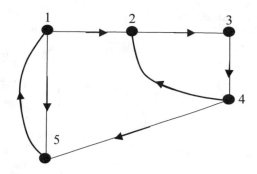

direction must be taken into account. So, there is really only one way to go from vertex 1 to vertex 5. Therefore, the first row of A looks like "0 1 0 0 1". The second row of A applies to connections that start out from vertex 2 and looks like "0 0 1 0 0". The entire matrix A looks like

$$A = \begin{pmatrix} 0 & 1 & 0 & 0 & 1 \\ 0 & 0 & 1 & 0 & 0 \\ 0 & 0 & 0 & 1 & 0 \\ 0 & 1 & 0 & 0 & 1 \\ 1 & 0 & 0 & 0 & 0 \end{pmatrix}. \qquad (14.3)$$

The next step in numerical loop enumeration is to compute A^n for logic paths n "steps" (edges) long. The diagonal elements of this matrix count logic paths that start and end at vertex i. These paths include undesirable redundant paths as well as desired simple logic loops without switchbacks or sub-loops (loops within loops). A redundant path is one that allows logic to flow over the same edge more than once, like a nine-step path connecting vertex 1 and vertex 4 (i.e., $1 \rightarrow 2 \rightarrow 3$ $4 \rightarrow 2 \rightarrow 3 \rightarrow 4 \rightarrow 2 \rightarrow 3 \rightarrow 4$; nine steps, edges, or arrows, long). For the FCG in Fig. 14.3, the analyst is most concerned with simple paths (loops) that start at vertex 3, are three or five edges long, and return to vertex 3. There are no loops that are 4 edges long and involve vertex 3. Vertex 3 is the analyst's focus because it marks a safety-critical decision. Therefore, the analyst will want to inspect the diagonal $(3, 3)$ matrix elements of A^3 and A^5. Since it is not possible to form a redundant loop with only three edges, and since there are no redundant paths 5 edges long that start and end at vertex 3, redundancy can be ignored.[*] MATLAB [9] can be used to calculate A^3 and A^5 by the following simple code:

$$A = [01001; 00100; 00010; 01001; 10000]$$
$$Acube = A \wedge 3 \qquad (14.4)$$
$$Afifth = A \wedge 5$$

Therefore,

$$A^3 = \begin{pmatrix} 0 & 1 & 0 & 1 & 1 \\ 0 & 1 & 0 & 0 & 1 \\ 1 & 0 & \boxed{1} & 0 & 0 \\ 0 & 1 & 0 & 1 & 1 \\ 1 & 0 & 1 & 0 & 0 \end{pmatrix}, \qquad (14.5)$$

and

$$A^5 = \begin{pmatrix} 1 & 1 & 1 & 1 & 1 \\ 0 & 1 & 0 & 1 & 1 \\ 1 & 1 & \boxed{1} & 0 & 1 \\ 1 & 1 & 1 & 1 & 1 \\ 1 & 1 & 1 & 0 & 1 \end{pmatrix}. \qquad (14.6)$$

The boxed "1" in the matrix of Eq. 14.5 tells the analyst that there is one logic loop 3 edges long that passes through vertex 3 (safety-critical Statement 3). The boxed "1" in the matrix of Eq. 14.6 tells the analyst that there is only one logic loop 5 edges long that passes through vertex 3. Therefore, only two simple logic loops, one 3 edges long and one 5 edges long, pass through Statement 3. Clearly, this is true by direct inspection of the simple FC in Fig. 14.1. For software that is considerably more complex, fail-safe loop counting by inspection would be difficult. As a practical matter, a loop counting procedure should directly follow algorithm verification by Inductive Assertion. This will give the analyst an initial "heads-up" if the number of simple loops passing through a safety-critical step does not match his/her expectations.

The next step in loop analysis is to identify each loop (i.e., identify the vertices it contains). In simple cases, this can be done by inspection, especially if the exact number of loops is known. However, computer search methods are available to extract all loops with certainty (see Chap. 11; [10]). Figure 14.4 displays the two safety-critical loops of concern.

When logic loops are separated in this fashion, it is easy for the analyst to pick out subtle logic flaws like M2, M3, M6, and M8. The effect of unusual or extreme conditions of the form R_2, $R_3 \rightarrow 0$ has already been discussed. At the other extreme, R_2, $R_3 \rightarrow \infty$, logic in both loops stops flowing. Why? If R_2 is very large, that implies that the midcourse correction is far too large. In that case, $\Delta_{k+1} \gg \Delta_k$ and the autodestruct command would be activated by Statement 3. In fact, in this case, Δ_{k+1} will *always* be larger than Δ_k and loop B will *never* be entered. This *overcorrection* is not what the designer wants either! The type of single-loop analysis just described can be partially automated so long as a computer knows what type of CFC devices lie between each pair of vertices, and the dangerous aspects of each CFC device. A typical interactive question sequence might look like that displayed below.

Enter loop vertices: 2, 3, 4, 2 ENTER
WARNING! An error in IF Statement 4 does not allow entry into this loop!

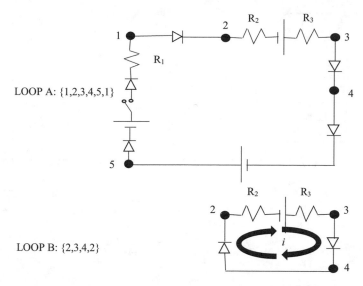

Fig. 14.4 The two simple logic loops that pass through safety-critical Statement 3 of Fig. 14.1. Loop A is 5 edges long, while loop B is 3 edges long. The notation { } encloses the vertices of the loop

Is loop {2, 3, 4, 2} extraneous code? (YES/NO): N̲ ENTER
WARNING! If R_2, R_3 = 0, a loop short circuit exists!
Is this what you want? (YES/NO): N̲ ENTER
WARNING! If R_2, R_3 → ∞, logic stops flowing in loops {2, 3, 4, 2} and {1, 2, 3, 4, 5, 1}
Is this what you want? (YES/NO): N̲ ENTER

14.4   Sorting and Searching

Until this point, little has been said about M7 errors. The reason, of course, is that sorting and searching problems were not encountered in the previous example, which focuses on the control software. However, nonnumerical algorithms like sorting and searching, although difficult to prove by inductive techniques, are nonetheless natural for, and susceptible to, loop analysis. Take, for example, the problem of a binary search. Consider a 2x(N − 1) integer array called ARRAY such that the first row of elements consists of sorted part numbers, while the second row consists of the corresponding flammability rating for each part (i.e. 0 = minimal hazard, 1, 2, 3, 4 = progressively more severe hazards). Furthermore, let $N = 2^n$, where n is a positive integer greater than 2. For definiteness set n = 3, then ARRAY has seven columns that might look like Eq. 14.7.

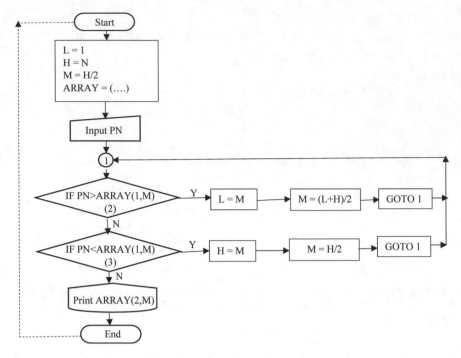

Fig. 14.5 FC for a binary search routine

$$ARRAY = \begin{pmatrix} 100 & 200 & 300 & 400 & 500 & 600 & 700 \\ 0 & 3 & 1 & 4 & 2 & 2 & 1 \end{pmatrix} \begin{matrix} \leftarrow \text{Part No} \\ \leftarrow \text{Flammability Rating} \end{matrix}$$

$$(14.7)$$

Figure 14.5 displays a FC for this binary search routine that supplies the flammability rating for any input part number. The code works as follows. A midpoint M is calculated from the high (H) and low (L) indices for the array, or some portion of the array. Then the input part number (PN) is tested to see if it is greater or less than element (1, M). In either case, a new midpoint is calculated and the test repeated until there is a match between the input (PN) and one of the array elements in the first row.

Figure 14.6a–c show that the FC of Fig. 14.5 can be neatly reduced to a CFC, FCG, and finally an NFC. The FCG shows that there are 3 return paths to vertex 1. One corresponds to a PN less than the array element it's being tested against. One corresponds to a PN greater than the array element. And, one corresponds to equality (the PN being searched for). There is nothing abnormal or alarming about this FCG, and the code will work as long as there are no human input (part number) errors (M9). This type of error cannot be detected by analysis. Therefore, in that case, the search will proceed without finding a match until eventually some non-integer arithmetic is encountered (M5), thereby producing an error message

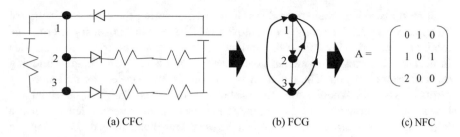

Fig. 14.6 The **a** CFC, **b** FCG, and **c** NFC for the FC in Fig. 14.5

that informs the operator of his/her mistake. So, loop analysis is a useful tool for debugging searching routines, even if Inductive Assertion is not, but it is subject to the same M9 errors of other software types.

14.5 The General Hierarchy of Corrections

It is clear from the previous analyses that loop analysis is a powerful tool that augments Inductive Assertion by removing many of its shortcomings; especially with regard to detection of M2, M3, M6, M7, and M8 errors. The pyramid of corrections in Fig. 14.7 summarizes the general sequence of debugging procedures for code that employs integer arithmetic so that roundoff errors (M5) do not occur.

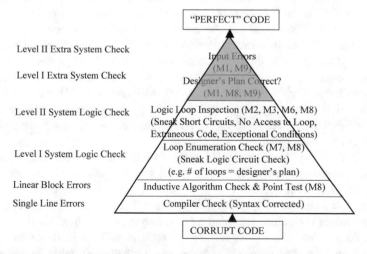

Fig. 14.7 The Hierarchy of Corrections. Each of the 4 lower layers indicates the kinds of errors that can be detected and corrected. Layers in gray indicate Myer errors that can occur and are attributable to human error. These are outside the scope of this chapter. M5 errors are not included because only integer arithmetic is allowed

It must always be remembered that human error during the process of software creation can never be *completely* removed (M8). Nor can human input errors be totally eliminated (M1, M9). If the designer does not have a clear and accurate idea in his mind with regard to his/her goals, mathematics is of limited help in bridging the understanding gap. The situation is not completely hopeless as evidenced by errors of type M8. These errors can exist at various levels. They can be trivial and low level; like a misplaced sign. Or, they can exist at an intermediate level such as a logic sneak short circuit. Or, they can be very high level indeed, involving the designer's understanding, or misunderstanding, of the physical process or system the designer wishes to model; in which case little can be done. It is also important to note that detection of an error like M8 at the intermediate level, may necessitate algorithm redesign and a return to inductive algorithm checks.

14.6   Conclusion

Designer *intent* is a difficult concept to quantify. Yet, if the designer's thinking is correct, it seems possible to build "perfect" software in the sense that the software accurately reflects the designer's plan (which may or may not be correct). However, no general proof of this conjecture has been provided to ensure that the sequence of procedures suggested in this chapter covers all possible cases—it probably doesn't. So, the answer to the initial question posed at the start of this chapter is this, "*Perfect code, in the absolute sense of the word perfect, is probably not achievable. However, if by perfect one excludes errors of the M1 and M9 type, as well as some M8 errors having to do with fundamental understanding of the programmer's problem, and if the code is restricted to integer arithmetic on a reliable machine, then perfect (provable) code can be written in at least some cases.*" So, here, a distinction is being made between *absolute perfection* and *practical perfection* that excludes shortcomings of the software creation and operation process that are due to human error, floating-point arithmetic errors, and machine hardware and radiation-induced "soft" errors. Although M1, M9, and some M8 errors are human in origin and can never be eliminated with complete confidence, their occurrence can be mitigated by peer review [1]. While machine errors can be mitigated by redundancy, voting procedures, and shielding.

Having given at least a qualified answer to the central question of software analysis, it seems appropriate to end this third, and last, chapter on software with a few big philosophical questions. Is it possible, using a combination of Inductive Assertion and loop analysis, to develop software that generates perfect code given input and output conditions specified by the analyst? The realization of such a goal may be far off into the future, and so no attempt will be made to answer this question here. But, if one piece of software can write another, what does that imply for reproduction as a hallmark of life? And, is such code generating software true artificial intelligence?

† According to Webster's, a *chimera* is, in general, "an often fantastic combination of incongruous parts," in art "a grotesque animal form in painting or sculpture compounded from parts of different real or imaginary animals," or in biology "an individual, organ, or part consisting of tissues of diverse genetic constitution." In the context of system safety, a chimera will refer to *a complex system composed of some combination of hardware, firmware, and software.*

‡ It should be noted that there is some debate in the system safety community about whether a mishap of this type should be considered a system safety issue. One school of thought is that once a missile exceeds safe separation distance "it's in the air, it's there, and we don't care." But, others in the community consider acquiring unintended targets to be a serious system safety issue that should be addressed.

* In fact, whenever the number of redundant paths is small, it is usually easier to get the number of simple paths or loops from the appropriate element of A^n by subtracting off redundant paths detected by inspection, rather than to calculate the elements of the Fundamental Matrix F_n discussed in Chap. 10.

14.7   Problem

Given the CFC of Fig. 14.2, and the expression for A^5 in Eq. 14.6, notice that 22 of the 25 matrix elements are 1's. What does that mean? Draw the paths. Which paths (loops) are redundant? Which paths (loops) are simple? A^5 has three zeros. Why?

References

1. Fetzer, J. (1988). Program verification: The very idea. *Communications of the ACM, 31,* 1048–1063.
2. Richardson, M. (1947). *College algebra* (pp. 316–320). New York: Prentice Hall.
3. Nance, D. W., & Naps, T. L. (1995). *Computer science,* 3rd edn., West Publishing Co., Minneapolis/St. Paul MN, pp. 56, 323–324, 820–825, 706–708.
4. Myer, G. (1976). *Software reliability: Principle and practices* (pp. 319–320). New York: Wiley.
5. Zito, R. R. (2012). 'Sneak Circuits' and related system safety electrical problems—I: Matrix methods. In *30th ISSC Proceedings* (pp. 6–10). GA, Aug: Atlanta.
6. Zito, R. R. (2015, 24–28 August). New flowcharts for a mathematical analysis of logic—Software II. In: *33rd ISSC Proceedings*, San Diego, CA.
7. Allen, F. E. (1970). Control flow analysis. *SIGPLAN Notices, 5*(7), 1–19.
8. Prosser, R. T. (1959). Applications of boolean matrices to the analysis of flow diagrams. In: *Proceedings of the Eastern Joint IRE-AIEE-ACM Computer Conference, Dec. 1–3,* pp. 133–138.
9. Hanselman, D., & Littlefield, B. (1998). *Mastering MATLAB 5*. Upper Saddle River, NJ: Prentice Hall.
10. Zito, R. R. (2012). *'Sneak Circuits' and related system safety electrical problems-II: Computer search methods, 30th ISSC Proceedings* (pp. 6–10). GA, Aug: Atlanta.

Part IV
Dangerous Goods

Chapter 15
Design Phase Elimination of Beryllium

15.1 Background

The author's interest in beryllium metal and its applications and risks began when he was working on an aerospace test program. During one design meeting, the author, who was then working as a system safety engineer, raised a question as to why certain parts were being made out of beryllium. The program's lead test engineer involved in this experimental effort began to shout, "Because it's light! Don't you understand?!" Well, as the author's supervisor at that time pointed out, "People who say things of that kind do so because they don't really know the answer themselves." And, as it turns out, she was quite correct. The true answer to this seemingly simple question is not as obvious as one might think. This investigation will commence by comparing the application virtues of just two metals (beryllium vs. magnesium). It will be discovered that even this simple "two-dimensional" problem will yield some surprising results. Later in this chapter, methods will be presented to extend the solution of this simple problem to optimizing materials selection when many different metals, alloys, or materials are candidates for use. This last more complex problem will prove to be an interesting exercise in matrix algebra and hyperspace geometry.

15.2 Graphical Solution to the "Two-Dimensional" Problem

To provide a concrete example, suppose an aerospace engineer wishes to construct a space experiment like the one in Fig. 15.1a. It has an upper and lower plate supported by 10 rods of square cross section. The total weight of these rods is limited to 500 nt, and all the rods together must have a strength of at least 185×10^4 nt. Furthermore, the upper and lower plate are separated by 1 m and a

© Springer Nature Switzerland AG 2020
R. R. Zito, *Mathematical Foundations of System Safety Engineering*,
https://doi.org/10.1007/978-3-030-26241-9_15

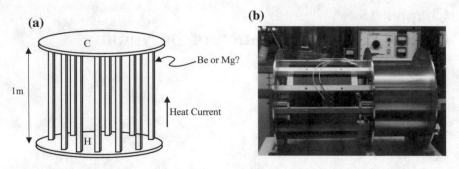

Fig. 15.1 a Model space experiment design. A hot plate is marked by an "H". A cold plate is marked by a "C". The temperature difference is 100 K^0. What should the vertical support members be made of to satisfy design requirements of weight, strength, and heat conduction? **b** CONCOP experiment, the inspiration for **a**

temperature difference of 100 K^0. A heat current from the lower to the upper plate is required to be at least 125 W (or 125 j/s, or 125 nt. m/s). If the reader thinks this example is contrived—don't! A very similar experiment, the Containerless Coating Process (CONCOP; Fig. 15.1b), flew on STS-60 (Feb. 3, 1994); only there, the thermal gradient was radial rather than longitudinal (Fig. 15.1b). The aerospace engineer now asks, "Can the desired strength and heat conduction be achieved without violating weight requirements? Should beryllium or magnesium be used? Or, should a combination of the two metals be used? Beryllium is stronger, but magnesium is lighter. What should be done?" The system safety engineer asks, "Can beryllium be eliminated altogether? And, if not, what is the least amount of beryllium needed to satisfy program requirements?" To answer all these questions, the first step will be to build a table of materials and requirements versus typical properties (Table 15.1) [1–3].

The second step in the analysis is to reduce the information in Table 15.1 to three inequalities. Let X equal the volume of magnesium used in m^3, and Y equal the volume of beryllium used in m^3. If g is the acceleration of gravity (9.8 m/s^2), then the weight requirements yield the inequality

Table 15.1 Materials and requirements versus typical properties

Properties and costs	Materials and requirements		
	Mg	Be	Requirements
Density (kg/m^3)	1.74×10^3	1.85×10^3	Wt. of supports $\leq$ 500 nt
Tensile strength (Pa)	185×10^6	370×10^6	$\geq 185 \times 10^4$ nt
Thermal conductivity (W/mK^0)	156	200	$\geq$ 125 W
Fabrication cost ($/$m^3$ of finished parts)	5×10^5	5×10^7	

$$1.74 \times 10^3 \, X \, g + 1.85 \times 10^3 \, Y \, g \leq 500 \quad \text{or} \quad \boxed{34.10 \, X + 36.26 \, Y \leq 1} \, . \tag{15.1}$$

Let L equal the plate separation (1 m), so that X/L is the total cross-sectional area of all magnesium support rods, and Y/L is the cross-sectional area of all beryllium support rods. Then the strength requirements yield the inequality

$$185 \times 10^6 \, X/L + 370 \times 10^6 \, Y/L \geq 185 \times 10^4 \quad \text{or} \quad \boxed{X + 2Y \geq 0.01} \, . \tag{15.2}$$

Let ΔT equal the temperature difference between the plates (100 K^0). Since the total thermal current in watts is just the product of the thermal conductivity of Mg times the cross-sectional area of the Mg rods divided by L, plus the thermal conductivity of Be times the cross-sectional area of the Be rods divided by L, the thermal conductivity heat sink requirements yield the inequality

$$156 \, X \, \Delta T / \, L^2 + 200 \, Y \, \Delta T / \, L^2 \geq 125 \quad \text{or} \quad \boxed{124 \, X + 160 \, Y \geq 1} \, . \tag{15.3}$$

Figure 15.2 is a plot of these last three boxed inequalities, where X and Y can never be negative (i.e., X, Y $\geq$ 0). Notice that the axes are labeled for Mg and Be use, both measured in cubic meters. It is for this reason that this Linear Programming problem is referred to as a "two-dimensional" problem. The shaded areas mark forbidden (X, Y) pairs, while the unshaded irregular pentagonal region contains (X, Y) pairs that satisfy all the design requirements of the system. The unshaded region is called the *Feasibility Set*. Notice that the Feasibility Set includes a subset of solutions that lie along the X-axis and use only magnesium metal (i.e. Y = 0). Furthermore, the solution of the minimum cost will be the one that uses the least metal and requires the least machining. This *optimal solution* is marked by a black dot in Fig. 15.2. In fact, a *Cost Function* C (or *Objective Function*) can be written down for this system using the information in the last row of Table 15.1 (Eq. 15.4). Notice that the cost of finished Be parts is about 100 times the cost of finished magnesium parts due to the health hazards involved and the intrinsic cost of the bulk metal.

$$\boxed{C(\$) = 5 \times 10^5 \, X + 5 \times 10^7 \, Y} \tag{15.4}$$

It is no accident that C (expressed in dollars) has its minimum value at a vertex along the polygonal boundary of the Feasibility Set. In fact, the *Fundamental Theorem of Linear Programming* states, "*The maximum (or minimum) value of the objective function is achieved at one of the vertices of the feasibility set.*" Finally, in addition to cutting costs, and reducing or eliminating beryllium from the design phase of a project, the Linear Programming method also supplies design information as well. Figure 15.2 shows that the optimal design solution requires 0.01 m^3 of magnesium, at a total fabrication cost of $(5 \times 10^5)(0.01)$, or \$5000, by Eq. 15.4. Since there are 10 support rods in the example above (Fig. 15.1a) each rod must

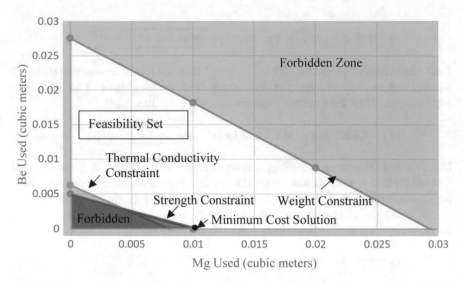

Fig. 15.2 The graphical solution to the "two-dimensional" Be/Mg materials choice problem under the triple constraints of weight, strength, and thermal conductivity

contain 0.001 m^3 of metal. And, since each rod is 1 m long, the cross-sectional area of each rod must be 0.001 m^2. Therefore, each rod must be 3.2 cm (or 1¼ in.) on a side. This type of information can now be fed back to the aerospace design team from system safety. As a check, the calculated rod dimensions and density of magnesium can be used to calculate the total weight of all support rods, with the result that the weight requirements in Table 15.1 are easily satisfied. The same type of direct calculation can be used to check that the other requirements are satisfied as well. In summary, all three constraints can be satisfied without the use of beryllium, and with the cost being minimized as well. A dangerous good has been completely eliminated from this design.

Further insight into the "two-dimensional" Be/Mg problem is gained if some of the parameters in Table 15.1 are changed. Suppose the design team wishes to conduct a higher heat current through the experiment, say at least 540 W. What would Fig. 15.2 look like then? Figure 15.3 is the result. Now, the Feasibility Set has shrunk to almost nothing. And, its shape has changed from a pentagon to a very small and narrow triangular zone whose base lies along the Be axis. In this case, only Be supports will be satisfactory. Notice that, in this case, the decision to use Be was based on the ability of Be to conduct 540 W. The fact that Be has a low density (i.e. is a "lightweight material") is irrelevant! One last example will complete this discussion on Geometrical methods. Suppose it is required to conduct 1000 W. What would Fig. 15.2 look like then? Figure 15.4 shows the result. The Feasibility Set has now vanished. Therefore, it is impossible to build this system with either Be or Mg and still meet *all* the project requirements. The design team must now relax requirement, or radically redesign the experiment.

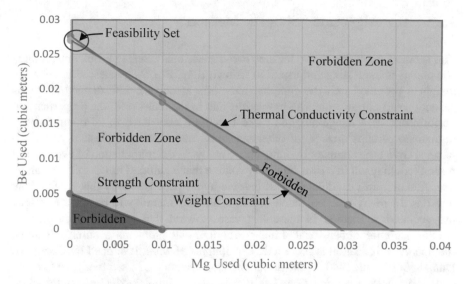

Fig. 15.3 If a minimum thermal current of 540 W is required, the Feasibility Set will shrink to a small triangular zone along the Be axis which has been marked by a circle. In this case, only Be supports will be satisfactory

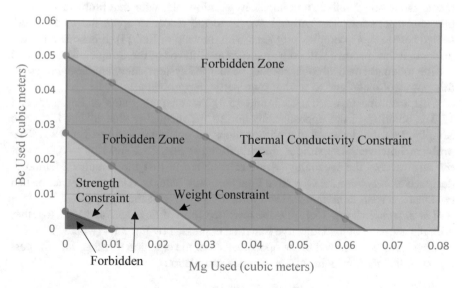

Fig. 15.4 If the minimum heat current is 1000 W, the Feasibility Set vanishes. It is impossible to build this system with either Be or Mg and still meet design requirements

15.3   The Algebraic "Simplex Method"

The geometric solution above for the "two-dimensional" Be/Mg problem with three constraints is interesting and useful in its own right. But, its greatest virtue was the insight it provided into the complex trade-offs associated with dangerous goods reduction and elimination. The problem is that in the real world, the properties and performance of a dangerous good like beryllium may have to be compared too numerous, perhaps dozens, of other substitute metals, alloys, plastics, and composites. Given the complexity of the above analysis, the introduction of even a third option[†] could result in a Feasibility Set volume that is difficult to visualize, let alone use. And, a fourth candidate would involve hyperspace geometry that is out of the question for many practical studies; although some attempt will be made to rescue the geometrical method in these last cases toward the end of this chapter. In complex, higher dimensional, Linear Programming problems a purely algebraic method of optimization is preferred. The *Simples Method* developed by George B. Dantzig in the late 1940s is the dominant algebraic method used today, and is well suited to machine computation. The details of the Simplex Method cannot all be presented here, so some results will simply be stated without proof, knowing that many good texts on Linear Programming exist [4]. In this chapter the Simplex Method will be reapplied to the "two-dimensional" Be/Mg problem above since the reader is already familiar, and hopefully comfortable, with that problem from the previous discussion. The author leaves it to the reader to apply the method presented here to more complex problems with the help of Ref. [4]. It is hoped that the discussion that follows will convey the general flavor of the Simplex Method, and exhibit the mathematical tools needed, so that the system safety engineer can apply this powerful technique to his/her own particular problem.

First, examine Eqs. 15.1, 15.2, and 15.3. The inequality sign of Eqs. 15.2 and 15.3 is pointing in the opposite direction to the inequality sign in Eq. 15.1. The Simplex Method demands that all equations be put in the form of Eq. 15.1. This problem, however, is superficial since multiplying Eqs. 15.2 and 15.3 by a factor of -1 will reverse the inequality. Another problem is that the Simplex Method is designed to *maximize* the Objective Function. But, here, the Objective Function is a *cost* that we wish to *minimize*. Therefore, the *negative* of the right side of Eq. 15.4 will be *maximized*! Eq. 15.1, plus the rewritten form of Eqs. 15.2 and 15.3, together with the negative of the Objective Function expressed by Eq. 15.4 (where the X and Y terms have been moved to the opposite side of the equation together with the cost C), form the starting point of our simplex calculations.

$$
\begin{aligned}
34.10X + 36.26Y &\leq 1 \\
-X - 2Y &\leq -0.01 \\
-124X - 160Y &\leq -1 \\
5 \times 10^5 X + 5 \times 10^7 Y + C &= 0
\end{aligned}
\tag{15.5}
$$

Next, the inequalities will have to be turned into equalities by the introduction of new parameters called *slack variables,* because they take up the "slack" between the left and right side of the inequalities. In this problem, the slack variables will be denoted by u, v, and w, such that

$$
\begin{aligned}
34.10X + 36.26Y + u &= 1 \\
-X - 2Y + v &= -0.01 \\
-124X - 160Y + w &= -1 \\
5 \times 10^5 X + 5 \times 10^7 Y + C &= 0.
\end{aligned}
\tag{15.6}
$$

This system of linear equations has an infinite number of solutions, because it has only four equations in six unknowns. The same was true for the graphical method discussed above because the Feasibility Set contained an infinite number of (X, Y) pairs. However, the Fundamental Theorem of Linear Programming says that only the (X, Y) pairs marking the vertices are important for maximizing (or minimizing) the computation of the Objective Function (cost function C). Similarly, only certain *Particular Solutions* to system 6 will be important. But, how are these particular solutions generated? The answer to that question brings the analyst to step 3, the reduction of Eq. 15.6 to a matrix called the *Simplex Tableau* (matrix 15.7 below). This matrix is composed of the coefficients of all variables (both original and slack). For clarity, each column of matrix 15.7 is labeled with a variable (both original and slack), while the rows (equations) are only labeled with the slack variable they contain (note that C is being treated as if it were a slack variable, although this is technically not true). Traditionally, the Simplex Tableau also contains a horizontal and vertical line to separate the last row and the last column, respectively. This is done because these numerical values have special properties that will be of particular use.

$$
\begin{array}{c}
 \\
u \\
v \\
w \\
C
\end{array}
\begin{array}{c}
X \quad\; Y \quad\;\; u \quad v \quad w \quad C \\
\left(\begin{array}{cccccc|c}
34.10 & 36.26 & 1 & 0 & 0 & 0 & 1 \\
-1 & -2 & 0 & 1 & 0 & 0 & -0.01 \\
-124 & -160 & 0 & 0 & 1 & 0 & -1 \\
5 \times 10^5 & 5 \times 10^7 & 0 & 0 & 0 & 1 & 0
\end{array}\right)
\end{array}
\tag{15.7}
$$

Now, this last matrix (Tableau) is not quite in what a mathematician would call *standard form*. And, Linear Programming calculations demand standard form (fourth step). The problem is that two of the constants to the right of the vertical line are negative. The simplest thing to do is just to multiply rows v and w by −1.

$$
\begin{array}{c}
 \\
u \\
v \\
w \\
C
\end{array}
\begin{array}{c}
X \quad\; Y \quad\;\; u \quad v \quad w \quad C \\
\left(\begin{array}{cccccc|c}
34.10 & 36.26 & 1 & 0 & 0 & 0 & 1 \\
1 & 2 & 0 & -1 & 0 & 0 & 0.01 \\
124 & 160 & 0 & 0 & -1 & 0 & 1 \\
5 \times 10^5 & 5 \times 10^7 & 0 & 0 & 0 & 1 & 0
\end{array}\right)
\end{array}
\tag{15.8}
$$

There is no harm in doing this since the matrix 15.7 just represents the system of linear Eqs. 15.6. Therefore, multiplying any row by a constant is always permissible, and will not change the solution to the system. Now, the analyst wishes to know if it is possible to build the experiment in Fig. 15.1 without any beryllium (i.e. Y = 0), using only magnesium (X). As will be shown, this question is equivalent to requiring that the entry in the second row of the first column be a 1 (which it is) while all other entries in the first column are zero. Recalling again that the matrix (tableau) 15.8 is just a system of linear equations, multiplying (or dividing) any row by a constant, interchanging row, or adding (subtracting) one row to (from) another, does not change the solution to the system. Specifically, multiply the second row of matrix 15.8 by 34.1 and subtract it from the first row, then multiply the second row by 124 and subtract it from the third row, finally multiply the second row by 5×10^5 and subtract it from the fourth row. This process, called *matrix reduction* (Step 5), yields matrix 15.9. Readers unfamiliar with matrix reduction should review Chap. 1 and Ref. [5].

$$
\begin{array}{c}
 \\
u \\
v \\
w \\
C
\end{array}
\left(
\begin{array}{cccccc|c}
X & Y & u & v & w & C & \\
0 & -31.94 & 1 & 34.10 & 0 & 0 & 0.659 \\
1 & 2 & 0 & -1 & 0 & 0 & 0.01 \\
0 & -88 & 0 & 124 & -1 & 0 & -0.24 \\
0 & 49 \times 10^6 & 0 & 5 \times 10^5 & 0 & 1 & -5000
\end{array}
\right)
\tag{15.9}
$$

Multiplying the third row by −1 to put matrix 15.9 in standard form, and interchanging columns X and v for clarity yields matrix 15.10.

$$
\begin{array}{c}
 \\
u \\
v \\
w \\
C
\end{array}
\left(
\begin{array}{cccccc|c}
v & Y & u & X & w & C & \\
34.10 & -31.94 & 1 & 0 & 0 & 0 & 0.659 \\
-1 & 2 & 0 & 1 & 0 & 0 & 0.01 \\
-124 & 88 & 0 & 0 & 1 & 0 & 0.24 \\
5 \times 10^5 & 49 \times 10^6 & 0 & 0 & 0 & 1 & -5000
\end{array}
\right)
\tag{15.10}
$$

Note that columns can always be interchanged because that just means the order of the variables in Eq. 15.6 has been changed. The necessary calculations are now complete. All that is required is the interpretation of matrix 15.10 (Step 6). Since the analyst is interested in solutions that do not require beryllium, set Y = 0. Therefore, it makes no difference what coefficients are contained within the rectangle under the column label "Y" of matrix 15.10. Also, v is a slack variable that can be picked however the analyst desires, so long as v ≥ 0. If the selection v = 0 is made, the entries (coefficients) within the rectangle under the column label "v" of matrix 15.10 are irrelevant, and matrix 15.10 becomes a well-defined system of 4 equations in 4 unknowns. Therefore, the equation containing the slack variable v (row v) tells the analyst that X = 0.01. Therefore, (X, Y) = (0.01, 0) is a *Particular*

Solution (when Y and v are 0), and the cost associated with this particular solution is C = −5000. The negative sign reminds the analyst that he/she has recast the original problem as a *maximization* problem to fit *standard form*. For the *minimization* problem, the analyst is interested in –C = \$5,000. This, of course, is exactly the same answer arrived at by the graphical method. But, not a bit of geometry was used! Furthermore, the fact that all the entries below the horizontal line (except the last) are positive, means that the value of C is *optimized*! So, the value of –C is as small as it will ever get. If some of these bottom entries had been negative, then the value of C would have been nonoptimal. In that case, the tableau could be transformed by the usual row operations to remove the most negative value first. This process would then be continued until all negative values are removed. So, the Simplex Method *guides* the analyst toward an optimal solution. Finally, as a check, it should be noted that if the Linear Programing calculations have been carried out correctly, no variable (original or slack) can be less than zero, except for C. Hence, in linear system (matrix) 15.10, v = 0, Y = 0, u = 0.659, X = 0.01, and w = 0.24, all positive as expected.

In fact, every number in Simplex Tableau 15.10 has some significance for the materials selection problem. For example, these figures can be used to evaluate the benefits, or liabilities, of making small changes to the constraints. This is called *marginal analysis*. Suppose the strength constraint is relaxed so that Eq. 15.2 becomes X + 2Y ≥ 0.009. The right side of this constraint has been changed by an amount $\Delta = -0.001$. In that case, the second line of Eqs. 15.6 becomes

$$-X - 2Y + v = -0.009 \tag{15.11}$$

With a new change $\Delta' = +0.001$ on the right of Eq. 15.11. In matrix (tableau) 15.10 look for the figure below the horizontal line in the v column. It is 5×10^5. *Marginal analysis* tells the analyst that the *change* in C will be $5 \times 10^5 \, \Delta'$, or 500. Therefore, the change in − C is − \$500. That is to say, the cost of construction will go *down* by \$500 because the design of the experiment no longer has to be as strong as it was! This is just one more example of how system safety can make significant contributions to the design process using tools that are unfamiliar to many design engineers.

There is so much to say about Linear Programming that it cannot all be addressed, or proved, here. Therefore, unfortunately, this discussion will have to be terminated at this point. However, the real power of the Simplex Method comes from the fact that it can be extended to the comparison of any number of materials without error, since all the calculations performed in this section would typically be carried out by computer. Nevertheless, the graphical method can be partially saved in simpler cases by new visualization tools. That's what will be discussed next.

15.4   Geometrical Methods Revisited: CAD, 3D Printing, and Steriography

In this section, the two-dimensional graphical method will be extended to three and four dimensions so that the performance of three or four materials can be compared visually. The two-dimensional Feasibility Set in Fig. 15.2 was an irregular pentagon. Suppose a third material was to be compared to beryllium and magnesium, and let the volume used of this new material be denoted by Z. Now the Feasibility set will be a volume. How many faces and how many vertices would such a volume possess if all three constraints on weight, strength, and heat conduction are in place? If the two-dimensional feasibility sets in the Y-Z and X-Z planes were similar to that in Fig. 15.2 for the X-Y plane, so that all faces fit together to form the "water-tight" closed solid depicted in Fig. 15.5 below, then the Feasibility Set would have 7 faces and 10 vertices. Each of these vertices would have to be tested to see which one minimizes C. That vertex, call it (X_m, Y_m, Z_m), would be the solution to the 3D problem. However, this is the most optimistic case. In general, the faces will intersect in a very complex way forming unexpected vertices like the one marked by the red dot in Fig. 15.5. Here is where Computer-Aided Design (CAD) comes into play to clarify shapes [6]. Viewpoints can be changed and solids can be rotated. Finally, the CAD file can be sent to a 3D printer for the production of a plastic model (Fig. 15.5—right).

The use of 3D-printing to visualize four-dimensional geometry is even more interesting. A point in Euclidian 4-space has four Cartesian coordinates instead of 2 for a point in a plane, or three for a point in 3-space. And, a line is still determined by any two points while a plane is determined by three. Consider the following

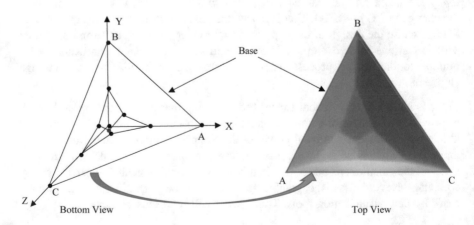

Fig. 15.5 The simplest solid (a septahedron) generated from three pentagons of the type in Fig. 15.2 (one pentagon in each perpendicular plane). The septahedron can be thought of as a three-sided pyramid whose edges have been cut into as the apex is approached

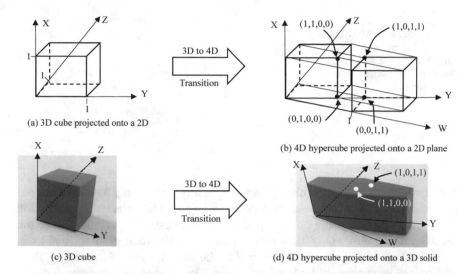

Fig. 15.6 **a** 3D cubic Feasibility Set projected onto a 2D space. **b** 4D cubic Feasibility Set projected onto a 2D space. Below the wire-mesh drawings are the corresponding solids created by 3D-printing (**c** and **d** respectively). In mesh B note how each of the six faces of the starting cube (in heavy lines and dashes) carves out a rhombohedral distorted "cube" (containing thin lines). These 6 distorted "cubes", plus the "initial" and "final" cubes, total 8 "cells". It is no accident that these 8 "cells" correspond in number to the 8 constraints of inequalities 12 (*two* for each of the four *double* inequalities in 4D)

simplest 3D and 4D constraint systems, where W represents the volume of material consumed by a given design for a fourth material.

$$3D(\text{Cube}) : 0 \leq X \leq 1,\ 0 \leq Y \leq 1,\ 0 \leq Z \leq 1$$
$$4D(\text{Hypercube}) : 0 \leq X \leq 1,\ 0 \leq Y \leq 1,\ 0 \leq Z \leq 1,\ 0 \leq W \leq 1 \qquad (15.12)$$

In the 3D case the Feasibility Set is a cube, while in the 4D case the Feasibility Set is a *hypercube*, also called a *tesseract* [7]. Just as a cube can be traced out by moving a square along the third dimension, a hypercube can be traced out by moving a cube along a fourth dimension. In Fig. 15.6a, a cube is projected onto the flat plane of this page. In Fig. 15.6b a hypercube is projected onto the flat plane of this page. The initial and final positions of the moving cube are drawn in heavy solid lines, while the thin lines indicate the direction of movement (4th dimension). Dashed lines are hidden lines. Below these projections onto 2D space are the three -dimensional forms created by 3D printing. When the hypercube is projected onto a three-dimensional space, a hexagonal right prism results. This may seem counter-intuitive. However, if it is remembered that each edge of a 3D cube is parallel to one of three axes, and if this idea is carried over into four dimensions, it is clear that Fig. 15.6b must result. Note that the axis system in Fig. 15.6a is nothing more than a comfortable illusion. Only two of the axes are actually perpendicular.

Fig. 15.7 A regular
dodecahedral Feasibility Set
in three dimensions (X, Y,
and Z). Each variable being
the volume consumed of a
given material

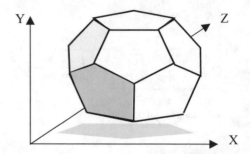

The third axis is at some arbitrary angle. Similarly, in 6B only two axes are perpendicular. The other two axes are at arbitrary angles. The tesseract has 16 vertices [7], 8 from the initial cube position and 8 from the final cube position. Notice that the 3D printing shows only 12 vertices because 2 of the vertices are internal to the top face (Fig. 15.6b), and 2 are internal to the bottom face, of the shape. These vertices can be marked on the faces with a pen. The value of any Objective Function (e.g., Cost Function, C) must be tested at all 16 vertexes. From Fig. 15.6b, and also from inequalities 15.12, it is clear that these vertices are (X, Y, Z, W) = (0, 0, 0, 0), (1, 0, 0, 0), (0, 1, 0, 0), (0, 0, 1, 0), (0, 0, 0, 1), (1, 1, 0, 0), (1, 0, 1, 0), (1, 0, 0, 1), (0, 0, 1, 1), (0, 1, 1, 0), (0, 1, 0, 1), (1, 1, 1, 0), (0, 1, 1, 1), (1, 0, 1, 1), (1, 1, 0, 1), (1, 1, 1, 1). Therefore, this 4D problem can be completely understood. Other, more complex, 4D problems can also be understood. There is, of course, no reason to stop this process at four dimensions. However, visualization becomes increasingly difficult at higher dimensions.

Next, a considerably more complex 3D problem and its 4D analog will be discussed. Suppose the 3D Feasibility Set for a given project looks like the dodecahedron in Fig. 15.7. What does such a set imply? A dodecahedron has 12 faces, and each planar face represents a three-dimensional constraint on the materials variables X, Y, and Z.

Notice that the Feasibility Set never contacts the X–Z plane, but lies above it. Therefore, in contrast to the Be/Mg problem of the previous section, an algebraic simplex calculation will fail to yield a solution for Y = 0. Therefore, even a superficial geometrical examination of a complicated Feasibility Set can yield some important information that can explain the seemingly aberrant behavior of some Simplex Method calculations. How does this 3D problem translate into a 4D Linear Programming problem involving materials variables X, Y, Z, and W? This time the 4D Feasibility Set cannot be constructed simply by translating the 3D figure along a new axis (W). That technique worked for the tesseract because a translated point yields a line, a perpendicularly translated line yields a square, a perpendicularly translated square yields a cube, and finally a perpendicularly translated cube yields a tesseract. However, those simple symmetries fall apart when one tries to generate a pentagon from a line, or a dodecahedron from a pentagon. And, it will not be possible to generate the 4D analog of the dodecahedron, called a *120-cell*, by simple translation either! Nevertheless, there are definite construction procedures

Fig. 15.8 A simplified (partial) 4D 120-cell projected onto 2D space. Notice all the pentagons in this figure. The complete 120-cell has 600 vertices, 1200 edges, and forms a complicated closed figure with a doughnut hole [7]

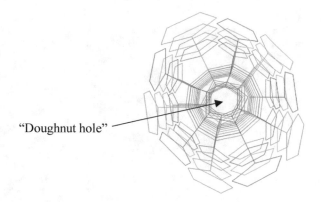

"Doughnut hole"

for generating the 120-cell [7], each "cell" (no longer cubic) representing a constraint (in analogy with the tesseract of Fig. 15.6) involving the four variables X, Y, Z, and W. Now one may ask, "When will an engineer ever have to deal with 120 constraints?" A system containing 15 interrelated parts with 8 materials constraints on each, or any system with 30 parts and four constraints on each, etc., qualifies. Each constraint is a linear equation in the variables X, Y, Z, and W, and one or more of these variables may represent the volume consumed of a dangerous good that the analyst wishes to minimize or eliminate. Figure 15.8 is a color coded, radially expanded, partial construction of a regular 120-cell projected onto a 2D surface (the plane of this page) from a slightly off-center viewpoint. Only 320 vertices and 400 edges have been drawn. The complete figure, containing 600 vertices and 1200 edges [7], is very complex and hard to understand because of line crossings that occur when the 120-cell is projected onto a plane. These crossings are not vertices. Few real-world engineering problems will have the beautiful symmetry of the 120-cell. Figure 15.8 gives the analyst some important *global* information. This 4D Feasibility Set has a "doughnut hole" at its center!

The geometry of the doughnut hole can be better understood if there was a way to get inside this cobweb of lines so that, by changing his/her perspective, the safety engineer can eliminate false vertices. Then, vertices of the Feasibility Set can be selected (as was done with Fig. 15.2) and tested for optimization of the Objective Function. There is, in fact, a way to do just that. Many universities have a *Stereography Facility* called a *CAVE* (Computer Analysis and Visualization Environment; Fig. 15.9). When not in use, the "CAVE" is just an empty white room (screens) fitted with projectors (Fig. 15.9a). When active, the analyst has to wear the usual glasses to sense the 3D effect (Fig. 15.9b) which, although technically different from holography, gives the same result (Fig. 15.9b). Each "lens" of these glasses is electronically controlled so that alternately one "lens" goes black while the other is transparent. And, each eye sees the same image from a slightly different perspective. Hence, the double lines in Fig. 15.9b. The cycle time is very fast so that there is no sense of "flicker". The illusion is so powerful that it is irresistible.

(a)   **(b)**

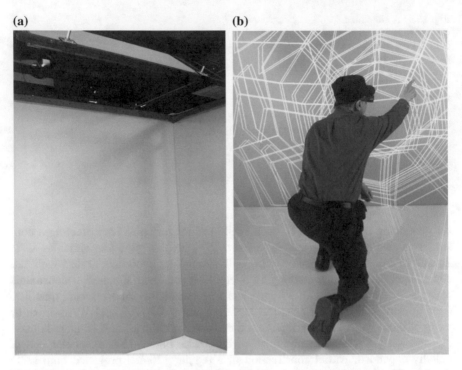

Fig. 15.9 a Inactive Cave Facility. **b** Looking for solutions in the fourth dimension. The author is immersed in a stereo environment while wearing 3D glasses. He thinks that he is touching a hyperspace vertex as he crawls through the four-dimensional doughnut hole. But alas, it's all a physiological and electronically induced optical illusion!

Images are projected on the walls, ceiling, and floor, so that the system safety engineer gets the sense of walking into the four-dimensional hyperspace doughnut hole. Then he/she can look around, select, and test, various vertices for optimization, moving in the direction that gives the most promising result or minimizes a dangerous good (as was done for the two-dimensional problem visualized by planar Fig. 15.2). Only the real 4D vertices will be present. However, some right angles will be lost and appear acute or obtuse. That is the price that must be paid for working in the fourth dimension. Figure 15.10a is a close-up of the doughnut hole, which is easier to understand if the pentagons are color coded so as to distinguish line crossings seen in projection from true vertices (Fig. 15.10b).

It should be clear to the reader at this point that hyperspace geometry is a complex business, thereby highlighting the utility of the algebraic Simplex Method. Pólya's Hyperspace Random Walk Theorem (1921) gives the analyst some idea of how the difficulties grow with dimension n. In one or two dimensions, a random walk will eventually bring the walker back to his starting point after a sufficiently large number of steps. However, a random walker in three or more dimensions has only a 1/2n chance of returning back to the starting point. In three dimensions there

(a)   **(b)**

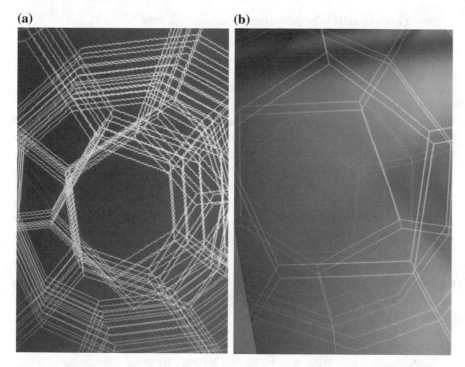

Fig. 15.10 a Close-up of the "doughnut hole". **b** Close-up of internal pentagons which may be color coded

is a 34% chance of returning home (think of the difficulties involved in unscrambling Rubik's Cube by random 3D rotations of its parts). However, in four dimensions the probability of return is only 12.5%. In fact, it can be shown that if a higher dimensional random walker does not return home after his first few steps, he is probably lost in hyperspace forever [8]. Nevertheless, graphical methods do provide some information not available by purely numerical methods. For example, notice how vertices seem to collect in concentric shells around the doughnut hole in Fig. 15.8. This kind of *global* information about *all* vertices is usually not available by "number crunching" alone. When vertices collect in small regions of hyperspace, it may be possible to eliminate a whole subset of vertices if a sample evaluation of the Objective Function at one representative vertex of the subset, with coordinates (X, Y, Z, W), yields a value that the analyst suspects is far from optimal. Graphical methods may also help in interpreting Simplex Method calculations. So, simplex calculations and graphical methods can be made to work together.

15.5   The Health Consequences of Beryllium

The principle health risk from beryllium is from the dust [9], which may cause acute pneumonitis (irritation of the respiratory tract), berylliosis (respiratory disease), contact dermatitis (irritation of the skin), or conjunctivitis (irritation of eye membranes).

The symptoms of acute pneumonitis (acute irritation of the respiratory tract) are difficult or painful breathing (dyspnea), chest pain, bronchial spasms, fever, cyanosis (bluish coloration of the skin), cough, blood-tinged sputum, and nasal discharge. Also, the right side of the heart may fail due to increased arterial resistance. These symptoms may occur two to three weeks after exposure of 1–20 days. In the case of chronic poisoning, pulmonary granulomatosis (pulmonary growths with a granular character; berylliosis) develops, with weight loss, and marked dyspnea beginning from 3 months to as much as 11 years after first exposure. Berylliosis may steadily worsen, or it may involve exacerbations and remissions. As with acute pneumonitis, failure of the right side of the heart may occur as a result of increased pulmonary resistance. Approximately 2% of people with berylliosis die of the disease. For the remainder of patients, symptoms may be relieved by hormones (prednisone or corticosteroid) for varying lengths of time, but there is no cure. The Threshold Limit Value (TLV) in the air is a minute 0.002 mg/cu m. Realistically, no beryllium is allowed in the air.

Acute contact poisoning can be caused by cuts from beryllium contaminated objects. Such injuries form slow healing deep ulcerations. Acute dermatitis simulating first- and second-degree burns can be caused by contact with beryllium dust. Chronic dermatitis appears as an eczematous (oozing), maculopapular (having raised lesions), erythematous, vesicular (having sacks containing fluid or gas), rash in a large percentage of workers exposed to beryllium dust. And, a patch test with dilute beryllium solutions shows a positive reaction. Treatment consists of surgical removal of beryllium-contaminated areas of skin.

Beryllium dust contact with the eye causes acute conjunctivitis with corneal maculae (raised lesions) and diffuse erythema. Treatment consists of removal from further exposure and thorough washing of skin and eyes—very painful. An anesthetic ointment is applied to reduce pain.

Since beryllium is in the second column of the periodic table it has a +2 valence like magnesium. And it is known that beryllium will inhibit certain magnesium activated enzymes. But, the exact relationship between this effect and the observed pathologic changes caused by beryllium is unknown.

15.6 Conclusion

Throughout this chapter, the interaction between system safety and the design engineering team has been stressed. All too frequently system safety has been looked upon by design engineers as a source of "interference". Or, worse still, as unnecessary personnel who "don't understand". Here, it has been shown how input from system safety not only eliminates serious health risk, but helps engineering to optimize their design and keep costs down. As discussed above, design engineering should not trivialize the risks posed by beryllium metal. System safety engineering understands these risks, as well as the benefits and trade-offs associated with the use of beryllium, very well—and better than design engineering. The time for the system safety engineer to be part of the design team is long overdue!

† Calcium metal (Ca) is chemically stable in oxygen-free environments like space, is more refractory than magnesium, has a thermal conductivity that is just as high as Be, and has a lower density than either Be or Mg. Although the atomic weight of Ca is 4.44 times that of Be, its atomic volume is also 5.98 times larger, giving Ca a lower density. To the author's knowledge, Ca metal has never been used as a structural material in space applications.

15.7 Problem

Suppose it was desired to increase the thermal conductivity requirement of the experiment in Fig. 15.1 by 10%.

(a) Do a Marginal Analysis to decide how such a change would impact the cost of the experiment.
(b) What do these results imply?
(c) Calculate the left side of inequality 15.3 for the original design, where $(X, Y) = (0.01, 0)$.
(d) Is your answer to part b justified?
(e) In the context of this problem, what does the word *small* mean when speaking about changes in constraints as discussed in the next to the last paragraph of Sect. 15.3 of this chapter?

References

1. Wikipedia. (2016). *Mechanical Properties*. Retrieved June 19, 2016, from https://en.wikipedia. org/wiki.Magnesium, https://en.wikipedia.org/wiki.Beryllium.
2. Anon. (2016). *Investing News*. Retrieved June 19, 2016, from InvestingNews.com/daily/ resource-investing/critical-metals.

3. Anon. (2016). Be Resources Inc. Retrieved June 20, 2016, from Beresources.ca/aboutberyllium/berylliumprices.htm.
4. Goldstein, L., Schneider, D., & Siegel, M. (1998). *Finite mathematics and its applications* (6th ed., pp. 107–187). Upper Saddle River, NJ: Prentice-Hall.
5. Shields, P. C. (1968). *Elementary linear algebra* (pp. 1–13). NY: Worth Publishers.
6. Elliot, S. D., Leigh, R. W., Matthews, B. (1995). *autoCAD: A concise guide to commands and features for release 13 for Windows* (4th ed., p. 646). Chapel Hill, NC: Ventana Press.
7. Coxeter, H. S. M. (1961). *Introduction to geometry* (pp. 396–414). New York: Wiley.
8. Pickover, C. A. (2009). *The math book* (pp. 344, 452). Sterling, New York.
9. Dreisbach, R. H. (1977). *Handbook of poisoning* (pp. 214–216). Los Altos, CA: Lange Medical Publications.

Chapter 16
Accelerated Age Testing of Explosives and Propellants

16.1 Introduction

It is of paramount importance for the users of explosives and propellants, as well as for the safety engineers that work with these dangerous materials, to understand their behavior as a function of time. Energetic materials (many of which have been listed in Appendix E) can change their characteristics after extended storage times in, perhaps, nonideal conditions such as high humidity or burial. The two most fundamental questions about old explosives and propellants from the standpoint of the user and safety engineer, respectively, are

(1) can the energetic material still deliver the same thrust (propellants), or specific impulse (explosives), and
(2) is the energetic material still safe [1]?

In the first case, the specific impulse (or impulse per unit mass of explosive) is measured by multiplying the force delivered to a target of area $\mathscr{A}$ at a specified distance $\mathscr{D}$ by the small time interval dt needed for the explosive shock wave to pass, and then dividing by $\mathscr{A}$ [2]. In the second case, the safety engineer may, for example, need to determine if the structural strength of a propellant has been compromised by aging, because structural failure during combustion might cause sufficiently severe deflagration in the damaged zone to result in catastrophic bursting of a rocket motor assembly [3].

These are very complex questions, and their answer will depend on the composition of the explosive or propellant under study. Because energetic materials are often stored for many years, testing of new formulations or designs by natural aging is not practical. Instead, aging is accelerated by short-term storage at temperatures that are typically 25°–75° above the anticipated storage temperature. It is believed that exposure of explosives and propellants to these elevated temperatures for relatively short times is chemically equivalent to natural storage for a much longer time period at lower temperatures. The artificially aged energetics can then be tested

© Springer Nature Switzerland AG 2020
R. R. Zito, *Mathematical Foundations of System Safety Engineering*,
https://doi.org/10.1007/978-3-030-26241-9_16

for thrust, or specific impulse, and safety. And, these results should be equivalent to results obtained with the naturally aged material.

The rationale for this procedure lies in the Arrhenius relation [4],

$$k = Ae^{-Ea/RT}. \tag{16.1}$$

In this equation, k is a reaction rate in moles/s for a chemical process like combustion, and Ea is the activation energy for that process ($\sim 10^4$ cal/mole). Strictly speaking, Ea should be measured at absolute zero, but the temperature dependence of Ea is so small that it is greatly overshadowed by the uncertainty of measuring Ea at room temperature. R is the Ideal Gas Constant (1.987 cal/K mol, or 8.31 J/K mol), T is the Kelvin temperature, and A is the pre-exponential factor that is assumed to be independent of temperature. This last assumption is good over a small temperature range, but 25 °C is not really small enough and, as will be demonstrated, the effects of the temperature dependence of A will begin to show themselves. Furthermore, the Arrhenius relation (Eq. 16.1) is really only valid for *simple energetics* composed of a single phase (like nitrocellulose) or a homogeneous mixture of fuel (or explosive) and oxidizer. The introduction of a plastic binder greatly complicates analysis. *In this chapter phase homogeneity restraints, as well as temperature restraints on A, will be relaxed. Equivalent baking times (with error bars) for these more general conditions will also be calculated.*

16.2 Simple Energetics with a Temperature Independent A

In order to successfully employ accelerated testing to answer the two central questions of the introduction, it is necessary to determine the proper equivalent accelerated aging time t_2, for any elevated temperature T_2, that corresponds to storage for a given time t_1 at an anticipated temperature T_1. Or, equivalently, the value of the ratio t_2/t_1 is desired. If a fuel is slowly consumed by chemical reaction with an oxidant, or the environment, at a rate that is *three* times faster at T_2 than at T_1, then aging at T_2 need only be *one-third* as long to achieve the same result. That is to say, t_1/t_2 is inversely proportional to the ratio of reaction rates at temperatures T_2 and T_1, respectively. Therefore

$$t_1/t_2 = k_2/k_1 = [A\exp(-Ea/RT_2)]/[A\exp(-Ea/RT_1)] = \exp(Ea/RT_1 - Ea/RT_2)$$
$$= \exp([Ea/R][1/T_1 - 1/T_2]) = \exp([Ea/(RT_1T_2)][T_2 - T_1]). \tag{16.2}$$

It is crucially important at this point to realize that the temperature-independent pre-exponential factor A has been cancelled from both the numerator and denominator of Eq. 16.2. This is very fortunate because the reaction rate for propellants and explosives is either too fast or too slow (depending on whether ignition has occurred or not) to determine A directly! As an example, consider a propellant reacting with its oxidant (e.g., air) during storage at 25 °C (298 K). A system safety engineer would like to know what the properties of this propellant are after, say, 10 years. Well, he/she isn't going to wait that long, so the analyst opts for accelerated testing at 120 °F (49 °C or 322 K). Quite warm, but not enough to be dangerous. How long will the engineer have to bake the test sample? Using Ea and R from the introduction, and the Kelvin temperatures the engineer plans to use, Eq. 16.2 yields $t_1/t_2 = 3.52$. That is to say, each day in the oven is equivalent to 3.52 days of natural aging. So, the analyst will have to bake the propellant for 10/3.52 years (i.e. 2.84 years or 34 months). This is still a long time. But, if the engineer wishes to shorten the procedure, an increase in oven temperature will be required. Equation 16.2 has been the principal tool used in accelerated testing of explosives and propellants. However, this simple picture can be complicated by a variety of factors, including side reactions and the temperature dependence of A.

16.3   Simple Energetics with a Temperature-Dependent A

Generally speaking, A will not be independent of temperature T, so that the first line of Eq. 16.2 will become

$$
\begin{aligned}
t_1/t_2 = k_2/k_1 &= [A(T_2)\exp(-Ea/RT_2)]/[A(T_1)\exp(-Ea/RT_1)] \\
&= [A(T_2)/A(T_1)]\exp(Ea/RT_1 - Ea/RT_2).
\end{aligned}
\tag{16.3}
$$

Several models can be selected for A(T), but the most common is a power law of the form of Eq. 16.4 below [4]. And, there are theoretical reasons for believing this power law; it's not just a fit to data.

$$
A = aT^m,
\tag{16.4}
$$

where both "a" and m are temperature independent empirical constants. Usually, the exponent m is difficult to determine experimentally because the exponential dependence on $1/T$ is usually much stronger than the dependence of A on T^m. This fact, combined with experimental errors in the determination of k, limit the engineer's ability to determine the functional dependence of the pre-exponential factor on T. However, studies with reacting gasses [4] suggest a value of about ½. Although, values of m may range as high as 2 [4]. Substituting Eq. 16.4 into Eq. 16.3 yields

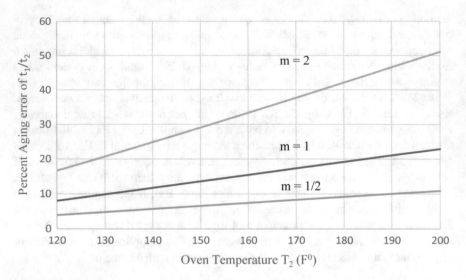

Fig. 16.1 Aging error versus oven temperature as a function of m. The storage temperature T_1 was set at 25 °C (298 K)

$$t_1/t_2 = k_2/k_1 = \left[T_2^m \exp(-Ea/RT_2)\right] / \left[T_1^m \exp(-Ea/RT_1)\right]$$
$$= (T_2/T_1)^m \exp(Ea/RT_1 - Ea/RT_2)$$
$$= (T_2/T_1)^m \exp([Ea/R][1/T_1 - 1/T_2]) = (T_2/T_1)^m \exp([Ea/(RT_1T_2)][T_2 - T_1]).$$

$$(16.5)$$

Figure 16.1 shows the accelerated aging error (i.e., the percent error in t_1/t_2) incurred by ignoring the temperature dependence of A, as a function of m and oven temperature T_2 where

$$\%error = \{(T_2/T_1)^m \exp([Ea/(RT_1T_2)][T_2 - T_1])$$
$$- \exp([Ea/(RT_1T_2)][T_2 - T_1])\} / \exp([Ea/(RT_1T_2)][T_2 - T_1]) \times 100$$
$$= ([T_2/298]^m - 1) \times 100.$$

$$(16.6)$$

At m = ½ and T_2 = 120 °F (322 K) the error is only about 4%. However, the percent error grows rapidly with temperature so that just below the boiling point of water (200 °F) the error has already reached 11%. And, the error is even worse for larger values of m. For m = 2 at 200 °F, the error in t_1/t_2 is 51%. The space trapped between the curves for m = 1/2 and m = 2 then *gives a range for the % error in t_1/t_2*.

16.4 Composite Energetics

a. *Binder Chemistry*: Most modern explosives and propellants are plastic composites. For example, fuel and oxidizer particles may be dispersed uniformly
   throughout a polyurethane matrix (binder). The total composite being called a
   grain. The polyurethane binder is typically made by mixing a dihydroxy
   compound (a liquid) and a diisocyanate (also a liquid) together. As the functional groups (end groups) of each component react (Fig. 16.2), a polymer chain
   of polyurethane plastic is formed, where R and R' are organic chains/rings
   selected to give the final polyurethane the appropriate mechanical properties and
   setup time [5].

The two starting liquids are doped with fuel and oxidizer particles, and equal
volumes are mixed at room temperature for about 20 s. "Setup" takes about 15 min
(depending on R and R'), and most curing is completed in 24–48 h. These figures,
of course, are very much dependent on the R and R' selected for the reaction
components. In any case, some curing will continue into the early stages of aging. It
is important to have an equal number of moles of each reactant (usually equal liquid
volumes) to within ±1%, otherwise unreacted liquid will remain that can, for
example, interfere with binding of a final composite to a rocket motor casing. Extra
dihydroxy is particularly problematic because it will remain in the polyurethane
matrix (binder) as bubbles of liquid. By contrast, an excess of diisocyanate will
react with growing polyurethane chains by a mechanism involving attack of a
polyurethane nitrogen atom, possessing a lone pair of electrons in –NHCOO–
groups, by an electropositive cyanate carbon atom. (i.e., C of OCN– in Fig. 16.3).
Once the cyanate nitrogen atom extracts the polyurethane hydrogen, one cyanate
carbon/nitrogen bond (of a double bond) is broken and the polyurethane nitrogen is
ready to donate an electron to the cyanate carbon hybrid orbital. Thereby forming a
new carbon/nitrogen sp^3 bond and, therefore, a side chain. The –NCO terminus of
this side chain may then bond to another polyurethane strand. The additional
cross-linkages will change the binder's properties, making it less elastic.

   In addition to these formation reactions, liquid additives may be introduced into
the reactive components of the polyurethane that remain as a dispersed liquid in the
grain (final composite solid). These liquid additives may possibly cause de-wetting
of the solid components (e.g., rocket motor casing) from the polyurethane. This, in
turn, may cause structural strength problems for the grain [6].

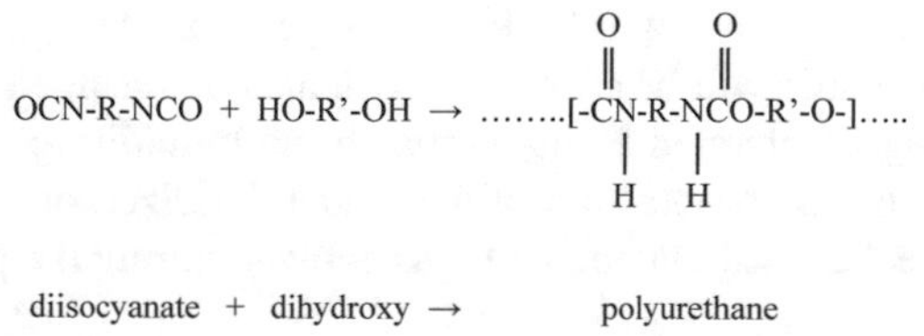

$$OCN\text{-}R\text{-}NCO + HO\text{-}R'\text{-}OH \rightarrow \ \ldots\ldots\ldots[\text{-}\overset{\overset{O}{\parallel}}{C}N\text{-}R\text{-}N\overset{\overset{O}{\parallel}}{C}O\text{-}R'\text{-}O\text{-}]\ldots\ldots \qquad (16.7)$$
$$\phantom{OCN\text{-}R\text{-}NCO + HO\text{-}R'\text{-}OH \rightarrow \ \ldots\ldots\ldots[\text{-}C}\underset{H}{|} \ \ \underset{H}{|}$$

diisocyanate + dihydroxy →   polyurethane

Fig. 16.2 The chemistry of formation of a polyurethane binder

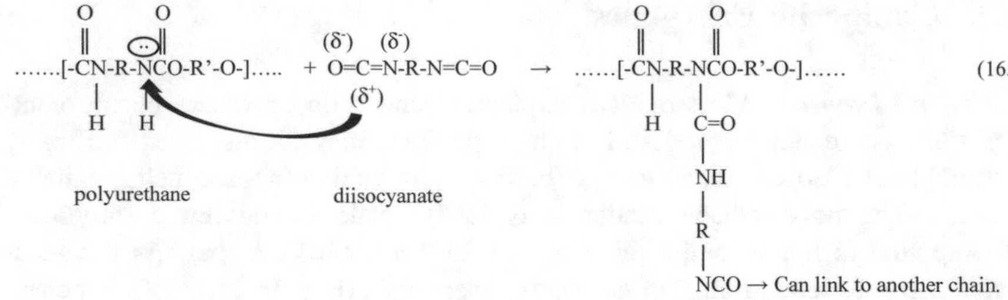

$$
\begin{array}{c}
\overset{\text{O}}{\underset{}{\parallel}}\quad\overset{\text{O}}{\underset{}{\parallel}} \\
......[\text{-CN-R-NCO-R'-O-}]..... \\
\underset{\text{H}}{\mid}\ \underset{\text{H}}{\uparrow}
\end{array}
\quad
\begin{array}{c}
(\delta^-)\ \ (\delta^-) \\
+\ \text{O=C=N-R-N=C=O} \\
(\delta^+)
\end{array}
\quad \rightarrow \quad
\begin{array}{c}
\overset{\text{O}}{\underset{}{\parallel}}\quad\overset{\text{O}}{\underset{}{\parallel}} \\
......[\text{-CN-R-NCO-R'-O-}]...... \\
\underset{\text{H}}{\mid}\ \underset{\text{C=O}}{\mid} \\
\mid \\
\text{NH} \\
\mid \\
\text{R} \\
\mid \\
\text{NCO} \rightarrow \text{Can link to another chain.}
\end{array}
\qquad (16.8)
$$

polyurethane   diisocyanate

Fig. 16.3 Growth of side chains in polyurethane due to excess diisocyanate in the original mixture

Although polymers are generally thought to be impervious to water and oxygen, they will, in fact, react with both. Figure 16.4a–c describe the mechanisms of these reactions for polyurethane.

Hydrolytic chain scission and oxidative crosslinking come from atmospheric water vapor and oxygen, and are introduced into the polyurethane by gaseous diffusion during aging. These reactions cause weakening and strengthening, respectively. Furthermore, water vapor introduced into the binder reactants during mixing produces CO_2 gas as a byproduct, causing bubbles in the grain (Fig. 16.5) that can lead to structural weakness if excessive. Water vapor absorbed by dihydroxy compounds during storage will also react with diisocyanate during mixing to create additional CO_2 gas bubbles (Fig. 16.4d). Finally, the reaction of diisocyanate with water vapor during storage will cause it to homopolymerize to polyurea and become useless for making polyurethane (Fig. 16.4d). During homo-polymerization CO_2 gas is released that can cause the metal or plastic isocyanate storage container to swell. Any such swelling is a signal to the system safety engineer that the diisocyanate should be disposed of. Hence, it is important to store reactants, and conduct mixing operations, in a bone dry environment and to degas the mixture before the polyurethane binder gels. Changes in material properties of the binder due to the oxygen and water vapor reactions in Fig. 16.4a through d manifest themselves in changes of the elastic modulus E (in MPa), the tensile strength S (in MPa), and the elongation e (measured by the length change in %) of the grain. The "elongation" here refers to elongation during tensile testing. Unacceptable changes may result in composite cracking, an abnormal burn, and possible explosion of a rocket motor.

In the calculations that follow, it will be assumed that a composite grain is dry (i.e., that the polyurethane components have fully reacted and that there are no liquid additives in the grain). Only hydrolytic chain scission and oxidative cross-linking of the polymer binder will be allowed. It will also be assumed that the humidity is controlled during accelerated aging to match the humidity conditions during normal storage. No other decomposition of the binder, oxidizer, or fuel, will be allowed. Finally, it will be assumed that the CO_2 gas evolved during the polymer

(a)

$$\ldots\ldots[-CN\text{-}R\text{-}N\text{-}C\text{-}O\text{-}R'\text{-}O\text{-}]\ldots\ldots + H_2O \rightarrow \ldots[-CN\text{-}R\text{-}N\text{-}C\text{-}OH + HO\text{-}R'\text{-}O\text{-}]\ldots \qquad (16.9)$$

(structure with O, O(δ^-) double bonds; H, H(δ^+); δ^- on H_2O; H, H)

(b)

Free radical end bonds to neighboring chain

O_2 CH_3 HN-C=O-R'...   H-N-C=O

$\rightarrow$ 1e⁻ M 1e⁻ $H_2CO\text{-}O\text{-}H$ HN-C=O-R'...   H-N-C=O

$\rightarrow$ $H_2C\text{-}O$ •   HN-C=O-R'...   H-N-C=O   +   OH •   Free Radical $\qquad (16.10)$

(c)

peroxide bridge

$$\ldots\text{-}CH = CH + O_2 \rightarrow \ldots\left[CH_2 - CH\right]O\text{-}O\left[CH_2 - CH\right] \qquad (16.11)$$

(with X substituents below)

(d)

$$O=C=N\text{-}R\text{-}N=C=O + H_2O \rightarrow OCN\text{-}R\text{-}N\text{-}C\text{-}OH \rightarrow OCN\text{-}R\text{-}NH_2 + CO_2\uparrow \rightarrow \ldots[-NHCNH\text{-}]\ldots \qquad (16.12)$$

(δ^+) (δ^-) (O double bond, H)   polyurea

Fig. 16.4 **a** Hydrolytic chain scission. The electronegative oxygen atom of water attacks an electropositive carbon atom of polyurethane. **b** Oxidative cross-linking. Here, the R group is toluene (very common), one of whose labile hydrogens are attacked by O_2 to produce a hydroperoxide that decomposes to an oxygen free radical termination capable of bonding to a neighboring polyurethane stand. **c** Oxidative cross-linking. Here, a double bond within R or R' is attacked by O_2 to produce a peroxide bridge between two polyurethane strands. "X" is just the rest of the polymer chain. **d** Reaction of water with diisocyanate. Here, the OCN–R–NH$_2$ intermediate product can react with another intermediate molecule to produce polyuria. Note the production of CO_2 gas during reaction of isocyanate with water

formation reaction (Eq. 16.7) via the side reaction in Fig. 16.4d is completely removed from the grain before binder setup (gel formation).

The required accelerated aging time at elevated temperatures to produce samples for thrust and impulse testing can, of course, be calculated as before with a value of Ea that is appropriate to the *composite* propellant. *However, the material properties of a sample may age at a different rate than the chemical thrust and impulse properties. Furthermore, the calculation of E, S, and e for naturally aged propellant as a function of E, S, and e measured after accelerated aging at elevated temperatures is a feature of composites that must also be addressed.* Both of these problems will be discussed next.

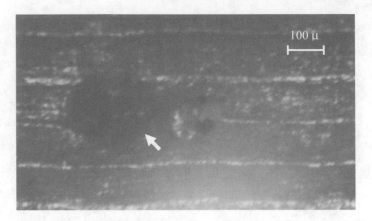

Fig. 16.5 Micrograph (30X) of a gas bubble in the interior of a polyurethane casting. The bubble (dark circular structure on the left) is 270 μm in diameter and was exposed by milling through the surface of the specimen. Occasional small bubbles are harmless, but large defects are dangerous

b. *Binder Mechanical Properties*: For propellants, three of the most important mechanical properties are the Young's Modulus E (i.e. the "spring constant" of the composite when subject to stress, or force per unit area), the tensile strength S of the composite (i.e., the maximum stress the grain can withstand before it tears), and the maximum elongation e of the composite (i.e., $\Delta l/l$ in % before failure). And, these mechanical properties of the composite will largely depend on the mean molecular weight (MMW) of the binder. Let MMW_i represent the starting MMW of N moles of binder before aging at some temperature T, corresponding to storage temperature for natural aging or an oven temperature for accelerated aging. After a time t, assume Δ_{O2} mol of the initial polyurethane are consumed by oxidative crosslinking to produce $\Delta_{O2}/2$ mol of cross-linked polyurethane with a mean molecular weight of $2(MMW_i)$. Similarly, assume Δ_{H_2O} moles of the initial polyurethane will be consumed by hydrolytic scission to produce $2(\Delta_{H_2O})$ moles of cut polyurethane with a mean molecular weight of $MMW_i/2$. What, then, is the mean molecular weight after time t, call it MMW_f? It is just the average

$$
\begin{aligned}
MMW_f &= [(N - \Delta_{O_2} - \Delta_{H_2O})(MMW_i) + (\Delta_{O_2}/2)(2MMW_i) \\
&\quad + 2(\Delta_{H_2O})(MMW_i/2)]/[N - \Delta_{O_2} - \Delta_{H_2O} + \Delta_{O_2}/2 + 2(\Delta_{H_2O})] \\
&= [N(MMW_i)]/[N - \Delta_{O_2}/2 + \Delta_{H_2O}] = (MMW_i)[1 - (\Delta_{O_2}/2N) + (\Delta_{H_2O}/N)]^{-1} \\
&\approx (MMW_i)[1 + (\Delta_{O_2}/2N) - (\Delta_{H_2O}/N)] \\
&= (MMW_i)(1 + \delta), \text{ where } \delta = (\Delta_{O_2}/2N) - (\Delta_{H_2O}/N).
\end{aligned}
$$

$$(16.13)$$

The change in the MMW_i per unit time is just $\delta/t = (1/N)(k_{O_2}/2 - k_{H_2O})$, where k_{O_2} and k_{H_2O} are the rate constants for oxygen cross-linking and hydrolytic scission, respectively. Since the change in E and S are both proportional to the change in the polyurethane MMW,

$$\Delta E/t = K(k_{O_2}/2 - k_{H_2O}), \text{ where K is a constant of proportionality.} \qquad (16.14)$$

Physically, Eq. 16.14 can be rationalized by noting that oxidative crosslinking increases the spring constant E of the binder, while hydrolytic scission breaks "springs" thereby reducing E. Similarly, $\Delta S/t = K'(k_{O_2}/2 - k_{H_2O})$. To a first approximation, it will be assumed that the constants of proportionality K and K' are both temperature independent so that they will cancel out of the numerator and denominator of the calculations that follow. The temperature independence of K is why E is almost constant for crystalline polymers (like polyurethane) of fixed MMW over a broad temperature range, well below the melting point [5]. Tensile testing takes place over such a short period of time that crosslinking and scission play no part in changing the MMW as the E versus T curve is generated.

c. *Calculation of Changes in E and S versus Temperature and Time (Constant A)*:
   In this subsection, the change in the Young's Modulus E and the tensile strength S due to temperature-dependent oxidative crosslinking and hydrolytic scission reactions will be addressed. Even without these chemical reactions, E and S will change slightly with temperature. However, those changes are small ($\sim 2\%$ over the temperature ranges of interest), especially for high molecular weight crystalline polymers (like polyurethane) far below the melting point of the plastic. By comparison, the chemical reactions of the aging process make large changes ($\sim 20\%$) in the mechanical properties of polyurethane. Hence, nonchemical temperature dependencies will be ignored. Since data is available on changes in E and s as a function of both temperature and time (Fig. 16.6; [6]), interest here will focus on the computation of the change in E (or S) as a function of time for two different temperatures. More specifically, the ratio of the slopes of the two E versus t lines will be computed, each straight line corresponding to a different temperature (T_1 and T_2).

Since $t_1 = t_2 = \Delta t = 150 - 30 = 120$ days, Eq. 16.14 implies

$$\begin{aligned}
[\Delta E(T_1, t_1)/t_1][\Delta E(T_2, t_2)/t_2] \\
= \Delta E(T_1, \Delta t)/\Delta E(T_2, \Delta t) \\
= \frac{A(O_2)\exp[-Ea(O_2)/RT_1] - A(H_2O)\exp[-Ea(H_2O)/RT_1]}{A(O_2)\exp[-Ea(O_2)/RT_2] - A(H_2O)\exp[-Ea(H_2O)/RT_2]},
\end{aligned} \qquad (16.15)$$

where $A(O_2)$ and $A(H_2O)$ are the pre-exponential factors for the oxygen crosslinking and hydrolytic scission reactions, respectively. And, where the factor of ½ associated with k_{O_2} has been absorbed into the pre-exponential factor $A(O_2)$ of the Arrhenius relation for k_{O_2}. Equation 16.15 would apply if the samples tested at

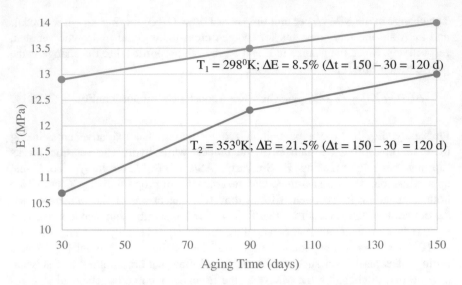

Fig. 16.6 E versus t graphs for samples at 25 °C (upper) and 80 °C (lower). The two curves are displaced because the initial material properties of the two samples were not identical. The sample at 353 K (80 °C) experienced a much greater percent change in E

each of the two temperatures had identical initial material properties. However, this was not the case. Therefore, the right side of Eq. 16.15 really applies to samples whose ΔE has been normalized by the value of E at 30 days (E_{30}), so that

$$\Delta E(T_1, \Delta t)/\Delta E(T_2, \Delta t) = (\Delta E(T_1, \Delta t)/E_{30})/(\Delta E(T_2, \Delta t)/E_{30})$$
$$= [\%\Delta E(T_1)]/[\%\Delta E(T_2)]. \qquad (16.16)$$

Now, Eq. 16.15 is problematic in several ways. First of all, the numerator and denominator on the right are the difference of two exponentials. So, unless the activation energies Ea(O_2) and Ea(H_2O) are known *very* well—usually *not* the case —the analyst will end up trying to take the difference between two very large and unstable numbers. It is unlikely that any reliable result can emerge from such a calculational procedure. Second the pre-exponential factors A(O_2) and A(H_2O) are unknown, and they don't cancel out of the numerator and denominator! So, evaluation of the right side of Eq. 16.15 is going to be challenging, but all is not lost. The first simplification is to define a ratio $\mathfrak{r}$ such that $\mathfrak{r} = A(H_2O)/A(O_2)$, or $A(H_2O) = \mathfrak{r}\, A(O_2)$. If this simple relationship is substituted into the right side of Eq. 16.15, A(H_2O) can be removed and A(O_2) completely cancels from the numerator and denominator. But, $\mathfrak{r}$ still remains an unknown. Nevertheless, the values of A(H_2O) and A(O_2) are no longer needed, only their ratio—an easier problem. Next, the mechanisms of hydrolytic scission and oxygen crosslinking are different, and so the activation energies are likely to be different. However, attack of a labile hydrogen during oxidative crosslinking is easy when R is toluene.

Therefore, the activation energy needed to form the intermediate structure is likely to be less for cross-linking than scission by an amount Δ, where $\Delta/Ea(H_2O)$ is *not* necessarily small. Combining Eqs. 16.15 and 16.16 with the substitutions just described yields

$$[\%\Delta E(T_1)]/[\%\Delta E(T_2)] = \frac{\exp\{-[Ea(H_2O) - \Delta]/RT_1\} - r\exp\{-Ea(H_2O)/RT_1\}}{\exp\{-[Ea(H_2O) - \Delta]/RT_2\} - r\exp\{-Ea(H_2O)/RT_2\}}.$$

(16.17)

If the numerator and denominator of Eq. 16.17 are both divided by $\exp\{-Ea(H_2O)/RT_1\}$, the result is

$$[\%\Delta E(T_1)]/[\%\Delta E(T_2)]$$
$$= \{\exp[\Delta/RT_1] - r\}/\{(\exp[-Ea(H_2O)/RT_2 + Ea(H_2O)/RT_1]\exp[\Delta/RT_2])$$
$$- \left(r\exp[-Ea(H_2O)/RT_2 + Ea(H_2O)/RT_1]\right)\}$$
$$= \{\exp[-Ea(H_2O)\Delta T/RT_1T_2]\}\{[\exp(\Delta/RT_1) - r]/[\exp(\Delta/RT_2) - r]\},$$

(16.18)

where $\Delta T = T_2 - T_1 = 353° - 298° = 55$ K. In the form of Eq. 16.18, [% $\Delta E(T_1)$]/[% $\Delta E(T_2)$] is not particularly sensitive to the selection of $Ea(H_2O)$ and Δ. Order of magnitude estimate will do. Setting $Ea(H_2O) = 10$ kJ/mole and $\Delta = 1$ kJ/mole yields the correct experimental ratio [6] for polyurethane, [% $\Delta E(T_1)$]/[% $\Delta E(T_2)$] = 0.395, provided $r = 1.95$. This value of r seems reasonable for the ratio of pre-exponential factors for two different reactions. *With these values of Ea(H_2O), Δ, and r, it is possible to use Eq. 16.15 for other temperatures in the range of 25–80 °C, and probably for temperatures a little outside this range as well.* As noted in subsection b above, [% $\Delta S(T_1)$]/[% $\Delta S(T_2)$] should also approximately obey Eq. 16.15. Experimentally, it has been found that this ratio equals 0.34 [6].

d. *Calculation of Equivalent Aging Times for Given Changes in E and S (Constant A)*: If equivalent aging times are required for E, at two different temperatures, then $\Delta E(T_1, t_1) = \Delta E(T_2, t_2)$. Therefore, from Eq. 16.15

$$[\Delta E(T_1, t_1)/t_1]/[\Delta E(T_2, t_2)/t_2]$$
$$= t_2/t_1$$
$$= \frac{A(O_2)\exp[-Ea(O_2)/RT_1] - A(H_2O)\exp[-Ea(H_2O)/RT_1]}{A(O_2)\exp[-Ea(O_2)/RT_2] - A(H_2O)\exp[-Ea(H_2O)/RT_2]}.$$

(16.19)

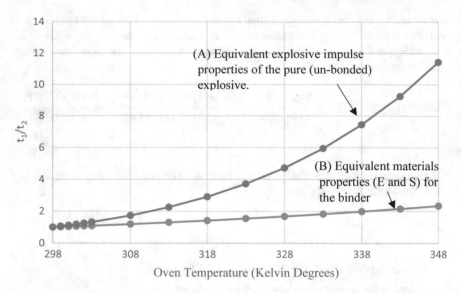

Fig. 16.7 The ratio of natural aging time to oven aging time (t_1/t_2) as a function of temperature (T_2) for equivalent (**A**) explosive impulse properties of the pure (un-bonder) explosive, and (**B**) materials properties (Young's Modulus E and tensile strength S) of a polyurethane binder

And, using the previous substitutions for $Ea(O_2)$ and $A(H_2O)$, and the estimated values for $Ea(H_2O)$ and Δ, together with the previously computed value of $\ast$, yields

$$t_1/t_2 = \{\exp[Ea(H_2O)\Delta T/RT_1T_2]\}\{[\exp(\Delta/RT_2) - \ast]./[\exp(\Delta/RT_1) - \ast]\}$$
$$= 1/0.395 = 2.53.$$

$$(16.20)$$

That is to say, each day in the oven is equal to 2.53 days of natural aging in so far as material properties of the binder are concerned. If Eq. 16.20 is reevaluated for a lower temperature $T_2 = 322$ K (49 °C or 120 °F), $t_1/t_2 = 1.57$. That is to say, each day in the oven at 322 K is equivalent to only 1.57 days of natural "shelf" aging. So, at this temperature, the material properties of the binder are aging much more slowly than the chemical properties of the pure (un-bonded) propellant as calculated from Eq. 16.2 (see Fig. 16.7). This is because the competition between oxidative crosslinking and hydrolytic scission only allows the polymer's mean molecular weight and mechanical properties to change slowly. Although this slow change appears linear in Fig. 16.7, plotting the curve for E and S by itself on an expanded t_1/t_2 scale shows that it has a slight curvature (Fig. 16.8). It is important to note

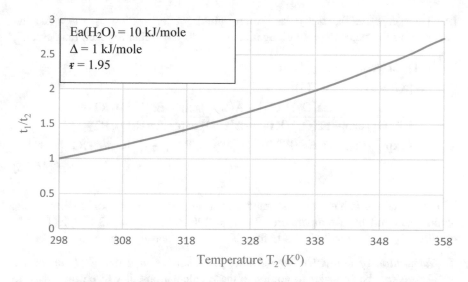

Fig. 16.8 The ratio of natural aging time to oven aging time (t_1/t_2) as a function of temperature (T_2) for equivalent changes in Young's Modulus E of a polyurethane binder. A similar relationship for aging times versus temperature is expected for the tensile strength S

that the choice of Ea(H_2O) and Δ is relatively unimportant. For example if Ea(H_2O) = 12 kJ/mole and Δ = 1 kJ/mole, the new material properties curve is virtually coincident with the old curve for E and S in Fig. 16.8, provided that now $\mathfrak{f}$ = 4.1. That is to say, sensitivity to the value of Ea(H_2O) and Δ has been lost, but the calculations are now sensitive to $\mathfrak{f}$. This, however, is not important to the analyst because $\mathfrak{f}$ is now just basically an adjustable parameter used to fit the available experimental data. It should also be noted that at T_2 = 298 K, t_1/t_2 = 1 in both Figs. 16.7 and 16.8, as it should be. Clearly, if the oven temperature is identical to the storage temperature, there is no accelerated aging (i.e., $t_1/t_2 \equiv 1$).

e. *Calculation of Changes in e versus Temperature and Time (Constant A)*: The maximum elongation before failure (e) is expressed as a percent change from the length of the original tensile test sample before stretching. And, e will decrease as the binder sample ages. Returning to basics, crosslinking will inhibit stretching, while molecular scission will favor it. In analogy with Eq. 16.14

$$\Delta e/t = K''(k_{H_2O} - k_{O_2}/2), \text{ where } K'' \text{ is a constant of proportionality.} \quad (16.21)$$

This equation has a form that is the negative of Eq. 16.14, so a negative slope for the change in e versus time is completely reasonable. Forming the ratio

$[\Delta e(T_1, t_1)/t_1]/[\Delta e(T_2, t_2)/t_2]$ as before yields, after setting $t_1 = t_2 = \Delta t$ and normalizing the percentage of elongation to the 30 day value for each temperature,

$$
\begin{aligned}
&[\Delta e(T_1, \Delta t)/e_{30}]/[\Delta e(T_2, \Delta t)/e_{30}] \\
&= [\%\Delta e(T_1)]/[\%\Delta e(T_2)] \\
&= \frac{-A(O_2)\exp[-Ea(O_2)/RT_1] + A(H_2O)\exp[-Ea(H_2O)/RT_1]}{-A(O_2)\exp[-Ea(O_2)/RT_2] + A(H_2O)\exp[-Ea(H_2O)/RT_2]} \\
&= \left\{\exp\left[-Ea(H_2O)\Delta T/RT_1 T_2\right]\right\}\left\{\left[\exp(\Delta/RT_1) - \tilde{r}\right]/\left[\exp(\Delta/RT_2) - \tilde{r}\right]\right\}
\end{aligned}
$$

$$(16.22)$$

as before; where it has been assumed that the constant of proportionality K'' is, again, independent of temperature. For $T_1 = 298$ K and $T_2 = 353$ K, $[\Delta e(T_1, \Delta t)/e_{30}]/[\Delta e(T_2, \Delta t)/e_{30}] = 0.395$ as before. The actual measured value [6] was 0.368.

f. *Calculation of Equivalent aging times for Changes in E and S (Temperature Dependent A)*: In order to complete these calculations, it will be necessary to establish a connection between the temperature independent coefficient "a" of Eq. 16.4 and the previous definition of $\tilde{r}$. For any given temperature T,

$$
\tilde{r} = A(H_2O)/A(O_2) = [a(H_2O)T^m]/[a(O_2)T^m] = a(H_2O)/a(O_2), \qquad (16.23)
$$

So long as m is the same for both the scission and cross-linking reactions. Therefore, $\tilde{r}$ as previously defined is also the ratio of $a(H_2O)$ to $a(O_2)$—very convenient. Therefore, after inversion, and making all necessary substitutions, Eq. 16.19 yields

$$
t_1/t_2 = (T_2/T_1)^m \frac{\exp\{-[Ea(H_2O) - \Delta]/RT_2\} - \tilde{r}\exp\{-Ea(H_2O))/RT_2\}}{\exp\{-[Ea(H_2O) - \Delta]/RT_1\} - \tilde{r}\exp\{-Ea(H_2O)/RT_1\}}.
$$

$$(16.24)$$

Letting the ratio of exponentials on the right of Eq. 16.24 equal $\mathcal{R}$, the expression for the percentage error in t_1/t_2 incurred by ignoring the temperature dependence of the pre-exponential factors is just

$$
\% \text{ error} = [(T_2/T_1)^m \mathcal{R} - \mathcal{R}] \times 100/\mathcal{R} = [(T_2/T_1)^m - 1] \times 100. \qquad (16.25)
$$

This is the same result as Eq. 16.6. Therefore, the graph of Fig. 16.1 still applies; so long as m is the same for both the scission and crosslinking reactions.

16.5   Some Experimental Observations on Gaseous Diffusion

The penetration of small gaseous molecules like O_2 and H_2O into polyurethane can be followed by observing its color change upon aging. Freshly gelled polyurethane is typically white, or a pale cream color. Upon aging, it "yellows", or becomes a tan color. This may occur in several ways, but is easiest to understand by examination of reaction D in Fig. 16.4. Notice how any unreacted diisocyanate may be attacked by water to produce the intermediate $OCN–R–NH_2$. When the R group is toluene (a methylated phenyl ring), and when O_2 is present (Fig. 16.4b), free radical reactions may produce the yellow dye (chromophore) aniline ($C_6H_5NH_2$). Figure 16.9 (left) shows a specimen of polyurethane, made from diphenyl methylene diisocyanate[†] and a hydroxyl-terminated polyether[†]. Therefore, production of a derivative of the chromophore aniline can be expected upon aging. The specimen had been naturally aged in the dark at 28 °C and 39% relative humidity for 24.0 years (8754 days), and had developed a deep tan color on its outer surface (location A of Fig. 16.9-left). This cylindrical casting was then cut apart to produce the flat region near the top of the specimen. Notice that some of the interior polyurethane has hardly changed from the original cream color of the specimen (location B of Fig. 16.9-right). Infrared spectra were then taken at locations A and B in Fig. 16.9. The result is shown in Fig. 16.10. *Pure aniline*, when trapped in an argon crystal at 5 K, shows an absorption band with a peak at a wavenumber of about 1594 cm^{-1}. And, *pure aniline gas* has a "band" composed of two closely spaced peaks at 1590 and

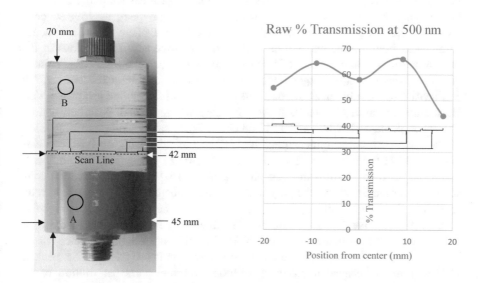

Fig. 16.9 (Left) Cylindrical polyurethane sample with embedded stainless steel tube. (Right) Transmission scan across a black and white photographic transparency of the "cut-away" of the cylindrical sample. The transmission of the clear transparency film at 5000 Å was 86.4%

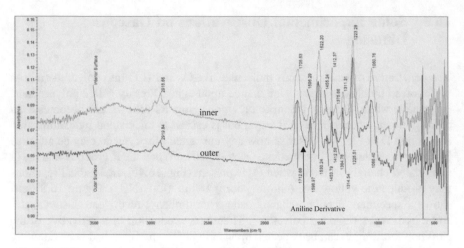

Fig. 16.10 The infrared spectrum of the tan outside surface versus the cream-colored interior of the polyurethane casting in Fig. 16.9 (left). The absorption band at 1641.6 cm^{-1} ($\lambda = 38.27$ µm) is probably the NH$_2$ scissor mode of an aniline derivative. The spectra for the outer and inner surfaces have been arbitrarily displaced by an absorbance of about 0.03 to facilitate comparison

1603 cm^{-1} (a mean of 1596.5 cm^{-1}). In fact, the position of an absorption peak for *pure aniline* can vary by ± 10 cm^{-1}, depending on the environment within which it is trapped. Figure 16.10 shows a bright absorption peak at 1598.97 cm^{-1}, a difference of only about 2.5 cm^{-1} from the aniline gas and 5 cm^{-1} from the aniline/Ar sample. However, it only takes a very small amount of a chromophore to color a matrix, and this peak is too prominent. Therefore, the peak at 1598.97 cm^{-1} probably represents some normal vibrational mode of a phenyl ring, a structure common to both aniline and polyurethane, so that agreement is fortuitous in this case. Much more characteristic of aniline is the nearby band of the –NH$_2$ "scissor" vibrational mode, that is unique to aniline, and has a peak at about 1620.65 cm^{-1} in aniline/Ar and a double peak at 1610 and 1619 cm^{-1} in the aniline gas phase. Very close examination of Fig. 16.10 shows that there is a very small peak between the absorption peaks at 1598.97 and 1712.68 cm^{-1}. This peak is hiding in the shoulder of the asymmetrical prominent peak at 1712.68 cm^{-1}, and its top has been estimated at 1641.6 cm^{-1} ($\lambda = 38.27$ µm), an error of 21 cm^{-1} from the aniline/Ar value and 22.6 cm^{-1} from the gas phase peak at 1619 cm^{-1}. However, it must always be remembered that the polyurethane environment is very different from the crystalline Ar or pure aniline gas phase. Furthermore, the chromophore is probably an aniline derivative, not *pure* aniline. Nevertheless, when the spectra from locations A and B are compared, the peak at 1641.6 cm^{-1} has almost vanished for the interior location that is well shielded from environmental gasses. The peak at 1641.6 is small because it can take as little as a few parts per million of the chromophore to color the polyurethane matrix. It was fortunate that this chromophore peak fell between two prominent absorption peaks of a typical polyurethane spectrum, or it would never have been detected! *Therefore, the chromophore is probably some derivative of aniline produced by water vapor and oxygen diffusion into the polyurethane matrix.*

Next, the flat surface was photographed and a black and white positive transparency was produced whose percent of transmission in the visible region of the spectrum can be measured from point to point. The scan from x = −18 mm to x = +18 mm utilized a 6 mm (1/4″) diameter circular window and a center wavelength of 5000 Å with a spread of ±100 Å (Fig. 16.9-right). Naturally, the part of the scan line that is closest to the polyurethane surface is darkest due to attack by atmospheric gasses. Surprisingly, however, the central portion of the scan (at x = 0) is also relatively dark.

The reason for this darkening is attacked by water vapor and oxygen along the interface between the central embedded stainless steel metal tube and the polyurethane. A close-up photograph (Fig. 16.11) of the top of the polyurethane casting at the interface reveals poor adhesion with pores where gasses can enter. Gas bubbles (probably a small amount of carbon dioxide generated by trace amounts of water contamination in the reactants) seem to have nucleated at the surface of the metal tube in much the same way as water vapor bubbles nucleate along the surface of a steel pot that is beginning to boil. Oxygen and water vapor have permeated this 4.5 cm diameter by 7.0 cm long polyurethane sample to varying degrees, depending on the location within the specimen. But, the zone of plastic halfway between the surface and the stainless steel tube is the least contaminated. Exposure to sunlight seems to accelerate darkening of polyurethane, possibly by additional UV free radical formation (Fig. 16.4b-right), followed by reaction with an isocyanate group (OCN–) leaving modified aniline molecules in the polyurethane matrix. Free radical producing ozone does something similar.

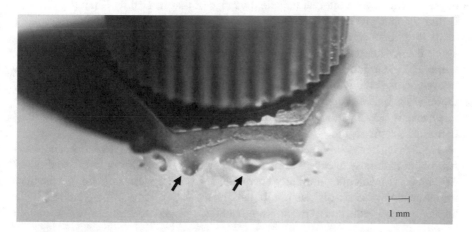

1 mm

Fig. 16.11 The close-up above shows mm scale openings (marked by arrows) along the interface between the stainless steel tube and the polyurethane. Oxygen and water vapor can enter the polyurethane casting along this interface

16.6   Conclusion

The calculations presented in this chapter seem to explain the observed chemical and physical facts about propellants, as they are known today, over the time intervals tested. And, it is also clear that separate specimens should be employed for studying chemical (explosive) properties and mechanical (material) properties during accelerated aging. But, as always, new questions and limitations arise. It has been assumed in the calculations for material properties that the naturally aged and oven aged (accelerated) samples have exactly the same geometry and are exposed to identical humidity and oxidation conditions. Although this may be true external to the polymer samples, the diffusion of oxygen and water into the grain is temperature dependent, and diffusion has not been taken into account other than experimentally in Sect. 16.5. However, over small temperature excursions, the diffusion rate is nearly constant since, by Frick's Law, the diffusion coefficient is only a linear function of temperature [4]. Furthermore, the rate of diffusion of water and oxygen will be different. Although, both molecules are small and differences in diffusion rate should, to some extent, be captured by the value off. The fact that equations like 16.15 represent a *ratio* of changes at two closely spaced temperatures should also mitigate errors.

Particular care was taken to only use E, S, and e data from specimens at least 30 days old to ensure that all curing was complete (i.e., to ensure that polymerization of the binder was complete). However, contrary to experience, perhaps this was not the case. Some additional curing may account for some of the increase in material strength (both E and S) over the time interval under study [6]. It is difficult to know the percent of polymerization without Gel Permeation Chromatography (GPC) tests. There is also some indication that E and S stop growing, and may even decrease, after a time interval outside the range of that studied by Khan [6]. Clearly, E and S for a given binder cannot grow forever. Perhaps excessive crosslinking leads to embrittlement. Perhaps fragile peroxide cross-linkages break down with aging due to thermal decomposition. Perhaps sites for cross-linkage are depleted while H_2O scission continues. Or, perhaps a secular equilibrium is established between crosslinking and scission that eventually causes the molecular weight and, therefore, material properties to level off as a function of time. Probably all these mechanisms play a part in capping off E and S. Clearly, more very long-term testing is needed in this regard.

Finally, it should be noted that this chapter involved many experimental tools from both materials science and analytical chemistry, in addition to the traditional experimental techniques normally encountered by the system safety engineer (like impulse testing, "cook-off" testing, bullet impact testing, drop testing, etc.). Tensile testing, optical reflectance/transmission studies, infrared spectroscopy, and visible light microscopy can aid the system safety engineer in diagnosing and correcting propellant problems at the macroscopic and molecular level. And, all of these techniques should be part of the system safety engineer's training.

†   Diphenyl methylene diisocyanate   Hydroxyl terminated poly ether

$$OCN-\bigcirc-\underset{\underset{H}{|}}{\overset{\overset{H}{|}}{C}}-\bigcirc-NCO \qquad\qquad HO-[CH_2O]-\ \cdots$$

16.7   Problem

Starting from the Arrhenius relation (Eq. 16.1), by what percent will the reaction rate increase by if T is raised by 10 °C from 25 °C (298 K)? What does this imply for the ratio of t_1/t_2 when studying the chemical properties of pure (un-bonded) explosives?

References

1. Anon. (2008, May 12). *WSESRB Interactive Safety Environment (WISE), Series C, Course 3*. Weapon System Explosives Safety Review Board (WSESRB). http://63.134.199.73/nw/lms/ LaunchMultiScoManifest.asp?courseid=60&addtomycourses=no&retake=undefined&.
2. Anon. (2012, July 30). *Department of Defense Ammunition and Explosives Hazard Classification Procedures (Joint Technical Bulletin), TB 700-2 NAVSEAINST 8020.8C TO 11A-1-47*. Washington, DC: Headquarters DoA, DoN, DoAF.
3. Sutton, G. P., & Ross, D. M. (1976). *Rocket propulsion elements* (4th ed., pp. 366, 358–361, 382–389). New York, NY: Wiley.
4. Alberty, R. A., & Silbey, R. J. (1997). *Physical chemistry* (2nd ed., pp. 640–641, 671–673, 681–687, 713). Wiley.
5. Stevens, M. P. (1999). *Polymer chemistry: An introduction* (pp. 378–382, 100–106). New York, NY: Oxford University Press.
6. Khan, M. B. (1993). Simulation of composite propellant aging. *Polymer-Plastics Technology and Engineering, 32*(5), 467–489 (published on-line 22 Sept. 2006).

Chapter 17
The Movement of Inorganic Cadmium Through the Environment

17.1 Background

Cadmium is a versatile metal that has found uses in the production of pigments and dyes, stabilizers in plastics, electrodes of nickel/cadmium alkaline batteries, printing, textiles, television phosphors, lasers, semiconductors, solar cells, photography, scintillation counters, neutron absorbers, florescent lamps, jewelry, engraving processes, pesticides, superphosphate fertilizers, alloys (typically with Cu, Ag, Ni, Au, Bi or Al), and even in dental amalgams. However, cadmium's most important use is as an electroplated coating on other metals. About 50% of all cadmium is used for this purpose. Cadmium plating has been used extensively as a corrosion inhibitor on a variety of steel surfaces (especially nuts and bolts, and parts used at sea) since the 1950s. Although its use is declining due to environmental concerns, the metal is still a popular coating. For public health reasons, it would be helpful to know what becomes of cadmium-plated materials. Millions of cadmium-plated screws and other components were used on military equipment of every kind. And, the "boneyard" at Davis-Monthan Air Force Base (DM AFB), Tucson, Arizona, contains such obsolete hardware as far as the eye can see (Fig. 17.1). If each fastener (nut, bolt, screw, lock washer, rivet, etc.) is coated with a tenth of a gram (10^{-4} kg) of cadmium, and if there are 10^5 fasteners per aircraft, and if ten thousand obsolete aircraft are stockpiled, then the total inventory of cadmium about is 100 metric tons!

Is the cadmium plating on various aircraft components still in the metallic state after decades of exposure to the elements? Does any cadmium reside in the soil as insoluble minerals? Or, has cadmium entered the water table as the mobile ion Cd^{2+}? These are some of the questions that will be answered in this chapter. Of course, it should be realized that the calculations presented here are also models for other heavy metals (e.g., mercury [Hg] that also has a valence of +2) thereby extending the utility of this chapter. On the other hand, it is also important to realize that heavy metals can react with organic compounds to form, for example, ions like

© Springer Nature Switzerland AG 2020
R. R. Zito, *Mathematical Foundations of System Safety Engineering*,
https://doi.org/10.1007/978-3-030-26241-9_17

Fig. 17.1 The obsolete military aircraft at Davis-Monthan Air Force base. Tons, perhaps 100 metric tons, of cadmium were used to coat the many millions of screws and other components used in these Cold War weapons

methyl mercury, CH_3Hg^+ (see Appendix F). This organo-metallic chemistry is not treated here, thereby limiting the utility of this chapter. However, if the chemical reactions involved are known, there is no reason why the methods described here cannot be applied. The keyword here is *known*. In the absence of known chemistry, the analyst is faced with a *known unknown*.

17.2   The Occurrence of Cadmium in Nature

The cadmium mineral system is, fortunately, quite simple. There are only three principal species; the carbonate ($CdCO_3$) called Otavite, the sulfide (CdS) called Greenochite (or Hawleyite), and the oxide (CdO) called Monteporrite [1]. The carbonate forms from the interaction of cadmium with the carbonate species CO_3^{2-}, HCO_3^-, and H_2CO_3. The ultimate source of these carbonate species is CO_2 in the atmosphere. And, some of this reservoir of gas can then dissolve in rainwater, surface freshwater, and seawater, to form an equilibrium mixture of the three basic species. Carbon dioxide can also be taken up by living organisms to form calcium carbonate (shells), which can eventually form limestone. And, solutions of limestone in water contain free chemically active CO_3^{2-} ions. Total environmental carbonates in solution are typically taken to have an activity (approximately the concentration) of 10^{-4} molal. The sulfide forms from the interaction of cadmium with the sulfur species S^{2-}, HS^-, and H_2S. Cadmium sulfide occurs in a hydrated form $[CdS(H_2O)_X]$ called Xanthocroite as well as an anhydrate. Sulfur is widespread in the environment, with a typical activity of 10^{-1} molal, and has volcanic activity as one of its primary sources, as can be verified by examination of the sulfur coating on rocks surrounding any caldera. It may seem strange to find the oxide of cadmium in nature, since the metal is used for corrosion protection. However, over geologic time even chemically resistant metals must succumb to attack by oxygen! Native cadmium metal has been found in the Vilyuy River basin in Siberia but is otherwise unknown in nature. The synthetically purified cadmium metal used by our civilization is only a meta-stable product that must eventually enter the environment by reacting with one of the first three elements in group VIB of the periodic table. In a later section, the chemical reactions that these species undergo will be considered in detail. But first, the chemistry of water must be understood, because water is the common medium for chemical reactions at or near the Earth's surface.

17.3   The Stability of Water

It may seem odd to discuss the stability of water in an engineering context but, in fact, it is of central importance since toxic cations cannot spread through the environment without liquid water, whose existence is more constrained than one might expect. Dry cadmium metal, or cadmium compounds, by themselves, are harmless, so long as they stay put. It is water that creates a problem! Therefore, much of the discussion in this chapter centers around the hydrolysis ("water splitting") equation [2].

$$2H_2O(1) \leftrightarrow 2H_2(g) + O_2(g)\ (20\,^{\circ}C) \tag{17.1}$$

By Le Chatelier's Principle, removal of the hydrogen or oxygen gas (g) product drives the equilibrium reaction to the right; thereby destroying liquid (l) water. But, how much hydrogen or oxygen must be removed for this to happen? The equilibrium constant (K_{eq}) for the hydrolysis of water is given by [2]

$$K_{eq} = 10^{-83.1} = P(H_2)^2 P(O_2)\ (20\,^{\circ}C), \tag{17.2}$$

where the hydrogen and oxygen gas pressures P are expressed in atmospheres (atm). Now, geologists seldom worry about oxygen pressures greater than 1 atm. The atmosphere is only 21% oxygen. Although, nearly pure oxygen is vented from some locations on Mt. Vesuvius at 1 atm of pressure. And, of course, artificial industrial processes may use pure oxygen. However, in a natural context, it is reasonable to set $P(O_2) = 1$ atm as an upper limit. In that case, $P(H_2) = \sqrt{(10^{-83.1}/1)} = 10^{-41.6}$ atm. If the partial pressure of hydrogen falls below this level, liquid water will decompose. But, what about the other limit, when the O_2 pressure is very low and the H_2 pressure is high. Natural gas wells do not vent pure H_2, but some bacteria release hydrogen as part of their metabolic processes, so that $P(H_2)$ may approach 1 atm in some soils. Therefore, it makes sense to set an upper limit of $P(H_2) = 1$ atm. In that case, $P(O_2) = 10^{-83.1}$ atm. That is to say, when $P(H_2) = 1$ atm, the oxygen pressure must be at least $10^{-83.1}$ atm for liquid water to exist. Otherwise, Eq. 17.1 will be driven to the right.

Now, one may ask, "Where do such low oxygen pressures exist?" Well, there are many places, both natural and artificial. In environments containing decomposing organic matter, biochemical reactions quickly remove oxygen, typically with a concomitant increase in CO_2 and hydrogen sulfide. Under these conditions, the water stability limit may be approached. To understand how effective decomposing organic matter is at consuming oxygen, one needs only to visit the National Museum of Ireland-Archaeology (Dublin). There, one can see the preserved bodies of Iron Age (600 BC) inhabitants who had fallen, or been thrown, or were buried, in the Irish bogs. Discovered in 2003, their soft fleshy tissues have been preserved in gruesome detail for over two and a half millennia [3]. Still older, are trees buried by fluvial activity. As branches and roots decay, anaerobic conditions quickly develop that arrest decay of trunks. The tree ring systems of original wood have been preserved for over 12,000 years [4]! Furthermore, observed mineral species suggest that regions of the Earth's crust must, at least at one time, been free of oxygen as well. The role of bacteria in mineral formation should not be underestimated! Industrially, many processes take place in sealed environments at zero, or near zero, oxygen pressure. For example, the melting and casting of propellants into solid rocket motors takes place in a completely anaerobic environment. Any minute amount of residual oxygen present during processing is quickly oxidized by the fuel itself [5]. Under these oxygen-free conditions, even liquid water cannot exist! Geologists (and even space scientists[†]) are concerned with the conditions under which water can exist because without this water, solutions of cations cannot form minerals like Greenochite and Otavite. Safety engineers worrying about the spread of cadmium through the environment must be concerned with exactly the same solution chemistry. However, environmental work takes place at or below $P(O_2) = 0.021$ atm (i.e., oxygen comprises 21% of the Earth's atmosphere).

It is difficult to measure a partial pressure, especially low partial pressures, in a mixture of gasses. Therefore, another method of detecting hydrogen and oxygen gas products is desired. Electrical oxidation potentials offer an excellent alternative for oxygen, while pH measurements (also ultimately electrical) offer an alternative for hydrogen. To this end, the oxidation/reduction reaction of Eq. 17.1 can be rewritten as two half-cell reactions. The first is

$$2H_2O(1) \leftrightarrow 4H^+(aq) + O_2(g) + 4e^-. \tag{17.3}$$

The second is

$$4H^+(aq) + 4e^- \leftrightarrow 2H_2(g), \tag{17.4}$$

where the argument (aq) means "aqueous". When Eqs. 17.3 and 17.4 are added together, Eq. 17.1 is recovered. Now, the Nernst equation [2] applies to any half-cell reaction and, for Eq. 17.3, can be written as

$$E = E^0 + (RT/n\mathcal{F}) \ln Q$$
$$= E^0 + \{(0.001987\,\text{kcal/K}^0)\,(298.15\,\text{K}^0)/[(n\,\text{moles})(23.06\,\text{kcal/volt} - \text{mole})]\}\ln Q$$
$$= E^0 + (0.025688/n)\ln Q = E^0 + (0.025688/n)(2.303)\log Q = E^0 + (0.05916/n)\log Q$$
$$= 1.23\,\text{volts} + (0.05916/4)\log\left([H^+]^4[O_2]/[H_2O]^2\right) = 1.23\,\text{v} + 0.01479\log\left([H^+]^4 P_{O2}/1\right).$$
$$(17.5)$$

R is the Ideal Gas Constant, T is the Kelvin temperature, $\mathcal{F}$ is Faraday's Constant, n is the number of moles of electrons in the half-cell reaction of Eq. 17.3 (i.e. 4), the factor of 2.303 converts base 10 logarithms to natural logarithms, Q is the reaction quotient where the concentration of pure water is unity ($[H_2O] = 1$) and the concentration of oxygen gas is its partial pressure ($[O_2] = P_{O2}$), and E^0 is the standard concentration half-cell potential (1.23 V) read from electrochemical tables or computed either from the equilibrium constant K_{eq} or the reaction change in the tabulated free energies of formation F^0 of reactants and products such that

$$E^0 = -(RT/n\mathcal{F})\ln K_{eq} = \Delta F^0/n\mathcal{F} = 1.23\,\text{V}. \qquad (17.6)$$

Note that a "+" sign has been used before the coefficient (RT/n) in Eq. 17.5 because the geochemical convention of placing reduced species on the *left* side of half-cell reactions (e.g. Eq. 17.3) has been adopted [2]. However, the "−" sign in Eq. 17.6 is also correct [2]. Using the properties of logarithms to simplify Eq. 17.5, and using the definition of pH (i.e. pH = − log {H^+}), yields

$$E = 1.23 + 0.01479\log P_{O2} - 0.05916\,\text{pH}. \qquad (17.7)$$

Substituting the three most important values for P_{O2} into this equation for the oxidation potential E yields

$$
\begin{aligned}
&E = 1.23 - 0.059\,\text{pH} &&@P_{O2} = 1\,\text{atm (upper limit)} \\
&E = 1.205 - 0.05916\,\text{pH} &&@P_{O2} = 0.21\,\text{atm (21\% of total atmospheric pressure)} \\
&E = -0.059\,\text{pH} &&@P_{O2} = 10^{-83.1}\,\text{atm (lower limit)}
\end{aligned}
$$
$$(17.8)$$

Figure 17.2 is a plot of these three equations. The value of E separating the highest and lowest lines is 1.23 volts, the theoretical voltage needed to split water into hydrogen and oxygen gas. Realistically, 1.5 volts are needed for hydrolysis in the laboratory to overcome heat loss [6]. Water can exist within the highest and lowest lines of Fig. 17.2. Notice that the second of Eq. 17.8 is almost co-linear with the first. This occurs because the expression for E (Eq. 17.7) is logarithmic for both O_2 and H^+. The same thing will happen at the lower limit of stability even if the oxygen level does not reach all the way down to $10^{-83.1}$ atm, as in the example environments discussed above. Several types of "natural water" have also been mapped

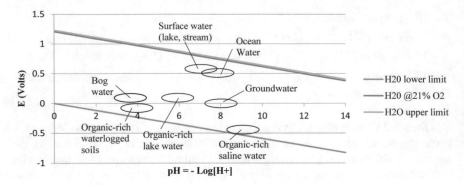

Fig. 17.2 Boundaries for the stability of water on the E versus pH diagram

onto Fig. 17.2. Notice how close "bog water", "organic-rich waterlogged soils", and especially "organic-rich saline waters" come to the lower limit for the existence of water, as discussed above.

All the "phases of Cd", that is to say, the various chemical species (compounds and ions) of concern in this chapter, must lie between the upper and lower bounds of water stability when plotted on the E versus pH diagram. Any compound or ion that can only exist outside the zone of water stability is of no concern because it cannot be mobilized by liquid water. Figure 17.2 is called a *Pourbaix* (Poor-bāy) *Diagram*, and it is the tool of choice for understanding chemical systems that involve both redox and acid/base reactions. The Pourbaix Diagram will be central to all that follow.

17.4   Cadmium Pourbaix Diagram

Figure 17.3 is the Pourbaix Diagram for the oxides, sulfides, and carbonates of cadmium ([7]—corrected). In this section, this diagram will be explained in quantitative detail. First of all, the upper and lower lines with a negative slope are just the water equations plotted in Fig. 17.2. These have already been discussed in detail. The safety analyst is only interested in chemical reactions that lie between these lines. Such as the carbonate reaction

$$CdCO_3(s) + 2H^+(aq) \leftrightarrow Cd^{2+}(aq) + H_2CO_3(aq). \tag{17.9}$$

The solid(s) cadmium carbonate ($CdCO_3$) in Eq. 17.9 exists as an insoluble precipitate in contact with water, while aqueous (aq) species are dissolved. This reaction does not involve a change in the oxidation state of any element. Therefore, the production of Cd^{2+} will be independent of oxidation potential (voltage E). However, Eq. 17.9 *is* pH dependent as is evident by the appearance of the H^+ ion on the left of

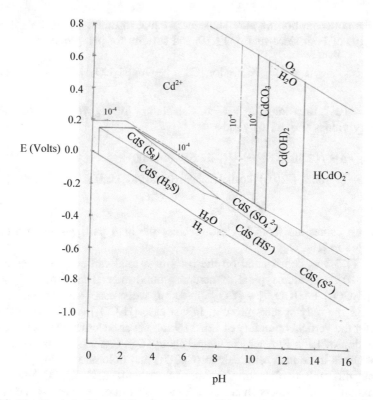

Fig. 17.3 The Pourbaix Diagram for the oxides, sulfides, and carbonates of cadmium [7]—recalculated and corrected. The total dissolved sulfur activity is 0.1 molal (approximately equal to the molarity concentration, M) and the *total* dissolved carbonate (CO_2 aq., $H_2.CO_3$, HCO_3^-, and CO_3^{2-}) activity is 10^{-4} molal, at 25 °C and 1 atm of pressure. The boundary for the ionic species (Cd^{2+}) applies to activity levels of 10^{-6} molal (solid vertical line) and 10^{-4} molal (dashed vertical line)

Eq. 17.9. The lower the pH, the greater the concentration of H^+, which will push the equilibrium to the right by Le Chatelier's Principle. On the other hand, an increase in pH means a decrease in the concentration of H^+, drawing the equilibrium to the left and reducing the concentration of Cd^{2+}. It is for this reason that there are two long vertical lines (chemical phase boundaries) at pH values of about 9 and 10. The more acidic (less basic) boundary (dashed) at pH = 9 corresponds to the higher Cd^{2+} ionic activity (approximately the same as concentration) of 10^{-4} molals. The less acidic (more basic) boundary (solid) at about pH = 10 corresponds to a lower Cd^{2+} activity of 10^{-6}. To understand where these pH values come from consider the equation for the equilibrium constant (K_{eq}) of reaction 17.9.

$$K_{eq} = \{[Cd^{2+}][H_2CO_3]\}/\{[CdCO_3][H^+]^2\}. \qquad (17.10)$$

Setting the concentration of pure substances equal to unity, taking logarithms (to the base 10) of both sides of Eq. 17.10, and solving for pH, yields

$$pH = \{\log K_{eq} - \log[Cd^{2+}] - \log[H_2CO_3]\}/2. \qquad (17.11)$$

If $[Cd^{2+}] = 10^{-6}$, then $\log[Cd^{2+}] = -6$. Calculating $\log K_{eq}$ from the difference in free energy values (Eq. 17.6) yields

$$\log K_{eq} = -\Delta F^0/\{2.303RT\} = -\{(-18.58 - 149.00)$$
$$- (-160.2 + 2(0))\}kcal/\{(2.303)(0.001987 \ kcal/K^0)(298.15K^0)\} = 5.4.$$
$$(17.12)$$

Free energy values are from [2]. Now comes the hard part, assigning a concentration value to $[H_2CO_3]$.

Figure 17.3 was constructed on the basis of a total carbonate species concentration of 10^{-4}, a value typical of natural groundwater. Therefore, $10^{-4} = [CO_2 (aq)] + [H_2CO_3] + [HCO_3^-] + [CO_3^{2-}]$. As a worst-case assumption, suppose H_2CO_3 was the only species present. In that case $[H_2CO_3] = 10^{-4}$ and pH = 7.7. The pH for the vertical boundary of Eq. 17.9 can never be less than this for a $[Cd^{2+}]$ concentration of 10^{-6}. But, why is the pH about 10 in Fig. 17.3? It is because deep groundwater is in contact with calcite ($CaCO_3$) and dolomite ($CaMg(CO_3)_2$), and these minerals will deplete groundwater of carbonic acid (H_2CO_3) so that its concentration is much lower than 10^{-4}. A great deal of effort was expended in the 1960s and 1970s determining the concentration of the various carbonate species at depth in groundwater [8], the result is that a concentration of H_2CO_3 will eventually reach about 2.5×10^{-9} molal. Substituting Log_{10} of this value (i.e. -8.6) into Eq. 17.11 yields a pH = 10, as expected.

The pure oxide of cadmium (CdO), when hydrated, forms a hydroxide which may be derived directly from the carbonate according to the reaction

$$CdCO_3(s) + 2H_2O \leftrightarrow Cd(OH)_2(s) + 2H^+(aq) + CO_3^{2-}. \qquad (17.13)$$

Again, there is no change of oxidation state for any element in Eq. 17.13. So, production of $Cd(OH)_2$ is independent of voltage E. However, the presence of the hydrogen ion on the right of Eq. 17.13 says that this reaction is directly pH dependent. Therefore, a vertical line separates the $CdCO_3$ phase from the $Cd(OH)_2$ phase at a pH of about 11 in Fig. 17.3. The equilibrium equation corresponding to Eq. 17.13 can be written in the form

$$pH = \{-\log K_{eq} + \log[CO_3^{2-}]\}/2. \qquad (17.14)$$

Again, for deep groundwater almost all carbonate species will be in the form CO_3^{2-} [8], so $[CO_3^{2-}] \approx 10^{-4}$. Calculating $\log K_{eq}$ from the difference in free energies yields -25.65, and substituting these values into Eq. 17.14 yields pH $= 10.8$. Essentially, the high concentration of basic hydroxyl ions at pH ≈ 11 will neutralize the acidic H^+ ions, thereby drawing the equilibrium to the right in Eq. 17.13, with the concomitant production of the $Cd(OH)_2$.

There is one more oxide to consider. It is formed according to the reaction

$$Cd(OH)_2(s) \leftrightarrow HCdO_2^-(aq) + H^+(aq). \qquad (17.15)$$

Again, there is no change in the oxidation state for any element in Eq. 17.15 (oxidation potential independence), but the presence of the H^+ ion indicates that this reaction is pH dependent. It takes a very basic environment to extract one hydrogen ion from cadmium hydroxide, so the vertical line that separates the $HCdO_2^-$ phase from the $Cd(OH)_2$ phase occurs at pH ≈ 13, as can be verified by repeating the pH calculations.

Finally, sulfide reactions will be considered. There is a vertical line separating the CdS phase from the Cd^{2+} phase at a very acidic pH of about 0.5. The CdS phase is also identified by H_2S in brackets. What does that mean? Examination of the reaction

$$CdS(s) + 2H^+(aq) \leftrightarrow Cd^{2+}(aq) + H_2S(aq) \qquad (17.16)$$

shows that it is oxidation potential independent, but pH dependent. Hence, the representation of this chemical phase change as a vertical line. Also, when CdS is the dominant phase, H_2S is a minor product. Hence, H_2S is enclosed in brackets after CdS. Now consider the reaction

$$8CdS(s) \leftrightarrow 8Cd^{2+}(aq) + S_8(s). \qquad (17.17)$$

This reaction is oxidation potential *dependent* because sulfur undergoes an oxidation from the -2 valence state to the zero valence state (i.e. there has been a loss of electrons). However, it is pH *independent* because no H^+ ions are involved in Eq. 17.17. Therefore, Eq. 17.17 is represented by a horizontal line at an oxidation potential of about $+0.15$ volts. This voltage is *positive* because sulfur was *oxidized*. Notice that the phase below this horizontal line is labeled CdS (S_8), meaning that S_8 is a minor constituent when CdS is the dominant chemical phase and reaction 17.17 is operative. So, there can be more than one CdS phase. In fact, there are five; depending on what the minor constituent is. Of these five phases, the one labeled CdS (SO_4^{2-}) is the most important because it separates insoluble CdS from aqueous Cd^{2+} according to the reaction

$$CdS(s) + 4H_2O \leftrightarrow Cd^{2+}(aq) + SO_4^{2-}(aq) + 8H^+(aq). \qquad (17.18)$$

Now, this reaction is interesting because it involves *both* oxidation potential dependence *and* pH dependence. So, this reaction will not be represented by a vertical or horizontal line on the E-pH diagram (Pourbaix Diagram). Equation 17.18

will be represented by a line with a slope (like the lines representing the decomposition of water). In Eq. 17.18, sulfur undergoes oxidation from the −2 state in CdS to the +6 state in $SO_4{}^{2-}$ (a change of 8 mol of electrons per mole of CdS). Applying the Nernst equation (Eq. 17.5) to the half-cell reaction in Eq. 17.18 yields

$$
\begin{aligned}
E &= E^0 + \{[(0.001987)(298)]/[(8)(23.06)]\}\left\{\ln\left([H^+]^8[SO_4^{2-}][Cd^{2+}]\right)\right\} \\
&= E^0 + (0.00321)\{8\ \ln[H^+] + \ln[SO_4^{2-}] + \ln[Cd^{2+}]\} \\
&= E^0 + (0.00321)\{\ln[SO_4^{2-}] + \ln[Cd^{2+}]\} + 8(0.00321)\ln[H^+] \\
&= (a\,constant) + 0.0257\ \ln[H^+] \\
&= (a\,constant) + (2.303)(0.0257)\log[H^+] = (a\,constant) \\
&\quad + 0.0592\log[H^+] = (a\,constant) - 0.0592\,pH.
\end{aligned}
$$

(17.19)

Comparing Eq. 17.19 with Eq. 17.7 shows that the slope of the water lines and the slope of Eq. 17.19 are the same; as is evident in Fig. 17.3. Similarly, the Nernst equation can be applied to the boundaries between the CdS phase and the $CdCO_3$, Cd $(OH)_2$, and $HCdO_2{}^-$ phases to produce three more sloped boundaries. The equation

$$CdS(s) + H_2CO_3(aq) + 4H_2O \leftrightarrow CdCO_3(s) + SO_4^{2-}(aq) + 10H^+(aq) \quad (17.20)$$

yields a Nernst equation of the form

$$E = (a\,constant) - 0.0590\,pH. \quad (17.21)$$

Notice the very slight change in slope. This is reflected in Fig. 17.3 (corrected from [7]). The equation

$$Cds(s) + 6H_2O \leftrightarrow Cd(OH)_2(s) + SO_4^{2-}(aq) + 10H^+(aq) \quad (17.22)$$

yields a Nernst equation having the same form and slope as Eq. 17.21. Therefore, the boundaries between CdS and $CdCO_3$, and CdS and $Cd(OH)_2$, are collinear. This phenomenon is reflected in Fig. 17.3. Finally, the equation

$$CdS(s) + 6\,H_2O \mapsto HCdO_2^-(aq) + SO_4^{2-}(aq) + 11\,H^+(aq) \quad (17.23)$$

yields a Nernst equation of the form

$$E = (a\,constant) - 0.0589\,pH. \quad (17.24)$$

Again, there is a *very* slight decrease in slope that is imperceptible in Fig. 17.3. All calculations and graphical constructions were made at 25 °C. So, Fig. 17.3 can be understood in intricate detail. And, this quantitative understanding will be required to solve the next two problems.

17.5   A Case of Cadmium Environmental Pollution

What exactly happens when rainwater, in chemical equilibrium with atmospheric CO_2, comes in contact with the pure meta-stable cadmium metal coating aircraft parts? To answer this question first consider the chemical reactions of CO_2 in pure neutral water.

$$CO_2(g) \leftrightarrow CO_2(aq) \qquad \text{(carbon dioxide dissolves in water)}$$
$$CO_2(aq) + H_2O \leftrightarrow H_2CO_3(aq) \qquad \text{(carbonic acid forms)}$$
$$H_2CO_3(aq) \leftrightarrow HCO_3^-(aq) + H^+(aq) \quad \text{(bicarbonate ion forms)}$$
$$HCO_3^-(aq) \leftrightarrow CO_3^{2-}(aq) + H^+(aq) \quad \text{(carbonate ion forms)}$$

$$(17.25)$$

These equations form a coupled system of chemical reactions whose equilibrium concentration of H^+ (and, therefore, pH) are very well known, so the calculations will not be presented here except by citation [9]. The "open-to-the-air" system equilibrium pH = 5.7. That is to say, rainwater is slightly acidic, and modern atmospheric pollution makes the acidity worse.

The next question is, "How does this slightly acidic rainwater solution attack the pure meta-stable cadmium metal?"

The calculation is performed as follows. Starting from the reaction of pure neutral meta-stable cadmium metal in the presence of the acidic hydrogen ion H^+ (or, more precisely the hydronium ion H_3O^+)

$$Cd^0 + 2H^+ \rightarrow Cd^{2+} + H_2. \qquad (17.26)$$

The equilibrium constant for this reaction is given by

$$K_{eq} = [Cd^{2+}][H_2]/\left([Cd^0][H^+]^2\right). \qquad (17.27)$$

Therefore,

$$\log K_{eq} = \log[Cd^{2+}] + \log[H_2] - \log[Cd^0] - 2\log[H^+]. \qquad (17.28)$$

Since the concentration (activity) of pure Cadmium metal at STP is unity, and since log 1 = 0, the term $- \log[Cd^0]$ may be set equal to zero. Similarly, evolved H_2 must push against the atmosphere at 1 atm so that the $\log_{10}$ of its concentration (measured as a partial pressure in atm) is given by $\log[H_2] = \log 1 = 0$. Finally, substituting the usual definition of pH (pH = $- \log[H^+]$), yields

$$\text{Log K}_{eq} = \log\left[Cd^{2+}\right] + 2\,pH. \qquad (17.29)$$

The value of log K_{eq} can be calculated from the difference in free energy between products and reactants of Eq. 17.26 at 25 °C, and its value is +13.6217 [2]. Therefore,

$$\boxed{13.6217 = \log\left[Cd^{2+}\right] + 2\,pH} \qquad (17.30)$$

This last equation is all-important and will serve as the starting point for all the calculations that follow in this section and the next. Setting pH = 5.7 for rainwater, and solving for $[Cd^{2+}]$ yields an equilibrium concentration of 158.5 mol/l in a closed system of Cd metal and acidic rainwater at 25 °C. Of course, such high concentrations will not be seen in practice since the reaction rate for Eq. 17.25 is *very* slow at the pH of rainwater; on the order of years before the first signs of "bloom" (oxidation) are noticed (Fig. 17.4).

Galvanic action between cadmium and steel will accelerate the corrosion process, while the passivating layer of oxide will slow it. In the end, however, rainstorm after rainstorm, year after year, decade after decade, cadmium metal will slowly be removed as the ion Cd^{2+} and will enter the soil below.

To understand what happens next, the Pourbaix Diagram of Fig. 17.3 needs to be consulted. First of all, Arizona soil is rich in carbonates, so deep groundwater in contact with soil will be basic at equilibrium. Values of groundwater pH typically run between 8 and 9 in the Tucson area. Second, deep groundwater is a low oxygen environment, but not a reducing environment since Arizona soils generally have a low content of organic matter. Therefore, $E \approx 0$. Under these conditions, the Pourbaix diagram indicates a Cd^{2+} concentration of $\sim 10^{-4}$ mol/l at pH = 9. The remainder of the cadmium is chemically pinned in place as an insoluble carbonate. Given a total initial mass of 10^5 kg of Cd metal, and making the *worst-case* assumption that all of this metal has been converted into its carbonate, a maximum of 10 kg of cadmium are mobile (because 10^5 kg $\times$ $[Cd^{2+}]/[CdCO_3] = 10^5 \times 10^{-4}/1 = 10$ kg). Since the area

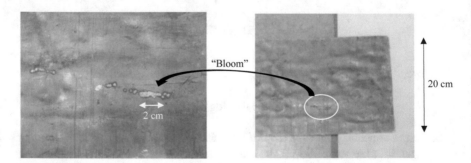

Fig. 17.4 "Bloom" on cadmium metal after 35 years of storage in air at 28 °C and $\sim 39\%$ relative humidity (no liquid water contact). The system safety engineer must be able to recognize bloom when he/she sees it so it can be removed and the surface decontaminated

of DM AFB is about 25 km^2 (or 25×10^6 m^2), the total number of kgs of Cd^{2+} washed into each m^2 of soil cannot exceed 0.4×10^{-6} kg/m^2 = 3.56×10^{-6} mol/m^2. If all this Cd^{2+} eventually turns up in the water table, and if the equivalent water table thickness is 1 m of pure water (actually it is about 20 m), then the final water table concentration will be about 3.56×10^{-9} mol/L. The *Maximum Contamination Level (MCL)* for Cd in drinking water is 5 μg/L = 4.5×10^{-8} mol/L = 8.1×10^{-10} mol fraction [10]. Therefore, the groundwater is safe by at least an order of magnitude. If the chemistry of the area had been different (say, more acidic), remediation might have required plowing humus (organic material) into the surface soil to reduce its oxidation potential and trap Cd^{2+} as an insoluble sulfide. Finally, one last mitigating factor should be mentioned. Groundwater is not a closed system. Groundwater flows and, in the Tucson area, the aquifer can recharge (flush itself out) every few months [11]; thereby preventing the buildup of toxic ions. In summary, the groundwater in the Tucson area will remain safe from cadmium pollution just so long as the desert remains alkaline. And, that will be the case for many millions of years.

17.6   A Case of Cadmium Poisoning

Amazingly, cadmium plating has been used for cooking utensils [12]. Acidic foods may directly attack the meta-stable metal via the reaction of Eq. 17.26. Furthermore, it can be shown‡ that the reaction rate is one million times faster at pH = 3 (fruit juice) than at pH = 6 (rainwater). Therefore, troublesome amounts of Cd^{2+} can be removed from cadmium coated utensils in minutes (instead of years). Solving Eq. 17.30 for pH, and setting [Cd^{2+}] = [MCL$_{Cd+2}$] = 4.5×10^{-8}, yields

$$pH = (13.6217 - \log[MCL_{Cd+2}])/2 = 10.47. \qquad (17.31)$$

Any solution with a pH less than 10.47 is potentially capable of dissolving enough cadmium metal to produce a poisonous solution given enough contact time. Also, notice that if the [MCL$_{Cd+2}$] is lowered in the future, a higher pH (less acidic environment) will suffice to extract dangerous levels of Cd^{2+}. Vinegar (5% acetic acid by volume) has a pH = 3.68. Cadmium-coated utensils immersed in salad dressing containing vinegar will quickly produce a cadmium ion concentration that exceeds the [MCL$_{Cd2+}$]. Repeated use of such utensils in this way will build up dangerous levels of cadmium in the human body. And, the health problems will be exacerbated in cadmium-sensitive individuals. Poisoning is even more of a problem with citrus fruit juice having a pH $\approx$ 2.87. Lemon juice is particularly bad, and it is a common ingredient in salad dressing. Aside from direct acidic attack, there are also other chemical routes by which the meta-stable metal may enter the food. Contact of cadmium metal with high sulfur foods like eggs, and the application of heat, is sufficient to produce a thin layer of sulfide, which will then dissociate to produce Cd^{2+} when the proper pH is reached in an oxygen-rich kitchen environment (see Fig. 17.3).

Earthenware vessels (approximately a mixture of 25% clay [kaolin plus traces of metal oxides], 28% kaolin [$Al_2Si_2O_5(OH)_4$], 32% quartz [SiO_2], and 15% feldspar [$KAlSi_3O_8$]) containing traces of cadmium have also been used for food. Acidic foods should never be stored or prepared in such vessels. Since the abundance of cadmium in the Earth's crust typically varies from 0.01 to 0.5 ppm by weight [1], and since the abundance of cadmium in soils and shale is 0.5 and 0.3 ppm by weight, respectively, a concentration of 0.5 ppm seems a conservative value to assume for the abundance of cadmium in earthenware that will not underestimate the dangers involved. Most of this in situ cadmium will exist as sulfides and carbonates that are ready to release Cd^{2+} under mildly acid, oxygen-rich conditions.[*] In fact, the examination of the Pourbaix Diagram (Fig. 17.3) shows that any pH ≤ 10 is potentially able to do this under the oxygen-rich conditions found in a kitchen. If earthenware is thought of as a solid solution of species, the mole fraction [2] of bound Cd^{2+} (denoted by $N_{Cd+2(s)}$) is given by

$$N_{Cd+2(s)} \approx \left(\text{ moles of } Cd^{2+} \text{ in compound} \right) / (\text{moles of clay} + \text{moles of kaolin} + \text{moles of quartz} + \text{moles of feldspar})$$
$$\approx (0.5 \times 10^{-6}/112.4)/([0.25/258] + [0.28/258] + [0.32/60] + [0.15/278]) = 5.55 \times 10^{-7}.$$

$$(17.32)$$

where masses have been translated into moles by dividing by the Gram Molecular Weight, and the nominal composition of earthenware above has been used. This mole fraction is about three orders of magnitude above the drinking water [MCL_{Cd+2}] expressed as a mole fraction. So, dangerous amounts of cadmium can slowly be extracted from the vessel to ultimately enter stored food and the consumer.

It should be noted that there was a trend 20 or 30 years ago to sell cookware made of exotic metals. Today, most metal cookware is made of aluminum, copper, or stainless steel, often with a ceramic or polymeric "non-stick" coating, and often colored. However, the long-term health effects of cooking food in such vessels are uncertain, especially if abrasives are used during cleaning. As far as metal cookware is concerned, only the old-fashioned iron pots and pans have been proven to be absolutely safe by both modern science and use since ancient times. With regard to ceramic cookware, pure clear uncolored Pyrex is safe.

17.7   Biochemical and Clinical Findings

Figure 17.3 is of central importance for any Programmatic Environmental Safety and Health Evaluation (PESHE) because it defines the conditions under which cadmium is mobile in the environment and can be absorbed by living organisms. In this regard, the species Cd^{2+} is of primary concern because this ion is doubly charged, like the calcium ion. Furthermore, the atomic radius of cadmium (1.7 Å) is almost the same as the covalent radius of calcium (1.74 Å). Therefore, it is not surprising that cadmium should occasionally substitute for calcium with devastating consequences for organism viability due to the different electron configurations and chemistry of the two atoms. In Japan, an outbreak of extremely painful bone softening disease has been traced to cadmium, as might be expected [13]. The source of the cadmium poisoning was rice and soybeans grown in soil contaminated by airborne cadmium from a nearby lead/zinc smelter [13]; cadmium and zinc are often found in the same ores because they are both group IIB elements. Calcium ions are essential for the operation of nervous and muscular tissues [14]. And, calcium ions are involved in the second-messenger system employed by all individual cells to communicate with the rest of the human body via hormones [14]. Therefore, it is not surprising to find that cadmium is damaging to all cells of the body [15]. Pathological findings in cases of fatal cadmium ingestion are severe gastrointestinal inflammation, and liver and kidney damage [15]. As mentioned above, the MCL for cadmium in drinking water is 5 µg/L. The Threshold Limit Value (TLV) in air for cadmium oxide fumes is 0.1 mg/cu m., and for cadmium metals and dust the TLV is 0.2 mg/cu. m [15]. In fatal acute poisoning from inhalation of cadmium fumes, pathologic examination reveals inflammation of the pulmonary epithelium and pulmonary edema [15]. Pathologic examination in fatalities following prolonged exposure to cadmium fumes reveals emphysema [15]. Finally, it should be noted that cadmium substitution for calcium in the environment allows it to move freely. On Long Island, New York, a metal plating waste containing cadmium and chromium traveled about 3000 ft. in a shallow aquifer [13].

17.8 Summary

Occasionally, in his/her capacity as a dangerous goods expert, the systems safety engineer is called upon to prevent pollution and poisoning, or control mishaps that have already occurred. The purpose of this chapter was to exhibit the methods to be used in such circumstances. Two current problems involving cadmium pollution and poisoning were examined in detail. And, although these problems are interesting in their own right, they also serve as a model for similar problems involving other inorganic toxic substances (many of which are listed in Appendix F together with a few Pourbaix Diagrams in Appendix G). Little has been said about organic pollutants in this chapter. However, Appendix H lists 48 volatile and 93 semi-volatile organic compounds on the Delaware Department of Natural Resources and Environmental Control (DNREC) Target Compound List (TCL) [16]. The system safety engineer's first step when it comes to organic toxins and pollutants is to be aware of which compounds are the most notorious so that hazards can be mitigated from the start. It is a great tragedy that most organizations consider dangerous goods work to be little more than filing the right paperwork with the appropriate government agencies to *"avert blame"*. The result of this attitude is often an expensive mishap. The loss of resources, the explosion of batteries mishandled because of ignorance about their internal chemistry, and the contamination of groundwater, are just a few of the consequences of having inexperienced personnel handle dangerous goods. Hopefully, this chapter has shown that the management of dangerous goods involves a considerable interdisciplinary knowledge of geology, biology, biochemistry, chemistry, and mathematics, and is more than just *secretarial work!*

† As an Interesting aside, at $P(O_2) = 0.21$ atm, the corresponding $P(H_2) = \sqrt{(10^{-83.1}/0.21)} = 10^{-41.2}$ atm. This last partial pressure is a minimum for hydrogen if liquid water is to exist. Below this pressure, there is only oxygen gas polluted with a trace amount of hydrogen. In fact, an assay of the Earth's atmosphere reveals 5.5×10^{-7} atm of H_2 gas [17]. Without this trace amount of hydrogen in our atmosphere, liquid water could not exist. The partial pressure of hydrogen gas would simply be insufficient to hold water molecules together. In that case, some of the Earth's water would have to decompose to restore the equilibrium H_2 partial pressure. If the Earth were too small to hang on to its tenuous hydrogen atmosphere, all of its water would eventually decompose, and Earth would become a dry lifeless world! The mean molecular velocity of H_2 at 20 °C is given by $v = \sqrt{(6kT/m)} = 2.7$ km/s, here T is the kelvin temperature, k is Boltzmann's constant, and m is the molecular mass of H_2. If this is set equal to the escape velocity of a planet equal to $\sqrt{(2GM/R)}$, where G is the gravitational constant, M is the mass of the planet, and R is its radius, then it is possible to solve for M. For a spherical body, $M = 4\pi R^3\rho/3$, where ρ is the body's density; typically about 2.7 gms/cc. Solving for R yields about 2,200 km. Therefore, no asteroid (maximum diameter 952 km), or moon smaller than our own Moon, can hold a hydrogen atmosphere at 20 °C, and no permanent liquid surface water can exist on such a body at the specified temperature. Of course, this prohibition on the existence of liquid water is in addition to the fact that liquid water must boil away to a vapor at the low gas pressures found on an asteroid's surface.

‡ The "velocity" v (measured in moles/sec) of the reaction in Eq. 17.26, as a function of pH, can be easily calculated by first noting that it can only proceed in the forward direction. The reason for this is that Eq. 17.26 is an electrochemically driven Redox reaction. Furthermore, once Cd^{2+} forms, it is rapidly, and very exothermically, complexed with water ($\Delta H = -1807$ kJ/mole) so that it is thermodynamically unfavorable to reform the native metal. It is for similar reasons that rust never goes back to pure iron. In that case, v is proportional to the concentration of cadmium and the square of the hydrogen ion concentration. Therefore,

$$v = k_{ab}\left[Cd^{2+}\right]\left[H^+\right]^2, \tag{17.33}$$

where k_{ab} is called the rate constant and the hydrogen ion concentration is squared because of the coefficient of 2 on the left of Eq. 17.26. Taking the logarithm to the base 10 of both sides of Eq. 17.33, substituting in the definition of pH (i.e. pH = $-\log[H^+]$), and defining a constant $C = \log(k_{ab})$, yields

$$\log v = C - 2\,pH. \tag{17.34}$$

Taking the antilog of both sides of Eq. 17.34 yields $v = C'/10^{2pH}$, where the new constant $C' = 10^C$. This last equation is the most useful form of v as a function of pH. If $v(3)$ is the reaction velocity at pH $= 3$, and if $v(6)$ is the reaction velocity at pH $= 6$, then the ratio of the two velocities is given by

$$v(3)/v(6) = (C'/10^6)/(C'/10^{12}) = 10^6, \text{ or } v(3) = 10^6\,v(6), \tag{17.35}$$

Therefore, the reaction velocity of Eq. 17.26 at pH $= 3$ is a million times faster than the velocity at pH $= 6$.

* Surprisingly, it has recently been discovered that exposure to light increases the solubility of cadmium red pigment in water [18]. This finding is significant in that such pigments are used in ceramics.

17.9 Problems

(1) The paragraph containing Eq. 17.15 states that the phase boundary of the Pourbaix diagram separating $Cd(OH)_2$ from $HCdO_2^-$ occurs at about pH $= 13$. Prove that this is so for $[HCdO_2^-] = 10^{-6}$.

(2) A surface settling pond with an oxidation potential of 0.2 volts contains hexavalent CrO_4^{2-} ions. Given the information in the Pourbaix Diagram of Fig. G.3 of Appendix G, what pH range will cause precipitation of hexavalent chromium in solution as solid chromite (essentially Cr_2O_3 + iron oxide) at the bottom of this pond, where it can be safely filtered away? Does this pH range overlap with the range to be expected for natural freshwater?

(3) Consider a solution of UO_2^{2+} ions.

  (a) What oxidation potential (in volts) is sufficiently reducing to bring these ions down as a solid precipitate in contact with water? (Hint: see Fig. G.8 of Appendix G)
  (b) What does reducing medium (solution) mean?
  (c) What does an oxidizing medium do?
  (d) The auto proteolysis equation says that in any solution a small amount of water must dissociate into hydrogen ions (H^+) and hydroxyl ions (OH^-).

$$H_2O \leftrightarrow H^+ + OH^- \text{ where } K_{eq} = 10^{-14}$$

  Where would the negative charge come from to neutralize UO_2^{2+} to form solid uraninite?
  (e) According to Le Chatelier's Principle, what would happen to the H^+ concentration if OH^- were removed from the solution? Qualitatively, is charge balance maintained in the oxygen/uranium/water system?

References

1. Fairbridge, R. W. (1972). *The encyclopedia of geochemistry and environmental science* (p. 99). NY: Van Nostrand.
2. Garrels, R. M., & Christ, C. L. (1965). *Solutions, minerals, and equilibria* (pp. 3–5, 16, 144–146, 403–428) Harper and Row, New York.
3. Condit, T., & Cooney, G., Kelly, E. P. (Eds.), *Kingship and sacrifice: Iron Age bog bodies and boundaries*, Archaeology Ireland with the National Museum of Ireland, Bray Co. Wicklow, Ireland, Heritage Guide No. 35, Sept. 2006.
4. Friedrich, M., et al. (2004). The 12,460-Year Hohenheim oak and pine tree-ring chronology from Central Europe-A Unique Annual Record for radiocarbon calibration and paleoenvironment reconstructions. *Radiocarbon, 46*(3), 1111–1122.
5. Zito, R. R., Accelerated testing of explosives and propellants—Dangerous goods. In *Proceedings of the 33rd International System Safety Conference*, Aug. 24–27, 2015, San Diego, CA.
6. Anon. *Water chemistry*. Retrieved 03 Aug 2012, from, http://www.science.uwaterloo.ca.
7. Schmitt, H. H. (Ed.), *Equilibrium diagrams for minerals* (pp. 163), Harvard, Boston Mass.
8. Freeze, R. A., & Cherry, J. A. (1979). *Groundwater* (pp. 108–112). Englewood Cliffs: NJ. Prentice–Hall.
9. Harris, D. C. (1995). *Quantitative chemical analysis* (4th ed., p. 271). N.Y.: Freeman.
10. Anon. *Primary maximum contaminant levels for drinking water*. Retrieved 22 March, 2012 from http://rules.sos.state.ga.us/docs/391/3/5/18.pdf.
11. Anon. Public Information and Conservation Department, City of Tucson Department of Water, 310 W. Alameda St., Tucson AZ 85701, (520) 791–4331.
12. IPCS INCHEM, *Cadmium*. Retrieved 16 March 2012, from http://www.inchem.org/documents/pims/chemical/cadmium.
13. Fetter, C. W. (1993). *Contaminant hydrology* (p. 280). Upper Saddle River, NJ: Prentice Hall.
14. Mathews, C.K., & Van Holde, K.E. (1996). *Biochemistry,* (pp. 264–266, 842–843, 2nd ed.), Benjamin, Menlo Park, CA.
15. Dreisbach, R. H. (1977). *Handbook of poisoning: Diagnosis and treatment* (9th ed., pp. 216–217). Los Altos, CA: Lang Medical Publications.
16. Anon. Delaware Department of Natural Resources and Environmental Control (DNREC) Target Compound List (TCL), http://www.dnrec.delaware.gov/dwhs/sirb/Documents/Target%20Compounds%20Lists.pdf. accessed May 27, 2017.
17. Anon. *Atmosphere of Earth,* Wikipedia, the free encyclopedia, http://en.wikipedia.org/wiki/Atmosphere_of_Earth, Mar. 8, 2012, p. 2 of 12.
18. Lockwood, D. (2017). Sunlight surprise raises cadmium pollution risk. *Chemical & Engineering News*. Retrieved June 22, 2017, from http://cen.acs.Org/articles/95/web/2017/06/Sunlight-surprise-raises-cadmium-pollution-risk, Web Date June 16, 2017.

Epilogue

Often, when an author writes a journal article, review, or book, it is instructive to end with the same questions and discussions that he/she began with. And, this book is no exception to that rule. A colleague of this author once complained, "These problems that we encounter in system safety are so varied and ill-defined that it is impossible to quantify them. You're really wasting your time even trying!" N. G. Leveson would agree to some extent. However, this author's response to such an indictment is, *"Unless it can be proven that quantification is impossible, it is never a waste of time to try because there is no payoff in NOT trying!"* This underlying philosophy of *tractability* has been adopted by the author in his mathematical quantification of system safety from 2011 to 2016. It was a difficult 7-year long journey, with occasional miss-steps along the way. But, in the end, the work was finished, as the author had promised his mentor Rene Fitzpatrick before he passed away.

When physicists, chemists, and other scientists are confronted with a complex and diverse collection of seemingly unrelated data, they usually suspect that there is some underlying order. So, a bewildering array of molecules (compounds) were shown to be composed of a set of 92 naturally occurring atoms (elements), and the atoms themselves were reduced to hadrons (protons and neutron) and leptons (electrons), with the hadrons finally being reduced to quarks. In system safety, a similar type of *underlying order* exists. Probability calculations and modeling can benefit from thinking of complex systems as systems of more elementary subsystems. Furthermore, many network problems involving, for example, electronics and software, have certain similarities, symmetries, differences, and dualisms that become evident when expressing these problems in a matrix format. These unsuspected relationships form an underlying order to a confusing tangle of seemingly unrelated problems when viewed superficially. And, it is the elucidation of this *underlying order* that is the primary theme of this book and its preceding published research.

The chart below shows how the solution to one system safety problem feeds into the solution of other system problems. The arrows indicate the direction of

© Springer Nature Switzerland AG 2020
R. R. Zito, *Mathematical Foundations of System Safety Engineering*,
https://doi.org/10.1007/978-3-030-26241-9

development and the feedback, or feed forward, of information from one model problem to another. Most nonhuman problems confronted by the analyst fall into four basic categories. There are general "Foundation" problems that apply to any type of system (electrical, mechanical, software, etc.). There are "Electronics" problems which cover such things as connector bent pin problems, sneak circuits, and the consequences of parts failures. There are "Software" problems (reliability prediction, debugging, cybersecurity, etc.). And, there are "Dangerous Goods" problems, which are the most diverse problems, and span the product life cycle. Typically, these last problems include handling and disposal of dangerous chemicals (explosives, propellants, pollutants, poisonous substances), handling of energetic parts (moving parts, hot surfaces, batteries, large compressed springs, unstable weights, etc.), and materials selection (elimination and replacement of dangerous materials, selecting materials with adequate strength and thermal conductivity, etc.), Almost all of these problems have been addressed in this book, at least in a representative way, by the various modules in the chart below; where each block is a paper in Refs. [1–17] and a chapter of this book. Only energetic parts, and cybersecurity (currently undergoing rapid evolution), have not been addressed in this edition. And, of course, fundamental to everything, either directly or indirectly, is a set of precise mathematical definitions (i.e., the block in the lower left corner of the chart below representing the Glossary of this book) that allow the system safety engineer to calculate system risks and probabilities of failure, and to make failure-safe detection and correction of hazards and defects.

Traditionally, failure databases have been little more than fault history logs. And, some of this data has even been reduced to failure rate curves. But, little has been done to use such curves to make accurate extrapolations to future system behavior or calculate probability caps on "Black Swan" events. Even less has been done to make early systematic failure-safe mathematical detection and correction of hazards before they become mishaps. *Mathematical prediction, detection, and correction of hazards* are the second major themes of this book.

Finally, at the deepest level, there is the realization that "accidents" (failures) are a natural consequence of entropy. There is an old Hindu saying, "all compound objects (a system) must eventually decay". And, so they must! Furthermore, the entropy of every system must increase as a function of time. *Unless* energy is expended (maintenance) to reverse entropy. The reason for this is that there is only one state for a system to operate properly. But, there exist many "broken" states. Entropy defines an "arrow of time". So there is a deep connection between entropy, failure, and time. Any system will fail given enough time! This *thermodynamic interpretation of failure* is the third, but most far reaching, theme of this book; with consequences in biological as well as mechanical, electronic, and software systems (Fig. 1).

If the author had to summarize these notions as a set of "assumption" (axioms), they might look something like the set below when combined with the two basic axioms of fallibility and noncompliance of chapter 1.

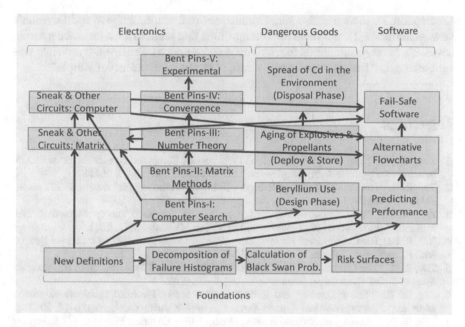

Fig. 1 The overall structure of system safety engineering. The arrows indicate the direction of development and the feedback, or feedforward, of information from one model problem to another

(1) *The Axiom of Human Fallibility*: Given enough time, a human operator will make a mistake with unit probability.

(2) *The Axiom of Human Noncompliance*: Given enough time, every organization will become system safety noncompliant with unit probability.

(3) *The Axiom of Tractability*: All system safety problem can be quantified to establish an upper bound on failure probability.

(4) *The Axiom of Underlying Order*: System safety problems are not just a collection of disconnected special cases, but are either intimately related or can be described by the use of common mathematical tools.

(5) *The Axiom of Prediction*: Statistical methods can be used to determine future failure rate behavior within limits.

(6) *The Axiom of Detection*: Most system safety hazards can be detected before they become mishaps. However, "Black Swan" events are undetectable a priori because they are unknown by definition.

(7) *The Axiom of Correction*: Most system safety hazards can be corrected before they become mishaps. However, "Black Swan" events cannot be corrected a priori because they are unknown by definition.

(8) *The Axiom of Entropy*: Failure is a thermodynamic process, and all systems tend toward disorder. Given sufficient time, all systems must eventually fail.

When these 8 axioms are compared to Leveson's 7 New Assumptions in Chapter there are some striking differences (c.f. axiom 5 above and Leveson's New

Assumption 3), there are also some similarities (c.f. axiom 2 above and Leveson's New Assumption 2), and there are some axioms that seem to have no counterpart in Leveson's New Assumptions (c.f. axioms 4 and 8 above). Perhaps Oscar Wilde was right after all, "The pure and simple truth is rarely pure and never simple."

References

1. Zito, R. R. "The Curious Bent Pin Problem—I: Computer Search Methods," *29th International Systems Safety Conference*, Las Vegas, NV, Aug. 8–12, 2011.
2. Zito, R. R. "The Curious Bent Pin Problem—II: Matrix Methods," *29th International Systems Safety Conference*, Las Vegas, NV, Aug. 8–12, 2011.
3. Zito, R. R. "The Curious Bent Pin Problem—III: Number Theory Methods," *29th International Systems Safety Conference*, Las Vegas, NV, Aug. 8–12, 2011.
4. Zito, R. R. "The Curious Bent Pin Problem—IV: Limit Methods," *30th International Systems Safety Conference*, Atlanta, GA, Aug. 6–10, 2012.
5. Zito, R. R. "The Curious Bent Pin Problem—V: Experimental Methods," *30th International Systems Safety Conference*, Atlanta, GA, Aug. 6–10, 2012.
6. Zito, R. R. "'Sneak Circuits' and Related System Safety Electrical Problems—I: Matrix Methods," *30th International Systems Safety Conference*, Atlanta, GA, Aug. 6–10, 2012.
7. Zito, R. R. " 'Sneak Circuits' and Related System Safety Electrical Problems—II: Computer Search Methods," *30th International Systems Safety Conference*, Atlanta, GA, Aug. 6–10, 2012.
8. Zito, R. R. "How Complex Systems Fail-I: Decomposition of the Failure Histogram," *Proceedings of the International Systems Safety Conference—2013*, Boston, Mass., Aug. 12–16, 2013.
9. Zito, R. R. "How Complex Systems Fail-II: Bounding the 'Black Swan' Probability," *Proceedings of the 34th International Systems Safety Conference—2016*, Orlando, FL, Aug. 8–12, 2016.
10. Zito, R. R. "How Complex Systems Fail-III: The System Risk Surface," *Proceedings of the 34th International Systems Safety Conference—2016*, Orlando, FL, Aug. 8–12, 2016.
11. Zito, R. R. "Predicting Software Performance- Software 1," *Proceedings of the 32nd International Systems Safety Conference—2014*, St. Louis, MO, Aug. 4–8, 2014.
12. Zito, R. R. "Alternative Flowcharts for a Mathematical Analysis of Logic- Software II," *Proceedings of the 33rd International Systems Safety Conference—2015*, San Diego, CA, Aug. 24–27, 2015.
13. Zito, R. R. "Fail-Safe Control Software- Software III," *Proceedings of the 33rd International Systems Safety Conference—2015*, San Diego, CA, Aug. 24–27, 2015.
14. Zito, R. R. "Accelerated Age Testing of Explosives and Propellants—Dangerous Goods I," *Proceedings of the 33rd International Systems Safety Conference—2015*, San Diego, CA, Aug. 24–27, 2015.
15. Zito, R. R. "The Movement of Inorganic Cadmium Through the Environment: Dangerous Goods II," *Proceedings of the 34th International Systems Safety Conference—2016*, Orlando, FL, Aug. 8–12, 2016.
16. Zito, R. R. "Design Phase Elimination of Beryllium: Dangerous Goods III," *Proceedings of the 35th International Systems Safety Conference—2017,* Albuquerque, NM, Aug, 21–25, 2017.
17. Zito, R. R. "New Definitions for a New Science," *Proceedings of the 35*th *International Systems Safety Conference—2017,* Albuquerque, NM, Aug, 21–25, 2017.

Appendix A
"Long Tailed" Distribution

$$\rho_{\text{global}}(t) = [(2\sqrt{\beta})/M]\Sigma_{j=0}^{\infty}\{[\beta^j/(j!)^2][\psi(j+1) - \ln\sqrt{\beta}]\},$$

where $\beta = (t/2M)^2$, M is the mode of the Rayleigh distribution of Rayleigh modes, and

$$\psi(j+1) = -c + \sum_{n=1}^{j}(1/n)$$

where c is the Euler–Mascheroni constant (0.57721....). Note that in the tables that follow the summation for $\rho_{\text{global}}(t)$ has been truncated after $j = 9$.

Probability density function

t	M				
	8	9	10	15	20
0	0.00000	0.00000	0.00000	0.00000	0.00000
1	0.03450	0.02868	0.02427	0.01259	0.00779
2	0.04817	0.04080	0.03505	0.01906	0.01214
3	0.05490	0.04728	0.04117	0.02337	0.01523
4	0.05778	0.05054	0.04458	0.02633	0.01753
5	0.05827	0.05176	0.04622	0.02837	0.01927
6	0.05724	0.05161	0.04665	0.02972	0.02059
7	0.05524	0.05055	0.04624	0.03055	0.02157
8	0.05263	0.04887	0.04523	0.03099	0.02229
9	0.04966	0.04678	0.04381	0.03110	0.02279
10	0.04650	0.04444	0.04210	0.03097	0.02311
11	0.04328	0.04197	0.04022	0.03064	0.02328

(continued)

© Springer Nature Switzerland AG 2020
R. R. Zito, *Mathematical Foundations of System Safety Engineering*,
https://doi.org/10.1007/978-3-030-26241-9

(continued)

Probability density function				
t	M			
12	0.04009	0.03943	0.03822	0.03015	0.02333
13	0.03698	0.03687	0.03618	0.02954	0.02327
14	0.03399	0.03440	0.03411	0.02884	0.02312
15	0.03115	0.03197	0.03207	0.02807	0.02290
16	0.02848	0.02964	0.03007	0.02724	0.02261
17	0.02597	0.02742	0.02814	0.02637	0.02228
18	0.02364	0.02531	0.02627	0.02548	0.02190
19	0.02148	0.02333	0.02448	0.02457	0.02149
20	0.01949	0.02146	0.02278	0.02366	0.02105
21	0.01766	0.01972	0.02117	0.02274	0.02059
22	0.01598	0.01809	0.01964	0.02183	0.02011
23	0.01444	0.01658	0.01821	0.02093	0.01961
24	0.01304	0.01518	0.01686	0.02005	0.01911
25	0.01176	0.01389	0.01559	0.01918	0.01860
26	0.01060	0.01269	0.01441	0.01834	0.01809
27	0.00955	0.01159	0.01330	0.01751	0.01757
28	0.00859	0.01057	0.01228	0.01671	0.01706
29	0.00773	0.00964	0.01132	0.01594	0.01654
30	0.00695	0.00879	0.01043	0.01519	0.01604
31	0.00625	0.00800	0.00961	0.01446	0.01553
32	0.00561	0.00729	0.00884	0.01377	0.01504
33	0.00504	0.00663	0.00813	0.01309	0.01455
34	0.00453	0.00603	0.00748	0.01245	0.01407
35	0.00407	0.00549	0.00687	0.01183	0.01360
36	0.00365	0.00499	0.00632	0.01124	0.01313
37	0.00329	0.00453	0.00580	0.01067	0.01268
38	0.00296	0.00412	0.00533	0.01013	0.01224
39	0.00266	0.00375	0.00489	0.00961	0.01181
40	0.00240	0.00341	0.00449	0.00911	0.01139
41	0.00216	0.00310	0.00412	0.00864	0.01098
42	0.00195	0.00282	0.00378	0.00818	0.01058
43	0.00176	0.00257	0.00347	0.00775	0.01020
44	0.00159	0.00234	0.00319	0.00734	0.00982
45	0.00143	0.00213	0.00292	0.00695	0.00946
46	0.00128	0.00194	0.00268	0.00658	0.00910
47	0.00112	0.00177	0.00247	0.00623	0.00876
48	0.00095	0.00162	0.00227	0.00589	0.00843
49	0.00075	0.00148	0.00208	0.00558	0.00811
50	0.00050	0.00135	0.00192	0.00527	0.00780

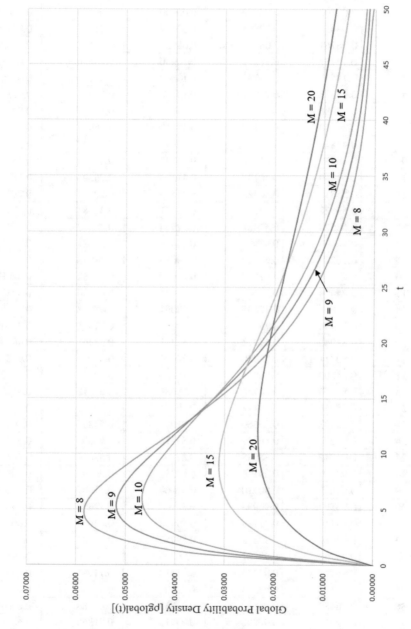

"Long Tailed" Probability Density Function
(Truncated after j = 9)

Cumulative distribution function

t	M				
	8	9	10	15	20
0	0.00000	0.00000	0.00000	0.00000	0.00000
1	0.03450	0.02868	0.02427	0.01257	0.00779
2	0.08267	0.06949	0.05933	0.03164	0.01992
3	0.13757	0.11677	0.10050	0.05500	0.03515
4	0.19535	0.16731	0.14508	0.08134	0.05261
5	0.25362	0.21907	0.19130	0.10971	0.07194
6	0.31086	0.27068	0.23795	0.13943	0.09253
7	0.36610	0.32123	0.28419	0.16998	0.11410
8	0.41873	0.37010	0.32942	0.20097	0.13639
9	0.46839	0.41688	0.37323	0.23207	0.15918
10	0.51489	0.46132	0.41533	0.26304	0.18229
11	0.55818	0.50329	0.45555	0.29367	0.20558
12	0.59827	0.54272	0.49377	0.32383	0.22890
13	0.63528	0.57961	0.52994	0.35337	0.25217
14	0.66924	0.61401	0.56405	0.38221	0.27529
15	0.70039	0.64510	0.59613	0.41028	0.29818
16	0.72887	0.67562	0.62620	0.43752	0.32080
17	0.75484	0.70304	0.65434	0.46390	0.34308
18	0.77848	0.72835	0.68061	0.48938	0.36498
19	0.79996	0.75168	0.70509	0.51395	0.38647
20	0.81945	0.77314	0.72787	0.53761	0.40752
21	0.83711	0.79286	0.74904	0.56035	0.42811
22	0.85309	0.81095	0.76868	0.58218	0.44822
23	0.86753	0.82753	0.78689	0.60312	0.46783
24	0.88056	0.84272	0.80374	0.62317	0.48695
25	0.89233	0.85660	0.81933	0.64235	0.50555
26	0.90293	0.86929	0.83374	0.66069	0.52363
27	0.91247	0.88088	0.84705	0.67820	0.54120
28	0.92107	0.89146	0.85932	0.69491	0.55826
29	0.92879	0.90110	0.87064	0.71085	0.57481
30	0.93574	0.90989	0.88107	0.72604	0.59084
31	0.94119	0.91789	0.89068	0.74050	0.60637
32	0.94760	0.92518	0.89952	0.75427	0.62141
33	0.95264	0.93181	0.90765	0.76736	0.63596
34	0.95717	0.93785	0.91513	0.77981	0.65003
35	0.96124	0.94333	0.92201	0.79164	0.66363
36	0.96489	0.94832	0.92832	0.80288	0.67676
37	0.96818	0.95286	0.93412	0.81355	0.68944
38	0.97113	0.95698	0.93945	0.82368	0.70168

(continued)

(continued)

Cumulative distribution function				
t	M			
39	0.97379	0.96073	0.94434	0.83328	0.71350
40	0.97619	0.96413	0.94883	0.84239	0.72489
41	0.97835	0.96723	0.95295	0.85103	0.73587
42	0.98030	0.97005	0.95673	0.85921	0.74645
43	0.98207	0.97261	0.96020	0.86697	0.75665
44	0.98366	0.97495	0.96339	0.87431	0.76647
45	0.98509	0.97708	0.96631	0.88126	0.77593
46	0.98637	0.97903	0.96900	0.88784	0.78503
47	0.98748	0.98080	0.97146	0.89407	0.79379
48	0.98843	0.98242	0.97373	0.89997	0.80222
49	0.98918	0.98390	0.97581	0.90554	0.81032
50	0.98968	0.98528	0.97773	0.91082	0.81812

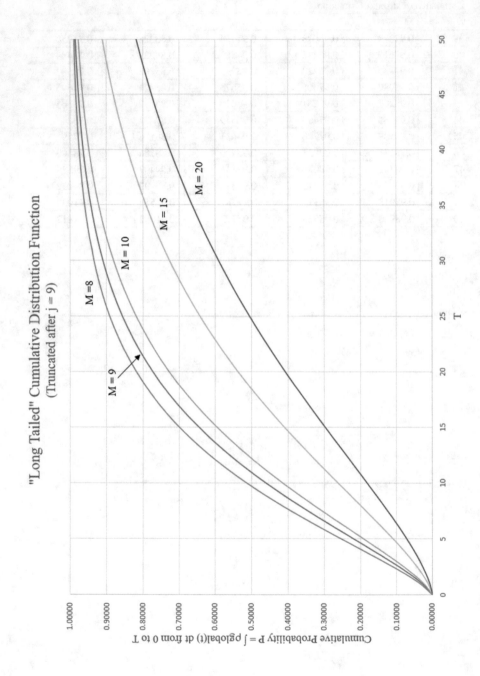

"Long Tailed" Cumulative Distribution Function
(Truncated after j = 9)

Appendix B
Chi-Square Probability Function[†]

$$P(\chi^2|v) = [2^{v/2}\Gamma(v/2)]^{-1} \int_0^{\chi^2} t^{v/2-1}e^{-t/2}dt \quad (0 \le \chi^2 < \infty),$$

where

(1) $P(\chi^2|v)$ = the probability that the quantity $X^2 = \Sigma X_i^2$ (Eq. 1.1, Chap. 1) is $\le$ χ^2 (by chance).
(2) v = a constant equal to the number of degrees of freedom of collected data. For $v > 100$, treat $\sqrt{2\chi^2} - \sqrt{2v-1}$ as a standard normal variable. It is for this reason that few χ^2 tables go beyond $v = 100$. This practice will be continued in the table that follows.
(3) Γ = the Gamma Function.
(4) Confidence Level = $Q(\chi^2|v) = 1 - P(\chi^2|v)$.
(5) The "Critical Values" in the following table are the values of χ^2 corresponding to frequently used confidence levels Q and v degrees of freedom.

† Formulas from the "Handbook of Mathematical Functions by M. Abramowitz and I. A. Stegun, Dover, New York, 1972, p. 940. The tables that follow are a combination of data from "Statistics, An Introduction", by D. A. Fraser, John Wiley, New York, 1958, and the "Handbook of Statistical Tables" by D. B. Owen, Addison-Wesley, Reading, Mass., 1962, both modified by the author with data truncated and rounded to the nearest third decimal place.

© Springer Nature Switzerland AG 2020
R. R. Zito, *Mathematical Foundations of System Safety Engineering*,
https://doi.org/10.1007/978-3-030-26241-9

ν	Confidence level					
	99.5%	97.5%	5%	2.5%	1%	0.5%
1	0.000	0.000	3.841	5.024	6.635	7.879
2	0.010	0.051	5.991	7.378	9.210	10.597
3	0.072	0.216	7.815	9.348	11.345	12.838
4	0.207	0.484	9.488	11.143	13.277	14.860
5	0.412	0.831	11.071	12.833	15.086	16.750
6	0.676	1.237	12.592	14.449	16.812	18.548
7	0.989	1.690	14.067	16.013	18.475	20.278
8	1.344	2.180	15.507	17.535	20.090	21.955
9	1.735	2.700	16.919	19.023	21.666	23.589
10	2.156	3.247	18.307	20.483	23.209	25.188
11	2.603	3.816	19.675	21.920	24.725	26.757
12	3.074	4.404	21.026	23.337	26.217	28.300
13	3.565	5.009	22.362	24.736	27.688	29.819
14	4.075	5.629	23.685	26.119	29.141	31.319
15	4.601	6.262	24.996	27.488	30.578	32.801
16	5.142	6.908	26.296	28.845	32.000	34.267
17	5.697	7.564	27.587	30.191	33.409	35.719
18	6.265	8.231	28.869	31.526	34.805	37.156
19	6.844	8.907	30.144	32.852	36.191	38.582
20	7.434	9.591	31.410	34.170	37.566	39.997
21	8.034	10.283	32.671	35.479	38.932	41.401
22	8.643	10.982	33.924	36.781	40.289	42.796
23	9.260	11.689	35.173	38.076	41.638	44.181
24	9.886	12.400	36.415	39.364	42.980	45.559
25	10.52	13.120	37.653	40.647	44.314	46.928
26	11.160	13.844	38.885	41.923	45.642	48.290
27	11.808	14.573	40.113	43.194	46.963	49.645
28	12.461	15.308	41.337	44.461	48.278	50.993
29	13.121	16.047	42.557	45.722	49.588	52.336
30	13.787	16.791	43.773	46.979	50.892	53.672
31	14.458	17.539	44.985	48.232	52.191	55.003
32	15.134	18.291	46.194	49.480	53.486	56.328
33	15.815	19.047	47.400	50.725	54.776	57.648
34	16.501	19.806	48.602	51.966	56.061	58.964
35	17.192	20.569	49.802	53.203	57.342	60.275
36	17.887	21.336	50.998	54.437	58.619	61.581
37	18.586	22.106	52.192	55.668	59.892	62.883
38	19.289	22.878	53.384	56.896	61.162	64.181
39	19.996	23.654	54.572	58.120	62.428	65.476
40	20.707	34.433	55.758	59.342	63.691	66.766

(continued)

(continued)

v	Confidence level					
	99.5%	97.5%	5%	2.5%	1%	0.5%
41	21.421	25.215	56.942	60.561	64.950	68.053
42	22.138	25.999	58.124	61.777	66.206	69.336
43	22.859	26.785	59.304	62.990	67.459	70.616
44	23.584	27.575	60.481	64.201	68.710	71.893
45	24.311	28.366	61.656	65.410	69.957	73.166
50	27.991	32.357	67.505	71.420	76.154	79.490
60	35.535	40.482	79.082	83.298	88.379	91.952
70	43.275	48.758	90.531	95.023	100.425	104.215
80	51.172	57.153	101.879	106.629	112.329	116.321
90	59.196	65.647	113.145	118.136	124.116	128.299
100	67.328	74.222	124.342	129.561	135.807	140.169

Appendix C
Truncated Pseudo-Weibull Distribution of Second Order (TPWD2)

$$\rho_{global}(t) \cong \left\{ (t/\sigma_m^2) \exp\left(-\tfrac{1}{2}[t/\sigma_m]^2\right) \right\} \left\{ 1 + \tfrac{1}{2}(s/\sigma_m)^2 [6 - 7(t/\sigma_m)^2 + (t/\sigma_m)^4] \right\}$$

σ_m =   the center (peak) of the narrow Gaussian distribution of Rayleigh modes.
s =   the standard deviation of the narrow Gaussian distribution of Rayleigh modes

Note: TPWD2 $\rightarrow$ Rayleigh pdf in the limit as $s/\sigma_m \rightarrow 0$.

Probability density function					
t	s				
	0.1	0.5	1	2	3
0	0.00000	0.00000	0.00000	0.00000	0.00000
1	0.00995	0.01002	0.01025	0.01113	0.01261
2	0.01961	0.01974	0.02016	0.02185	0.02465
3	0.02869	0.02887	0.02945	0.03176	0.03562
4	0.03693	0.03715	0.03783	0.04055	0.04508
5	0.04413	0.04436	0.04508	0.04793	0.05269
6	0.05013	0.05034	0.05102	0.05373	0.05826
7	0.05480	0.05498	0.05556	0.05787	0.06172
8	0.05810	0.05823	0.05865	0.06033	0.06314
9	0.06003	0.06010	0.06032	0.06121	0.06269
10	0.06065	0.06065	0.06065	0.06065	0.06065
11	0.06007	0.05999	0.05977	0.05886	0.05735
12	0.05840	0.05826	0.05782	0.05607	0.05314
13	0.05583	0.05563	0.05501	0.05252	0.04837

(continued)

© Springer Nature Switzerland AG 2020
R. R. Zito, *Mathematical Foundations of System Safety Engineering*,
https://doi.org/10.1007/978-3-030-26241-9

(continued)

Probability density function

t	s				
14	0.05253	0.05229	0.05152	0.04847	0.04337
15	0.04869	0.04841	0.04756	0.04413	0.03843
16	0.04447	0.04419	0.04329	0.03971	0.03374
17	0.04007	0.03978	0.03890	0.03537	0.02948
18	0.03561	0.03535	0.03452	0.03122	0.02571
19	0.03124	0.03101	0.03028	0.02735	0.02248
20	0.02706	0.02686	0.02626	0.02382	0.01976
21	0.02315	0.02300	0.02252	0.02064	0.01750
22	0.01956	0.01945	0.01913	0.01782	0.01564
23	0.01633	0.01627	0.01608	0.01534	0.01409
24	0.01347	0.01345	0.01340	0.01316	0.01278
25	0.01098	0.01100	0.01106	0.01127	0.01163
26	0.00885	0.00890	0.00905	0.00963	0.01060
27	0.00706	0.00712	0.00734	0.00820	0.00963
28	0.00556	0.00564	0.00591	0.00695	0.00870
29	0.00433	0.00442	0.00471	0.00587	0.00780
30	0.00334	0.00343	0.00373	0.00493	0.00693
31	0.00254	0.00264	0.00293	0.00412	0.00609
32	0.00192	0.00201	0.00229	0.00341	0.00528
33	0.00143	0.00151	0.00177	0.00280	0.00453
34	0.00105	0.00113	0.00136	0.00228	0.00382
35	0.00077	0.00083	0.00103	0.00184	0.00319
36	0.00055	0.00061	0.00078	0.00147	0.00262
37	0.00040	0.00044	0.00059	0.00116	0.00212
38	0.00028	0.00032	0.00044	0.00091	0.00170
39	0.00020	0.00023	0.00032	0.00070	0.00134
40	0.00014	0.00016	0.00023	0.00054	0.00104
41	0.00009	0.00011	0.00017	0.00041	0.00080
42	0.00006	0.00008	0.00012	0.00030	0.00060
43	0.00004	0.00005	0.00009	0.00022	0.00045
44	0.00003	0.00004	0.00006	0.00016	0.00033
45	0.00002	0.00002	0.00004	0.00012	0.00024
46	0.00001	0.00002	0.00003	0.00008	0.00017
47	0.00001	0.00001	0.00002	0.00006	0.00012
48	0.00000	0.00001	0.00001	0.00004	0.00009
49	0.00000	0.00000	0.00001	0.00003	0.00006
50	0.00000	0.00000	0.00001	0.00002	0.00004

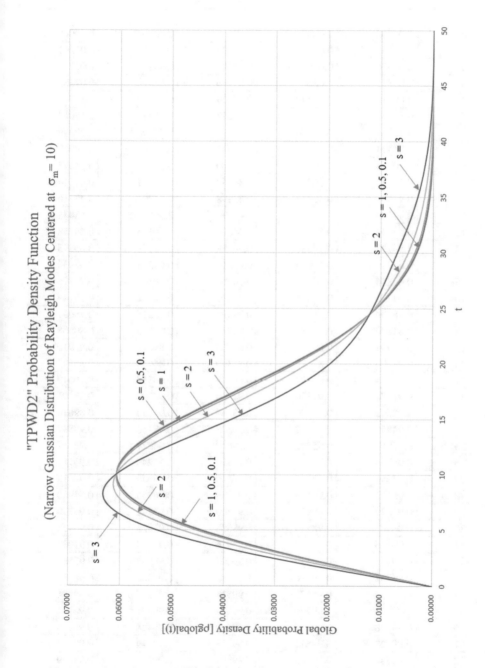

"TPWD2" Probability Density Function
(Narrow Gaussian Distribution of Rayleigh Modes Centered at $\sigma_m = 10$)

Cumulative distribution function

T	s				
	0.1	0.5	1	2	3
0	0.00000	0.00000	0.00000	0.00000	0.00000
1	0.00995	0.01002	0.01025	0.01113	0.01261
2	0.02956	0.02977	0.03041	0.03298	0.03726
3	0.05825	0.05864	0.05986	0.06474	0.07288
4	0.09518	0.09579	0.09769	0.10529	0.11795
5	0.13932	0.14016	0.14277	0.15322	0.17064
6	0.18944	0.19050	0.19379	0.20696	0.22890
7	0.24424	0.24548	0.24935	0.26482	0.29062
8	0.30234	0.30371	0.30800	0.32516	0.35375
9	0.36237	0.36381	0.36832	0.38637	0.41644
10	0.42302	0.42447	0.42898	0.44702	0.47710
11	0.48309	0.48446	0.48874	0.50588	0.53445
12	0.54149	0.54272	0.54657	0.56195	0.58758
13	0.59733	0.59836	0.60158	0.61447	0.63595
14	0.64986	0.65065	0.65310	0.66294	0.67933
15	0.69855	0.69906	0.70066	0.70707	0.71775
16	0.74302	0.74325	0.74395	0.74618	0.75149
17	0.78309	0.78303	0.78285	0.78215	0.78097
18	0.81870	0.81838	0.81737	0.81336	0.80668
19	0.84994	0.84938	0.84765	0.84072	0.82916
20	0.87700	0.87625	0.87390	0.86453	0.84892
21	0.90014	0.89924	0.89643	0.88518	0.86642
22	0.91970	0.91870	0.91556	0.90300	0.88206
23	0.93603	0.93496	0.93164	0.91833	0.89616
24	0.94950	0.94842	0.94503	0.93150	0.90894
25	0.96049	0.95942	0.95609	0.94277	0.92057
26	0.96934	0.96832	0.96514	0.95240	0.93117
27	0.97640	0.97544	0.97247	0.96059	0.94079
28	0.98195	0.98109	0.97838	0.96755	0.94950
29	0.98629	0.98551	0.98309	0.97342	0.95730
30	0.98962	0.98894	0.98683	0.97835	0.96423
31	0.99216	0.99158	0.98976	0.98247	0.97032
32	0.99408	0.99359	0.99205	0.98588	0.97560
33	0.99551	0.99510	0.99382	0.98868	0.98013
34	0.99656	0.99623	0.99517	0.99097	0.98396
35	0.99733	0.99706	0.99621	0.99281	0.98714
36	0.99788	0.99767	0.99699	0.99428	0.98976
37	0.99828	0.99811	0.99758	0.99544	0.99189
38	0.99856	0.99843	0.99801	0.99635	0.99359

(continued)

(continued)

Cumulative distribution function				
T	s			
39	0.99876	0.99865	0.99833	0.99705	0.99492
40	0.99889	0.99881	0.99857	0.99759	0.99596
41	0.99898	0.99892	0.99874	0.99800	0.99676
42	0.99905	0.99900	0.99886	0.99830	0.99736
43	0.99909	0.99905	0.99895	0.99852	0.99781
44	0.99912	0.99909	0.99900	0.99868	0.99814
45	0.99913	0.99911	0.99905	0.99880	0.99839
46	0.99915	0.99913	0.99908	0.99888	0.99856
47	0.99915	0.99914	0.99910	0.99894	0.99868
48	0.99916	0.99915	0.99912	0.99898	0.99877
49	0.99916	0.99915	0.99912	0.99901	0.99882
50	0.99916	0.99916	0.99913	0.99903	0.99886

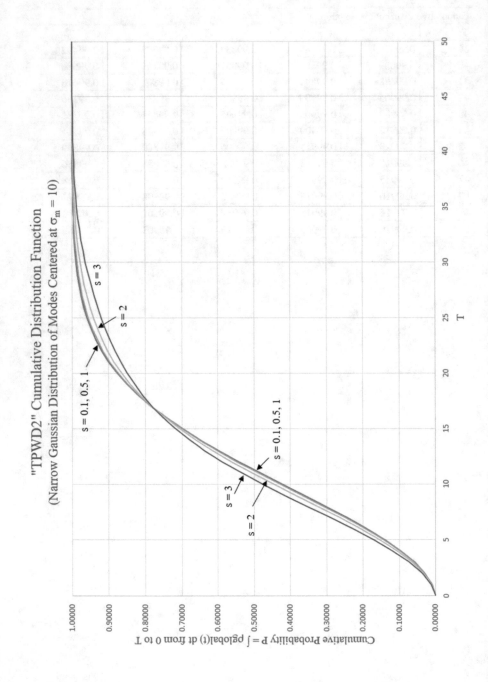

"TPWD2" Cumulative Distribution Function
(Narrow Gaussian Distribution of Modes Centered at $\sigma_m = 10$)

Appendix D
Sparse Matrices

The Adjacency Matrix for the array in Fig. 8.1 of Chap. 8 was constructed in Chap. 6 [1], and is displayed in Fig. D.1. It is called a *sparse matrix* because 70% of its entries are zeros. Furthermore, this particular sparse matrix is special because all its nonzero entries are clustered in a strip around the main diagonal. In this particular case, the strip is 3 sub-diagonals wide on each side of the main diagonal. Sparse matrices of this type are well known, and their properties have been studied for a long time [2–10]. So, there is a substantial body of information available to facilitate and simplify calculations. In addition, the diagonal elements of A are all zero, because no pin touches itself when bent away from its unbent position, and the nonzero entries are usually all 1's; further simplifying calculations. The matrix A is also symmetrical about the main diagonal ($A = A^T$). Also, the entries in the first row are simply the reverse of the entries in the last row, and the entries in the first column are simply the reverse of the entries in the last column. All of these properties are typical of *symmetrical* pin arrays.

Suppose it is desired to calculate the number of shortest conducting paths, formed during a bent pin event, between (1, 1) and (3, 3). These paths will all be 4 bent pins long. Therefore, an examination of the elements of A^4 is of interest. But, $A^4 = (A^2)(A^2)$, so the calculation of A^2 is an intermediate goal. What short cuts are possible given the sparsity and symmetry of A? Some thought, the definition of matrix multiplication [11], and all the special symmetries and numerical values of A discussed above, will justify the numerical algorithm that follows.

First, count the number of 1's in each row. That sum will be the value of the diagonal element for that same row in A^2. Next, fill the first row of A^2 as follows. Notice that there is a 1 in columns 2 and 4 of the first row of A. Look below these 1's and ask, "Are there any other rows with 1's in these same columns?" Yes, there is. Row 3 has a 1 in column 2. So, put a 1 in location (1, 3) of A^2. Row 5 has 1's in both columns 2 and 4. So, add these 1's and put an entry of 2 in location (1, 5) of A^2. Finally, row 7 has a 1 in column 4, so put a 1 in location (1, 7) of A^2. The rest of the entries in the first row of A^2 are zeros due to the sparsity of A. Now that the first row of A^2 has been calculated, the last row, first column, and last column of A^2 are

© Springer Nature Switzerland AG 2020
R. R. Zito, *Mathematical Foundations of System Safety Engineering*,
https://doi.org/10.1007/978-3-030-26241-9

Contact Pin

A=		(1,1)	(2,1)	(3,1)	(1,2)	(2,2)	(3,2)	(1,3)	(2,3)	(3,3)
	(1,1)	0	1	0	1	0	0	0	0	0
	(2,1)	1	0	1	0	1	0	0	0	0
	(3,1)	0	1	0	0	0	1	0	0	0
	(1,2)	1	0	0	0	1	0	1	0	0
	(2,2)	0	1	0	1	0	1	0	1	0
	(3,2)	0	0	1	0	1	0	0	0	1
	(1,3)	0	0	0	1	0	0	0	1	0
	(2,3)	0	0	0	0	1	0	1	0	1
	(3,3)	0	0	0	0	0	1	0	1	0

Bent Pin

Fig. D.1 The adjacency matrix for the array in Fig. 8.1 of Chap. 8. For each pin location on the left, that row has a 1 in each column labeled by a pin that it can touch if bent (nearest neighbor contacts only)

also known by symmetry. At this point the matrix A^2 looks as shown in Fig. D.2 (top). If needed, continue filling in the entries of A^2 in a similar manner, recalling that each entry found can be used to fill in 3 other elements of A^2. Now, A^2 (Fig. D.2—bottom) is still sparse (54% zeros), and still has symmetry across the main diagonal, but all the nonzero elements are no longer unity, and the nonzero entries are more scattered. Therefore, more sophisticated sparse matrix techniques then the one presented here must be used to calculate A^4. However, since only one element of A^4 is of concern, it is easier to just multiply the elements in the first row of A^2 by the elements in the last column of A^2 and sum the result to get the value of the desired element $A^4(1, 9) = 6$. In point of fact, for strictly stable connectors, the analyst will usually not have to worry about chains longer than 4 pins. In that case, all these calculations can be performed with nothing more than a paper and pencil. Software isn't even needed.

(1) Zito, R. R. "The Curious Bent Pin Problem—III: Number Theory Methods," *29th International Systems Safety Conference*, Las Vegas, NV, Aug. 8–12, 2011.

(2) Press, W. H., Flannery, B. P., Teukolsky, S. A., & W. T. Vetterling, *Numerical Recipes*, Cambridge U. Press, Cambridge, 1986, pp. 64–73.

(3) Golub, G. H. & C. F. Van Loan, *Matrix Computations*, Johns Hopkins U. Press, Baltimore, 1983, Ch. 5 and 6.

(4) Jacobs, D. A. H. (ed.), *The State of the Art in Numerical Analysis*, Academic Press, London, 1977, Ch. 1.3.

(5) Tewarson, R. P., *Sparse Matrices*, Academic Press, New York, 1973.

(6) Bunch, J.R. and D. J. Rose, (eds.), *Sparse Matrix Computations*, Academic Press, New York, 1976.

(7) Duff, I.S. and G. W. Stewart, (eds.), *Sparse Matrix Proceedings*, S.I.A.M., Philadelphia, 1979.

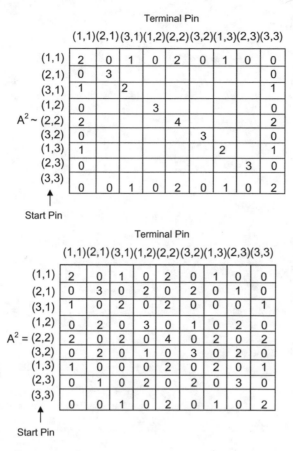

Fig. D.2 The computation of A^2

(8) Stoer, J., and R. Bulirsch, *Introduction to Numerical Analysis*, Springer, 1980, Ch. 8.

(9) Anon., *IMSL Library Reference Manual, Ed. 8*, IMSL Inc., 7500 Bellair Blvd, Houston, TX 77036.

(10) Anon., *NAG Fortran Library Manual, Mark 8*, NAG Central Office, 7 Banbury Road, Oxford OX26NN, U.K.

(11) Selby, S. M., *Standard Mathematical Tables, 7th ed.*, The Chemical Rubber Company, Cleveland, OH, 1968, pp. 108.

Appendix E
Short Table of Explosives, Propellants, Oxidizers, and Dangerously Flammable and Reactive Substances

This is a *partial* list of explosive substances across the entire periodic table. It must *not* be considered complete. Some of these compounds were intended to be sold as commercial or military explosives. However, many of these substances are simply compounds that are too unstable and unpredictable to be practical explosives. Hence, their on-site creation or storage, whether accidental or intentional, is to be avoided. The physical descriptions (color, state, etc.) of many of these unstable compounds have been included in this appendix to facilitate their immediate recognition by system safety. Still, other compounds are safe when handled properly, but very unsafe when handled or stored improperly. And, there are an enormous number of combinations of ordinarily safe substances, or at least controllable substances, that are unsafe when combined. Finally, there are so many organic compounds with strained molecular bonds that are capable of explosion, flammability, or high reactivity, that it is not possible to list them all here. A more complete compilation with thousands of entries can be found in the latest edition of the Hazardous Chemicals Desk Reference by Richard J. Lewis, Sr, (the 6th. ed. 2008, Wiley, NJ). However, it must always be remembered that new explosive compounds are always being discovered and devised. Only the most common energetic materials that the system safety engineer is likely to come in contact with, in the author's opinion, are listed here. In short, this list is no substitute for the careful examination of the Material Safety Data Sheet (MSDS) that is provided by suppliers of chemicals and other hazardous substances.

© Springer Nature Switzerland AG 2020
R. R. Zito, *Mathematical Foundations of System Safety Engineering*,
https://doi.org/10.1007/978-3-030-26241-9

Many Nitrogen Compounds are Explosive!!!!

Chemical Name	Formula	Combustion Products	Comments
Ammonium nitrate (s)	$NH_4NO_3 \rightarrow$	N_2O (g) + 2 H_2O (g)	Can detonate via another high explosive (TNT) or traces of acid and chlorine ion as catalysts. Nitrates, in general, are potentially explosive.
Trinitrotoluene (s) (TNT)	$2(C_7N_3O_6H_5) \rightarrow$	$3N_2$ + $5H_2O$ + 7CO + 7C or $3N_2$ + $5H_2$ + 12CO + 2C	High Explosive

NO₂ / H / H / O₂N / NO₂ / CH₃ "nitro" group

Note: s = solid; g = gas

Nitrogenous Explosives, Propellants, and Oxidizers

Chemical Name	Formula	Combustion Products	Comments
RDX (Research Department Explosive)	$C_3H_6N_6O_6$		High Explosive
Propane- 1,2,3 trinitrate "Nitroglycerine"	$C_3H_5N_3O_9$		Unstable colorless, oily, liquid. Will be discussed in detail later.

O₂N—N N—NO₂ / N / NO₂

Chemical Name	Formula	Combustion Products	Comments
Pentaerythrital tetranitrate (PETN)	$C_5H_8N_4O_{12}$		High Explosive
Cellulose nitrate ("Nitrocellulose") (Will be discussed in detail in the next section.)	Variable: $C_6H_9NO_7$ $C_6H_8N_2O_9$ $C_6H_7N_3O_{11}$		High Explosive and Propellant

O₃N / NO₃ / O₃N / NO₃

Chemical Name	Formula	Combustion Products	Comments
Mercury fulminate O-N≡C-Hg-C≡N-O	$Hg(ONC)_2$	Various: N_2, CO_2, HgO, CO, Hg, $Hg(CN)_2$, $Hg(OCN)_2$, $Hg(OCN)CN$	Will Detonate!
Lead Styphnate (lead 2,4,6 trinitroresorcinate)	$Pb\ C_6H\ N_3O_8$		Will Detonate!

Chemical Name	Formula	Combustion Products	Comments
Diazo dinitro phenol (DDNP)	$C_6H_2N_4O_5$		Will Detonate! (less sensitive to friction). Azides, in general, are potentially explosive.
Ammonium Chlorate	$2(NH_4\ ClO_3)$	(Heat) → $2NH_4Cl\ +\ 3\ O_2\,(g)$	Oxidizer for solid propellant
Ammonium Perchlorate	$2(NH_4\ ClO_4)$	(Heat) → $2NH_4\ Cl\ +\ 4\ O_2\ (g)$	Oxidizer (more stable than chlorates)

Chemical Name	Formula	Combustion Products	Comments
Unsymmetrical dimethyl hydrazine (UDMH)	$C_2H_8N_2$		Rocket fuel (hypergolic with NO_2, liquid oxygen [LOX], or HNO_3) PEL = 0.5 PPM
Hydrazoic Acid	$2(HN_3)$	$3N_2 + H_2$	Colorless, dangerously explosive liquid.
Sodium azide	$2(NaN_3)$	$2Na(\ )\ +3N_2(g)$	Azides in general are potentially explosive.
Nitronium perchlorate	NO_2ClO_4		Reacts violently with organic matter

Definition: Hypergolic fuel spontaneously ignites on contact with its oxidizer, or air.

Chemical Name	Formula	Combustion Products	Comments
Dinitrogen pentoxide	$2(N_2O_5)$	$4NO_2 + O_2$	Colorless unstable crystals
Nitrogen fluorodichloride	$NFCl_2$		Explosive
Tetraflurohydrazine	N_2F_4		Explosive reaction with hydrogen
Platinum amine nitrates and perchlorates	Various	Various	Detonate when heated or are impact sensitive.

Structure for Dinitrogen pentoxide: $O_2N{-}O{-}NO_2$ shown with NO_2^+ NO_3^-

Chemical Name	Formula	Combustion Products	Comments
Nitrogen trichloride	NCl_3		A pale yellow explosive photosensitive oil
Difluoroamine	HNF_2		A colorless explosive liquid
Chlorine nitrate	$ClNO_3$		Reacts explosively with organic mater
Fluorine nitrate	FNO_3		Intrinsically explosive
Nitrogen dioxide + nitric acid	$NO_2 + HNO_3$		Powerful oxidizing agent with aniline

Chemical Name	Formula	Combustion Products	Comments
Tetrsulfer tetranitride	S_4N_4	Probably N_2 + sulfur oxides + nitrogen oxides	Orange crystals that may be detonated by shock.
Anhydrous titanium nitrate	$Ti(NO_3)_4$ $\xrightarrow{\text{+ Organic compounds}}$	Nitrogen, nitrogen oxides, titanium oxide?	Inflammation or explosion.

Chemical Name	Formula	Combustion Products	Comments
Azido Ru^{3+} ammines	e.g. $[Ru(NH_3)_5N_3]^{2+} \rightarrow$	$[Ru(NH_3)_5N_2]^{2+} + \frac{1}{2}N_2$	All unstable and explosive.
Hydrazine	N_2H_4		colorless, oily, Fumming liquid or white crystals. Explodes on contact with many materials, including iron, stainless steel, and even glass. There is a long list of other compounds for which hydrazine will explode on contact.
Chlorine Azide	ClN_3		Extremely unstable explosive.

This list contains the more common and simpler nitrogen explosive, propellants, and oxidizers, and should not be considered complete!!!

Non-Nitrogenous Energetics (Non-transition Elements)

Chemical Name	Formula	Combustion Products	Comments
Lower aluminum alkyls	AlR_3 (R= organic alkyl)	$\xrightarrow{+6H_2O}$ $2Al(OH)_3 + 3H_2$	A reactive liquid that is hypergolic in air, exploding with water.
Salts of aluminum hydride	$M^+AlH_4^-$ (e.g. Li AlH_4)	$\xrightarrow{+4H_2O}$ $4H_2 + Al(OH)_3 + M^+OH^-$	Explosively hydrolyzed by water
Salts of gallium hydride	$M^+GaH_4^-$		
Hydrogen	H_2	$\xrightarrow[\text{+ignition source}]{\text{Air or oxygen}}$ H_2O	Explosive gas

$=AlH_4^-$

M = any metal atom

Chemical Name	Formula	Combustion Products	Comments
Boron triiodide	BI_3	$\xrightarrow{3H_2O}$ $3HI + B(OH)_3$	White solid below 43 C. Explosively hydrolyzed.
Beryllium alkyls	$R\text{-}Be\text{-}R$	$\xrightarrow{2H_2O}$ $Be(OH)_2 + H_2$	Hypergolic in air and explosively hydrolyzed.
Methylpotassium	KCH_3		Pyrophoric*
Hydrides (e.g. NaH, RbH, CsH, BaH, UH_3, Si_4H_{10}, La, and Lanthanide hydrides, etc.)	NaH	$\xrightarrow{H^+}$ $Na^+ + H_2(g)$	The hydrides in general are very reactive with air and water (hypergolic). UH_3 is also radioactive.

*Pyrophoric: will ignite spontaneously on contact with air.

Chemical Name	Formula	Combustion Products	Comments
White (allotrope of) phosphorous	P_4	air, O_2 → P_4O_{10} + 4 other trace oxides	Flammble in air
Phosphine	PH_3	+air, O_2 + ignition → $P_4 + H_2O$	Explosive and exceedingly poisonous
Diphosphine	P_2H_4	+air, O → Probably $P_4 + H_2O$	Spontaneously flammable
Phosphorous trichloride	PCl_3	H_2O → $H_2[HPO]_3$	Violently hydrolyzed by water to give phosphorous acid

Chemical Name	Formula	Combustion Products	Comments
Trimethylphosphine	$P(CH_3)_3$	air, O_2 → Probably $P_4 + H_2O + CO_2$	Spontaneously flammable in air
Hydrogen carbon phosphine	$HC{\equiv}P$	air, O_2 → Probably $P_4 + H_2O + CO_2$	Pyrophoric* substance
Ozone	O_3	Ignition → O_2	70% liquid phase (deep purple) is explosive as is the pure liquid (deep blue). The solid is black-violet. *Powerful oxidizing agent*, especially for organic compounds.

*Pyrophoric: will ignite spontaneously on contact with air.

Chemical Name	Formula	Combustion Products	Comments
Oxygen (Liquid; LOX) (pale blue)	O_2	Spark + H_2, CH_4, CO, or organics → $H_2O + CO_2$	Explodes violently with any organic compound if a spark is present. Oxidizer for rocket fuel
Oxygen difluoride (pale yellow poisonous gas! Will even react with Xe gas.)	OF_2	Spark +, H_2, CH_4, or CO →	Explodes violently. Rocket fuel oxidizer.
		Cl_2, Br_2, I_2 →	Explodes at room temperature
		Steam →	Explodes
Dioxygen difluoride	O_2F_2		Explodes on contact with many compounds.
Hydrogen (Liquid)	H_2	Spark + air	Explodes violently
Alkali metals	Li, Na, K, Rb, Cs	Water → Alkali metal hydroxides + H_2	Explodes violently due to auto-ignition of H_2

Chemical Name	Formula	Combustion Products	Comments
Hydrogen peroxide	H_2O_2 $\longrightarrow$	$H_2O + O_2$	Pale blue syrupy liquid when pure. *Strong oxidizer used with rocket fuel.*
Sodium oxide	Na_2O_2 $\xrightarrow{\text{Some metals (e.g. Fe)}}$	FeO_4^{2-} (or other metal oxides)	Violent reaction
Perchlorates (especially of organo-metallic ions)	$[\text{organo-metallic}]^+ClO_4^-$		Often dangerously explosive.
Silanes (Si hydrides)	e.g. Si_4H_{10}, Si_2H_6, SH_4, etc.		Spontaneously flammable in air. Rapidly hydrolyzed. React explosively with halogens.
Sulfur dust	S_8 $\xrightarrow{\text{Air + ignition}}$	Sulfur oxides	Explosive!

Chemical Name	Formula	Combustion Products	Comments
Thyonyl fluoride	$SOCl_2$ $\xrightarrow{\text{water}}$	$SO_2 + 2HCl$	
Thyonyl bromide	$SOBr_2$ $\xrightarrow{\text{water}}$	Similar to reaction above.	
Thyonyl fluorochloride	$SOFCl$ $\xrightarrow{\text{water}}$	Similar to reaction above.	All of these compounds react rapidly, and sometimes violently with water
Selenyl fluoride	$SeOF_2$ $\xrightarrow{\text{water}}$	Similar to reaction above.	
Selenyl chloride	$SeOCl_2$ $\xrightarrow{\text{water}}$	Similar to reaction above.	
Selenyl bromide	$SeOBr_2$ $\xrightarrow{\text{water}}$	Similar to reaction above.	
Clorosulfuric acid	$\xrightarrow{\text{water}}$		Explosively hydrolyzed by water.

Chemical Name	Formula	Combustion Products	Comments
Sulfur tetrafluoride	SF_4	Instantly hydrolyzed by water	Extremely reactive substance!
Chlorosulfuric acid	$HClSO_3$	Probably Cl_2, SO_2, and H_2SO_4	Colorless fuming liquid explosively hydrolyzed by water.
Fluorine gas	F_2	Various. Fluorine is the most chemically reactive of all elements and combines directly at ordinary or elevated temperatures with all elements other than oxygen and the lighter noble gases, often with extreme vigor. It also attacks many compounds.	Will vigorously attack organic compounds, breaking them down to fluorides; organic material will often inflame and burn in the gas, or *explode*! Phosphorus + fluorine gas will spontaneously explode! The same is true for some other non-metals.

Chemical Name	Formula	Combustion Products	Comments
Chlorine gas	Cl_2	Instantly hydrolyzed by water	Extremely reactive substance! Explodes on contact with many substances!
Chlorine trifluoride	ClF_3		Sponaneously flammable. A powerful oxidant which may reat violently with oxidizable materials. A rocket propellant. Explosive reaction with water.
Chlorine pentafluoride	ClF_5		Vigorous reaction with water or anhydrous nitric acid. Violent reaction on contact with metals.
Chlorites			Many chlorite salts are heat and impact sensitive.

Chemical Name	Formula	Combustion Products	Comments
Oxides of chlorine (See list below)	Various (see list below)	Various	All chlorine oxides are highly reactive and unstable, tending to explode under various conditions.
Dichloro trioxide	Cl_2O_3		Dark brown solid that explodes readily below 0^0C.
Dichloro hexoxide	Cl_2O_6		Red oily liquid. Reacts explosively with organic mater.
Chlorine monoxide	Cl_2O	Cl_2, O_2	Yellowish-red gas at room temperature. Explodes when heated or sparked.
Chlorine dioxide	ClO_2		Yellowish gas at 20^0C. Highly reactive and liable to explode very violently.
Dichlorine tetroxide	Cl_2O_4		Reacts explosively with organic mater.
Dichlorine heptoxide	Cl_2O_7		Take precautions against explosion.

Chemical Name	Formula	Combustion Products	Comments
Perbromic acid	$HBrO_4$		12M acid will explode on contact with tissue paper.
Interhalogens (see bromine trifluoride below)	Various	Mixture of halides	Reactive, corrosive, oxidizing substances that readily react vigorously or explosively with water or organic compunds.
Bromine trifluoride (Halogen fluorides in general)	BrF_3		Dangerously explosive with water.
Interhalogens	Various	Various	All reactive, corrosive, oxidizing substances.

Chemical Name	Formula		Combustion Products	Comments
Xenon trioxide	XeO_3		Probably Xe gas + O_2	Colorless crystals. *Explosive!* Will also react violently with alcohols.
Xenon tetroxide	XeO_4		Probably Xe gas + O_2	Colorless gas. *Explosive!*
Xenon Tetrafluoride	XeF_4	$\xrightarrow{\text{Water}}$ XeO_3		Will react violently with water to produce the trioxide (also explosive)!
Xenon hexafluoride	XeF_6	$\xrightarrow{\text{Water}}$ XeO_3		Will react violently with water to produce the trioxide (also explosive)!

Non-Nitrogenous Energetics (Non-transition Elements: General)

Chemical Name	Formula	Combustion Products	Comments
Peroxides of ethers	Various	Various	Peroxides of ethers are shock sensitive and explosive. Even an "empty" ether container should be considered explosive. It must always be remembered that ethers form peroxides on contact with oxygen (air). *Peroxides of ethers should be removed by a bomb squad!*
Hydrochloric acid	HCl	?	Not normally considered an explosive. However, there is one known case of a bottle of HCl spontaneously exploding. The reason is still unknown. In any case concentrated HCl should always be considered a reactive substance.

Chemical Name	Formula	Combustion Products	Comments
Acetylene gas	HC≡CH	$CO_2 + H_2O$	Shock sensitive and explosive!
Ethanol + Sulfuric acid	$CH_3CH_2OH + H_2SO_4$	CO_2 + Sulfur oxides	Very dangerous! The heat of reaction can cause a *major* ethanol fire. Can explode on contact with other organics.
Perchloric acid + wood	$HClO_4$ + organics	CO_2 + Sulfur oxides	Perchloric acid reacts explosively with organic material. The storage of perchloric acid on wood shelving will result in an explosion if the bottle should leak.

Non-Nitrogenous Energetics (Transition Elements: General)

Chemical Name	Formula	Combustion Products	Comments
Metal carbonyl	$M(CO)_n$ (typically n= 1→ 6) (M = a transition metal)	Probably metal oxide + CO_2	At least one type are known for all transition metals. In general they are very toxic and Ni carbonyl is flammable as well!
Transition metal ketone azides	$MC(O)N_3$ (M = a transition metal)	Probably metal oxide + N_2	Probably explosive

Non-Nitrogenous Energetics (Transition Elements: 1st Series)

Chemical Name	Formula	Combustion Products	Comments
Chromium trioxide	CrO_3	$\xrightarrow{\text{Organic substances}}$ Cr^{3+} + dehydrogenated organic	May explode.
Chromate esters	Various	$\longrightarrow$ Various	Highly explosive when pure
Dichromate ion	$Cr_2O_7^{2-}$	Various	Powerful oxidizing agent! (e.g. Potassium dichromate + powdered magnesium metal will explode upon ignition)
Lower alkyl zinc compounds	Various	Various	Liquids or low-melting solids that are spontaneously flammable.

Chemical Name	Formula	Combustion Products	Comments
Chromyl chloride	CrO_2Cl_2	$\xrightarrow{\text{Organic substances}}$ Various	Vigorously oxidizes organic matter.
Ti, V, Cr peroxo compounds	e.g. CrO_5	O_2 + other products	Unstable. Explosive or flammable in air.
[K, NH_4, or Tl] dichromate	[K, NH_4, or Tl] Cr_2O_7	$\xrightarrow{H_2O_2}$ $[CrO(O_2)_2OH]^-$ (Intermediate product)	Forms blue-violet violently explosive salts.
Hydrogen permanganate	$HMnO_4$	$\xrightarrow{\text{Organic substances}}$ Various	Violent oxidant for organic materials and itself decomposes violently above 3°C.

Chemical Name	Formula	Combustion Products	Comments
Dimanganese heptoxide	Mn_2O_7	?	Explosive oil.
Manganese monoxo trichloride	$MnOCl_3$	?	Green, volatile, explosive liquid.
Manganese trioxo chloride	MnO_3Cl	?	Green, volatile, explosive liquid.
Manganese dioxo dichloride	MnO_2Cl_2	?	Brown and unstable.
Manganese trioxo Fluoride	MnO_3F	?	Green liquid that explodes at room temperature.

Chemical Name	Formula	Combustion Products	Comments
Pure powdered iron (finely divided)	$Fe \xrightarrow{O_2}$	Iron oxides	Pyrophoric.[*] Also combines vigorously with Cl, F, Br, I, P, B, C, and Si.
Iron oxide (powder)	FeO	Higher iron oxides	Pyrophoric [*]
Powdered nickel	Ni	Nickel oxides	Pyrophoric [*]
Potassium nickel cyanide	$K_4[Ni(CN)_4]$	Various	Extremely reactive.
Potassium nickel dicarbon hydride	$K_4[Ni(C{\equiv}CH)_4]$	Various	Extremely reactive.

[*]Pyrophoric: will ignite spontaneously on contact with air.

Chemical Name	Formula	Combustion Products	Comments
Zinc metal dust (dust of many metals)			Dust may ignite spontaneously in dry air. Explosive with acids.
Vanadium tetrachloride	VCl_4 $\xrightarrow{water}$	Oxovanadium IV chloride	An oil violently that is hydrolyzed.

Non-Nitrogenous Energetics (Transition Elements: 2nd & 3rd Series)

Chemical Name	Formula	Combustion Products	Comments
Di(cyclopentadienyl) niobium hydride	$(h^5\text{-}C_2H_2)_2NbH$	?	Exceedinly reactive.
Dihydrogen Tungsten tetrasulfide	H_2WS_4		Unstable red solid. May release hydrogen gas which can explode.
Molybdenum oxide tetrachloride	$MoOCl_4$		In general highly reactive. Violently hydrolyzed by water.
An organo-molybdenum compound	$Na_2\{[Mo_6(OMe)_8](OMe)_6\}$ (Note: Me = methyl group)	?	Pyrophoric*

*Pyrophoric: will ignite spontaneously on contact with air.

Chemical Name	Formula	Combustion Products	Comments
Hexafluorides of platinum metals (Ru, Os, Rh, Ir, Pd, and Pt)	e.g. IrF_6 e.g. OsF_6	$\xrightarrow{water}$ HF, O_2, O_3, IrO_2(aq) $\xrightarrow{water}$ HF, OsO_4, OsF_6^-	React violently with liquid water
Platinum metal pentafluorides	$(Ru, Os, Rh, Ir, Pd, or Pt)F_5$		Very reactive hydrolyzable substances.
Platinum metal tetrafluorides	$(Ru, Os, Rh, Ir, Pd, or Pt)F_4$		Violently hydrolyzed by water.
Ruthenium tetroxide	RuO_4	$\longrightarrow$ RuO_2, O_2	Can explode.
Osmium tetroxide	OsO_4		Powerful oxidizing agents.
Tungsten oxide tetrachloride	$WOCl_4$		Scarlet crystals. Violently hydrolyzed.

Chemical Name	Formula	Combustion Products	Comments
Platinum hexafluoride	PtF_6		One of the most powerful oxidizing agents known.
Hexafluorides			All the hexafluorides are extraordinarily reactive and corrosive substances and normally must be handled in Ni or Monel apparatus although quartz can be used if necessary. However, Pt hexafluoride and Rh hexafluoride will react with glass (even when rigorously dry) at room temperature!

Chemical Name	Formula	Combustion Products		Comments
Disodium platinum hexahydroxide	$Na_2Pt(OH)_6$	Acetic acid + nitric acid	?	Such solutions have been known to explode.
Ta, Mo, W peroxo compounds				Discussed above (1st transition series). Explosive.
Cadmium metal dust				Ignites spontaneously in air. Also toxic!
Cadmium Phosphide	Cd_3P_2			Explosive reaction with nitric acid. Toxic.
Peroxo complexes of Nb, Ta, Mo, or W	Various			Explosive or flammable in air.

Non-Nitrogenous Energetics (Lanthanides)

Chemical Name	Formula	Combustion Products	Comments
Pure lanthanide metal	Ce, Pr, Nd, Pm,[†] Sm, Eu, Gd, Tb, Dy, Ho, Er, Tm, Yb, Lu $\xrightarrow{O_2}$	(Lanthanide)$_2O_3$ (except Ce which forms CeO_2)	All react directly with water to liberate hydrogen gas, itself explosive. All burn easily in air. All can be expected to explode if ignited in a finely divided state in air.
Lanthanide hexafluorides	(Lanthanide)F_6		Extraordinarily reactive and corrosive substance.
Lanthanide alkyl complexes	Poorly characterized	?	Pyrophoric[*]

*Pyrophoric: will ignite spontaneously on contact with air.
†Promethium (Pm) is intensely radioactive

Non-Nitrogenous Energetics (Actinides†)

Chemical Name	Formula	Combustion Products	Comments
Powdered uranium metal	U $\xrightarrow{O_2}$	Uranium oxides	Frequently pyrophoric.[*] Also radioactive and capable of sustaining a nuclear chain reaction when enriched. Chemically reactive. Combines directly with most elements.
Powdered thorium metal	Th $\xrightarrow{O_2}$	Thorium oxides	Pyrophoric[*] and radioactive.
Plutonium	Pu $\xrightarrow{O_2}$	Plutonium oxides	Pyrophoric,[*] radioactive, and fissile. Chemically reactive like U. Pu is so toxic that a 1 microgram dose is potentially lethal!

*Pyrophoric: will ignite spontaneously on contact with air.
†All the synthetic elements and their compounds are intensely radioactive and dangerous.

Chemical Name	Formula	Combustion Products	Comments
Uranium hexafluoride	UF_6		Extraordinarily reactive and corrosive substances. Rapidly hydrolyzes water.
Uranium hydride			See hydrides of non-transition elements.
Uranium pentachloride	UCl_5		Violently hydrolyzed by water
Uranium hexachloride	UCl_6		Violently hydrolyzed by water
Plutonium hexafluoride	PuF_6		Very much less stable than UF_6. Most system safety engineers will never need to work with this deadly compound.

†All the synthetic elements and their compounds are intensely radioactive and dangerous.

Appendix F
Short Table of Inorganic Toxins and Pollutants

This short table of toxic and polluting inorganic compounds and elements involves the entire Periodic Table. Some of the metals listed here (or more properly their ions) are also listed in the environmental "Delaware Department of Natural Resources and Environmental Control Target Analyte List (TAL)—Metals and Cyanide" (www.dnrec.deleware.gov/dwhs/sirb/Documents/Target%20Compound%20Lists.pdf, accessed May 27, 2017; also see Appendix H). However, this list is longer than the TAL because the system safety engineer may be confronted with toxins and pollutants in a laboratory, military, or commercial setting where release into the environment is strictly controlled, or not possible, but the exposure of personnel is still a risk. The TAL also includes metals like sodium, potassium, zinc, and iron that are not normally considered toxic and are, therefore, not listed here. All concentrated acids and bases should be considered dangerous because they are generally very *corrosive*. However, many in *dilute aqueous solutions* (e.g. acetic acid, hydrochloric acid, phosphoric acid, sodium hydroxide, etc.) are composed of ions that are either harmless, or even essential to life. That is to say, they are neither *intrinsically toxic* nor polluting in a certain specified pH range (sufficiently diluted). Substances of this kind are not included in this appendix. Although, HCl as a *gas* is very "toxic", because of its ability to form concentrated acid in the respiratory tract, and is included here. This list is necessarily *incomplete*, and is only included to give a general idea of the kinds of compounds that are toxic. It only includes the simplest and perhaps most notorious toxic substances. A more complete compilation with thousands of entries can be found in the latest edition of the Hazardous Chemicals Desk Reference by Richard J. Lewis, Sr, (the 6th. ed. 2008, Wiley, NJ). And, it must always be remembered that new compounds are always being devised. Therefore, the Material Safety Data Sheet (MSDS) supplied by vendors of chemicals should always be consulted as a primary source of information about toxic or polluting substances. The physical descriptions of the species in this appendix have often been included so that the system safety engineer can recognize a dangerous material when it is seen.

© Springer Nature Switzerland AG 2020
R. R. Zito, *Mathematical Foundations of System Safety Engineering*,
https://doi.org/10.1007/978-3-030-26241-9

Chemical name	Chemical formula	Comments
Nontransition elements		
Arsenic	As	Poisonous metal. See Appendix G
Arsenic compounds	Various	Very toxic by all routes (ingestion, inhalation, etc.), Especially arsenic oxide
Arsine	AsH_3	Extremely poisonous gas
Beryllium	Be	Inhalation of airborne dust may result in permanent disability and eventual death
Chlorine fluorides	ClF_3, ClF_5	Poisonous by inhalation. Also explosive
Chlorine gas	Cl_2	Toxic and corrosive
Cyanides (inorganic)	Various (e.g. CN^-, KCN, NaCN, HCN (gas), etc.)	Exceedingly poisonous via ingestion or inhalation Also, see Appendix H
Disulfur decafluoride	S_2F_{10}	Extremely poisonous, being similar to phosgene in its physiological action
Fluorine gas	F_2	Toxic and corrosive. A deadly substance
Fluorophosphates	Various	Exceedingly toxic neurotoxin
Hydrofluoric acid	HF(gas or aqueous solution)	Toxic. Corrosive. Will even attack glass
Hydrogen chloride (gas)	HCl (gas)	Toxic. Corrosive. Forms hydrochloric acid in water
Hydrazines	N_2H_4, and derivatives (e.g. UDMH)	Poisonous gases. Very toxic and carcinogenic
Interhalogens (general)	Various	Poisons. All reactive, sometimes explosive, corrosive, oxidizing substances
Lead & its compounds	Various	Poisonous. See Appendix G
Nitric acid	HNO_3 (red fuming)	Poison by inhalation. Corrosive. Explosive in combination with other materials
Organic compounds	Various and many	See Hazardous Chemicals Desk Reference cited above. Also, see Appendix H
Oxygen difluoride	OF_2	Pale yellow poisonous gas. Very explosive
Phosgene	CCl_2O	Colorless poison gas or volatile liquid
Phosphine	PH_3	Exceedingly poisonous and explosive
Polonium and its compounds	Various	Toxic and radioactive. compounds
Selenium dihydride	H_2Se	Extremely poisonous. The odor is revolting
Silane	Si_4	Poisonous gas

(continued)

(continued)

Chemical name	Chemical formula	Comments
Silanes (chemical family)	Various (Si_2H_6, Si_4H_{10}, etc.)	Poisonous
Stibine	SbH_3	Similar to arsine but less stable
Sulfur dihydride	H_2S	Extremely poisonous. Far exceeds the toxicity of hydrogen cyanide gas. The odor is revolting
Tellurium dihydride	H_2Te	Extremely poisonous. The odor is revolting
Thallous solutions	Tl^+	Exceedingly poisonous. In traces causes loss of hair
Unsymmetrical dimethyl Hydrazine	$N_2H_2(CH_3)_2$ (UDMH)	Poisonous gas or liquid. Very toxic. Used as rocket fuel
Transition elements		
Carbonyls	Various [e.g. $Ni(CO)_4$]	Extremely Poisonous! Used extensively in industry. Carbonyl derivatives of at least one type are known for all of the transition metals
Chromium trioxide	CrO_3 (hexavalent chromium; Cr^{6+})	Soluble in water, highly poisonous. See Appendix G
Hexafluorides (all)	MF_6 (e.g. PtF_6)	Extraordinarily reactive and corrosive substances (M = metal atom)
Mercury & its compounds	Hg, Various (e.g., CH_3Hg^+) (e.g., Methylmercury ion$^+$)	Very high toxicity, especially alkyls. But, therapeutic organo-metallic compounds are much less toxic and very useful. Some have antifungal properties. Also, see Appendix G
Osmium tetroxide	OsO_4	Toxic substance. A particular hazard to the eyes
Platinum compounds	Various	Toxicity varies from allergic reactions, to symptoms of intoxication, to mutagenesis, to carcinogenesis
Ruthenium tetroxide	RuO_4	Toxic substance
Technetium and its compounds	Tc, Various	Radioactive! Compounds
LANTHANIDES AND ACTINIDES:		
Hexafluorides (all)	MF_6 (e.g. UF_6, PuF_6) (M = metal atom)	All very toxic and corrosive substances

(continued)

(continued)

Chemical name	Chemical formula	Comments
Plutonium metal	Pu	Very toxic. One microgram may be lethal
Synthetic elements (all)	Ac, Pu, Am	Exceedingly radioactive and dangerous
Thorium & its compounds	Th, Various	Carcinogenic when ingested due to its radioactivity. See Appendix G
Uranium and its salts	U, Various	High chemical toxicity largely shown in kidney damage, may not be reversible. See Appendix G

† The methylmercury ion has become widespread in the environment and poses a risk to the fishing industry. In 1994 the United States Environmental Protection Agency and the U.S. Food and Drug Administration recommended "limiting or avoiding shark and swordfish". In 2001, tilefish and king mackerel were added to this list. Currently, marlin, orange roughy, and bigeye tuna have also been added. But, it is recommended to avoid tilefish only if it comes from the Gulf of Mexico. *Source* http://federalregister.gov/documents/2017/01/19/2017-01037/advice-about-eating-fish-from-the-environmental-protection-agency-and-food-and-drug-administration

Appendix G
Pourbaix Diagrams for Common Elemental Toxins and Pollutants

The source references for each diagram that was used, or modified, in this Appendix have been cited in the captions. The most important reference (cited below) for Pourbaix Diagrams is N. Takeno, *Atlas of Eh-Ph Diagrams: Intercomparison of Thermodynamic Databases*, Geological Survey of Japan, AIST website: https://www.gsj.jp/data/openfile/no0419/openfile419e.pdf, 2005. Not cited was the very useful compilation by T. S. Bowers, K. J. Jackson, and H. C. Helgeson entitled *Equilibrium Activity Diagrams for Coexisting Minerals and Aqueous Solutions at Pressures and Temperatures to* 5 kb *and* 600 °C, Springer, Berlin, 1984. Although these diagrams are in a different, but related, format to those presented in this text, the compilation by Bowers et al contains a wealth of information that is useful to the system safety engineer. Since published Pourbaix Diagrams are known to contain errors and omissions of phases, the safety engineer should always check each critical boundary by direct calculation, via the methods of Chapter 17, before use. Cross-checking important results with other sources is also essential (Figs. G.1, G.2, G.3, G.4, G.5, G.6, G.7, G.8 and G.9).

© Springer Nature Switzerland AG 2020
R. R. Zito, *Mathematical Foundations of System Safety Engineering*,
https://doi.org/10.1007/978-3-030-26241-9

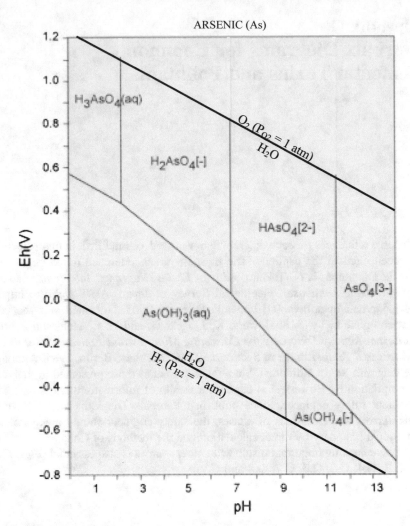

Fig. G.1 Pourbaix diagram for the arsenic/oxygen system at 25 °C and 1 atm of pressure (*Source* N. Takeno, *Atlas of Eh-Ph Diagrams: Intercomparison of Thermodynamic Databases*, Geological Survey of Japan, AIST website: https://www.gsj.jp/data/openfile/no0419/openfile419e.pdf), 2005, Fig. 12 on p. 31, accessed July 11, 2017). All species shown are water soluble, and the total As species concentration is 10^{-10} moles/L. Note that the species As(OH)$_3$ (aq) is sometimes written as H$_3$AsO$_3^0$ (aqueous) in some chemistry texts. Naturally occurring common arsenic minerals have low solubility at ordinary groundwater pH values: native arsenic (As), Orpiment (As$_2$S$_3$), Realgar (As$_4$S$_4$), Allemontite (AsSb), Arsenopyrite (FeAsS), Gersdorffite (NiAsS), Cobaltite (CoAsS), Chloanthite ([Ni,Co]As$_2$), Rammelsbergite/Pararammelsbergite (NiAs$_2$), Ulllmannite (NiSbS), Lollingite (FeAs$_2$), and Niccolite (NiAs). The complexing and compounding of arsenic with iron and sulfur is of particular significance and limits the mobility of arsenic in the environment

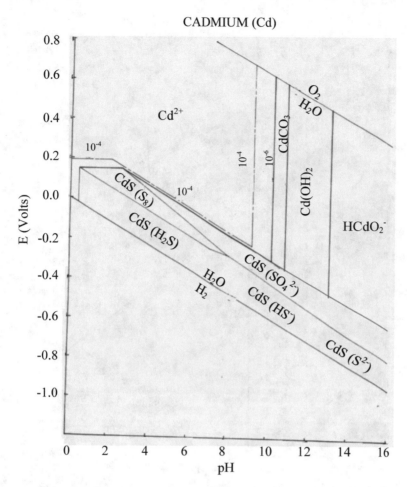

Fig. G.2 Pourbaix Diagram for the oxides, sulfides, and carbonates of cadmium (*Source* diagram constructed by M. B. Fraser (1961) in *Equilibrium Diagrams for Minerals*, Schmitt, H.H. (editor), Geological Club of Harvard, Boston Mass., 1962, p. 163; recalculated and corrected). The total dissolved sulfur activity is 0.1 molal (approximately equal to the molarity concentration, M) and the *total* dissolved carbonate (CO_2 aq., H_2, CO_3, HCO_3^-, and CO_3^{2-}) activity is 10^{-4} molal, at 25 °C and 1 atm of pressure. The boundary for the ionic species (Cd^{2+}) applies to activity levels of 10^{-6} molal (solid vertical line) and 10^{-4} molal (dashed vertical line)

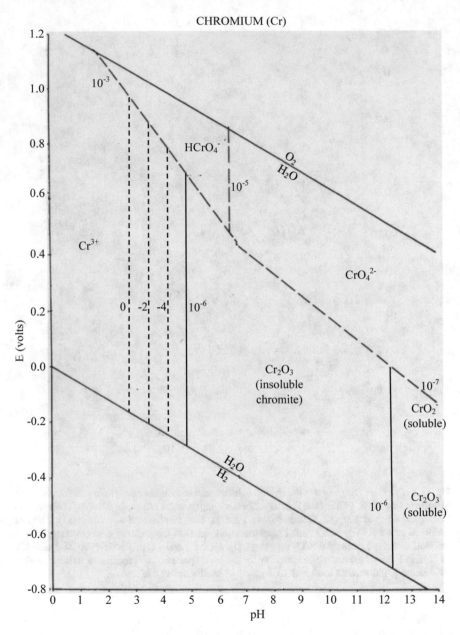

Fig. G.3 Pourbaix Diagram for the chromium/oxygen system. Species and boundaries are shown for 25 °C and 1 atm of pressure (*Source* diagram constructed by Cornelis Klein [Jan. 1961] in *Equilibrium Diagrams for Minerals*, Schmitt, H.H. (editor), Geological Club of Harvard, Boston Mass., 1962, p. 57 (relabeled for legibility), modified with data from George Mason University Neurological Engineering Laboratory (http://complex.gmu.edu/people/peixoto/subpages/ece590S08/ pourbaix.pdf, accessed July 17, 2017). The only common naturally occurring ore is insoluble chromite [(Fe,Mg)O·Cr$_2$O$_3$]. Dashed boundaries marked by −4, −2, and 0 correspond to Cr^{3+} molal concentrations of 10^{-4}, 10^{-2}, and 10^0 (or unity: i.e. all Cr^{3+}) respectively

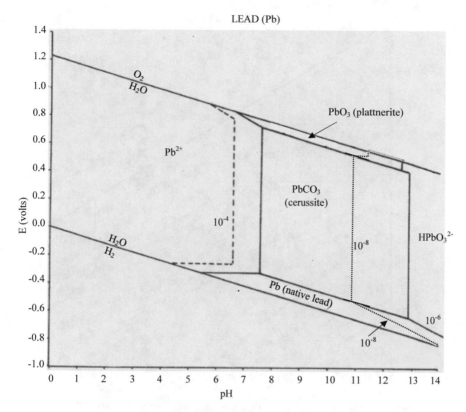

Fig. G.4 Pourbaix Diagram for lead/oxygen/carbonate system (i.e. the native lead/plattnerite/ cerussite system) at 25 °C and 1 atm total pressure (*Source* diagram constructed by Cornelis Klein [Jan. 1961] in *Equilibrium Diagrams for Minerals*, Schmitt, H. H. (editor), Geological Club of Harvard, Boston Mass., 1962, p. 191; relabeled for legibility and consistency). Boundaries of solids lie at a total ion activity ($\sim$concentration in molals) of 10^{-6} (solid lines). The dashed boundary (left) corresponds to an activity of 10^{-4} molals of Pb^{2+}, while the dotted boundary (right) corresponds to a concentration of 10^{-8} molals of $HPbO_3^{2-}$. Notice that Pb^{2+} exists over a broad range of oxidation potentials and pH values. And, some of these pH values match those found in natural drinking water and surface water (e.g. reservoirs), thereby allowing low levels of lead to exist in such water sources and be consumed by plants, animals, and humans; either directly or through the food chain

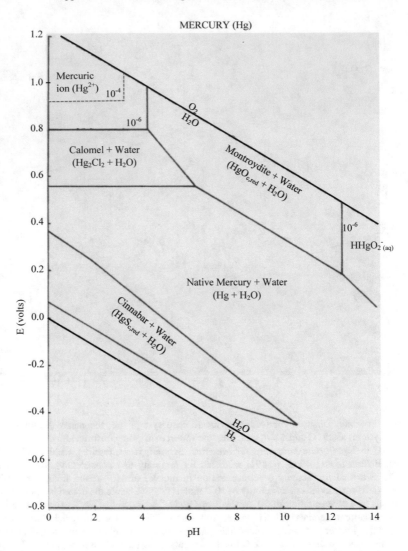

Fig. G.5 Pourbaix diagram for the mercury/sulfur/chlorine/oxygen/carbonate system (i.e. the native mercury/red cinnabar/calomel/montroybite/clearcreekite system) at 25 °C and 1 atm total pressure (*Source* diagram constructed by D. Symons (1961) in *Equilibrium Diagrams for Minerals*, Schmitt, H. H. (editor), Geological Club of Harvard, Boston Mass., 1962, p. 175; relabeled). The total dissolved sulfur species was 10^{-6} molal, total dissolved chlorine species was 10^{-5} molal, and the total dissolved carbonate species was 10^{-1} molal. Solid lines (boundaries) represent mercury ion concentrations of 10^{-6} molal, while the dotted line represents an Hg^{2+} concentration of 10^{-4} molal. Since no mobile mercury species exists at the pH range of drinking water, mercury metal is seldom a problem. However, at the pH of some acidic foods (e.g., salad dressing, lemon juice, vinegar, etc.) native mercury can enter solution. Finally, notice that no carbonate phase appears in this diagram because clearcreekite is a *very* rare mineral and difficult to form under natural conditions. Note: the subscripts "c", "red", and "aq" refer to crystalline, red, and aqueous species, respectively

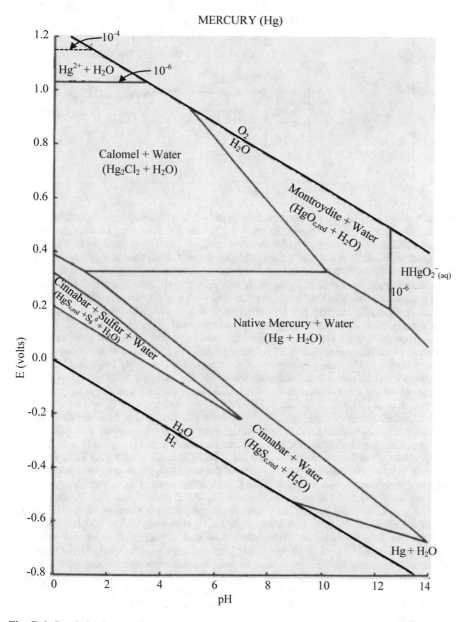

Fig. G.6 Pourbaix diagram for the mercury/sulfur/chlorine/oxygen/carbonate system at 25 °C and 1 atm total pressure (*Source*: diagram constructed by D. Symons (1961) in *Equilibrium Diagrams for Minerals*, Schmitt, H.H. (editor), Geological Club of Harvard, Boston Mass., 1962, p. 174; relabeled for legibility). The total dissolved sulfur species was 10^{-2} molal, the total dissolved chlorine species was 10^{-1} molal, and the total dissolved carbonate species was 10^{-1} molal. Solid lines (boundaries) represent mercury II ion concentrations of 10^{-6} molal, while the dotted line represents an Hg^{2+} concentration of 10^{-4} molal. Relative to the previous Fig. G.5, the great increase in sulfur and chlorine has caused the sulfide (cinnabar) and chloride (calomel) phases to increase in size, the $HHgO_2^-$ boundary to move negligibly to the left, and the Hg^{2+} region to shrink so it can only be reached by the lowest pH values. It is even less likely that native mercury will enter the solution in this system

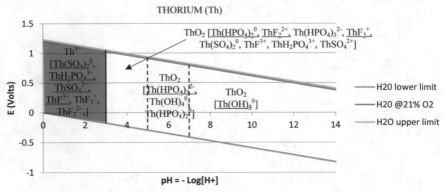

Fig. G.7 Pourbaix Diagram for the thorium/oxygen/hydrogen/fluorine/chlorine/phosphate/sulfate system at 25 °C and 1 atm pressure. The solid vertical line is the boundary between the sparingly soluble solid thorianite phase (ThO_2; the chemical analog of uraninite or pitchblende, UO_2) and the dangerous mobile ionic phase (Th^{4+}; shaded) in pure water at a total thorium species concentration of 10^{-10} moles/L (N. Takeno, *Atlas of Eh-Ph Diagrams: Intercomparison of Thermodynamic Databases*, Geological Survey of Japan, AIST website, 2005, pp. 258–9, accessed July 11, 2017). The principal naturally occurring ore of thorium is monazite, which contains thorianite. However, thorium oxide usually has very low solubility, *unless it is in an acidic environment*. The dashed vertical boundaries separate the inorganic species (enclosed in square brackets in descending concentration with the most dominant first) for "natural water" with total fluorine, chlorine, phosphate, and sulfate concentrations of 0.3 ppm, 10 ppm, 0.1 ppm, and 100 ppm, respectively (D. Langmuir and J.S. Herman, *Geochemica et Cosmochemicaa*, vol. 44, 1980, pp. 1753–1766). Underlined species have a peak concentration within the indicated pH range. This Pourbaix Diagram could be a reasonable model for surface water without much organic matter (e.g. Colorado river in Arizona), in which case the inorganic complexes of Th will be the primary species. Concentrations of inorganic thorium species are further reduced by adsorption into clay, which is nearly total above pH = 6.5. However, there are several ominous aspects to thorium chemistry. First, the mobility of thorium complexes formed with certain organic ligands, like citric acid (absolute pH = 1.13, citrus fruit juice pH = 2.87), is much greater than those formed by inorganic ligands. Also, much higher than normal thorium levels (~ 38 ppb) have been found in groundwater associated with uranium mining and milling in the U.S.A. and Canada (op. cit. Langmuir and Herman). Russian researchers (A. A. Drozdovskaya and Y. P. Mel'nik, *Geokhimiya*, vol. 4, 1968, pp. 151–167) have found comparably high values of up to 10 pbb in ground and mine waters. And, V. S. Dementyev and N. G. Syromyatnikov (*Geokhimiya*, vol. 2, 1965, pp. 211–218) have measured Th concentrations in clear (unfiltered) spring and well waters from granite aquifers as high as 40 ppb. These latter concentrations are two orders of magnitude higher than "normal". So, there may be some local thorium contamination problems associated with mining operations. Particles in unfiltered groundwater from granite aquifers may pose a much wider health risk. Finally, it should be noted that the concentration of Th in groundwater can increase by an order of magnitude with temperatures ranging from 25 to 91 °C (op. cit. Langmuir and Herman)

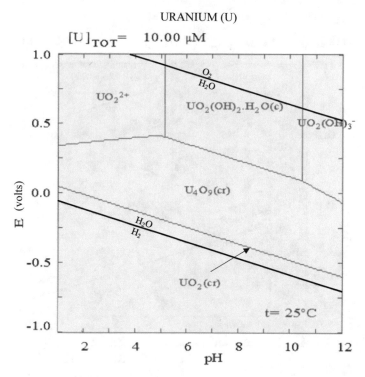

Fig. G.8 Pourbaix diagram for the uranium/oxygen/water system at 25 °C and 1 atm. pressure. Solid (crystalline) species are denoted by the suffix (cr). The boundary lines correspond to a total dissolved uranium concentration of 10^{-5} mol/L. (*Source* http://www.wow.com/wiki/Pourbaix_diagram, accessed May 19, 2017)

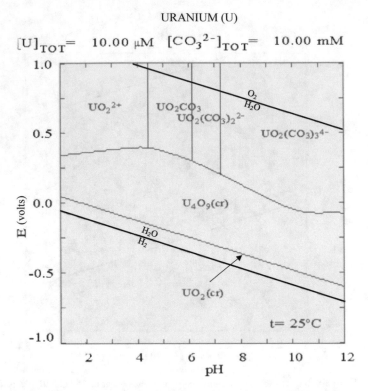

Fig. G.9 Pourbaix diagram for the uranium/oxygen/carbonate/water at 25 °C and 1 atm. pressure. Solid (crystalline) species are denoted by the suffix (cr). The boundary lines correspond to a total dissolved uranium concentration of 10^{-5} mol/L (*Source* http://www.wow.com/wiki/Pourbaix_ diagram, accessed May 19, 2017)

Appendix H
Delaware Department of Natural Resources and Environmental Control (DNREC) Target Compound List (TCL): Volatile and Semi-volatile Organic Compounds and the Target Analyte List (TAL): Metals and Cyanide

Author: T.L. Wilson, DNREC, Created 8/15/2012, Accessed July 18, 2017, http://www.dnrec.delaware.gov/dwhs/sirb/Documents/Target%20Compound%20Lists.pdf.

Target Compound List (TCL)
Volatiles

Dichlorodifluoromethane	1,1-Dichloroethane	4-Methyl-2-pentanone
Chloromethane	cis-1,2-Dichloroethene	Toluene
Vinyl Chloride	2-Butanone	trans-1,3-Dichloropropene
Bromomethane	Chloroform	1,1.2-Trichloroethane
Chloroethane	1,1,1-Trichloroethane	Tetrachloroethene
Trichlorofluoromethane	Cyclohexane	2-Hexanone
1,1-Dichloroethene	Carbon Tetrachloride	Dibromochloromethane
1,1.2-Trichloro-1,2,2-trifluoroethane	Benzene	1,2-Dibromoethane
Acetone	1,2-Dichloroethane	Chloro benzene
Carbon Disulfide	Trichloroethene	Ethylbenzene
Methyl Acetate	Methylcyclohexane	Xylenes (total)
Methylene Chloride	1,2-Dichloropropane	Styrene
trans-1,2-Dichloroethene	Bromodichloromethane	Bromoform
Methyl tert-Butyl Ether	cis-1,3-Dichloropropene	Isopropylbenzene
1,1,2,2-Tetrachloroethane	1,3-Dichlorobenzene	1,4-Dichlorobenzene
1,2-Dichlorobenzene	1,2-Dibromo-3-chloropropane	1,2,4-Trichlorobenzene

Target Compound List (TCL)
Semivolatiles

Benzaldehyde	Hexachlorocyclopentadiene	Hexachlorobenzene
Phenol	2,4,6-Trichlorophenol	Atrazine
bis-(2-Chloroethyl) ether	2,4, S-Trichlorophenol	Pentachlorophenol
2-Chlorophenol	1,1'-Biphenyl	Phenanthrene
2-Methylphenol	2-Chloronaphthalene	Anthracene
2,2'-oxybis (1-Chloropropane)	2-Nitroaniline	Carbazole
Acetophenone	Dimethylphthalate	Di-n-butylphthalate
4-Methylphenol	2,6-Dinitrotoluene	Fluoranthene
N-Nitroso-di-n-propylamine	Acenaphthylene	Pyrene
Hexachloroethane	3-Nitroaniline	Butyl benzyl phthalate
Nitrobenzene	Acenaphthene	3,3'-Dichlorobenzidine
Isophorone	2,4-Dinitrophenol	Benzo(a)anthracene
2-Nitrophenol	4-Nitrophenol	Chrysene
2,4-Dimethylphenol	Dibenzofuran	bis(2-Ethylhexyl) phthalate
bis (2-Chloroethoxy) methane	2,4-Dinitrotoluene	Di-n-octylphthalate
2,4-Dichlorophenol	Diethyl phthalate	Benzo(b)fluoranthene
Naphthalene	Fluorene	Benzo(k)fluoranthene
4-Chloroaniline	4-Chlorophenyl-phenyl ether	Benzo(a)pyrene
Hexachlorobutadiene	4-Nitroaniline	Indeno(1,2,3-cd) pyrene
Caprolactam	4,6-Dinitro-2-methylphenol	Dibenzo(a,h)anthracene
4-Chloro-3-methylphenol	N-Nitrosodiphenylamine	Benzo(g,h,i)perylene
2-Methylnaphthalene	4-Bromophenyl-phenyl ether	

Target Compound List (TCL)	
Semivolatiles	
alpha-BTC	4,4'DDT
beta-BHC	Methoxychlor
delta-BHC	Endrin ketone
gamma-C (Lindane)	Endrin aldehyde
Heptachlor	alpha-Chlordane
Aldrin	gamma-Chlordane
Heptachlor epoxide	Toxaphene
Endosulfan 1	Aroclor-1016
Dieldrin	Aroclor-1221
4,4'-DDE	Aroclor-1232
Endrin	Aroclor-1242
Endosulfan II	Aroclor-1248
4,4'DDD	Aroclor-1254
Endosulfan sulfate	Aroclor-1260
Target Analyte List (TAL) *Metals and Cyanide*	
Aluminum	Magnesium
Antimony	Manganese
Arsenic	Mercury
Barium	Nickel
Beryllium	Potassium
Cadmium	Selenium
Calcium	Silver
Chromium	Sodium
Cobalt	Thallium
Copper	Vanadium
Iron	Zinc
Lead	Cyanide

PERIODIC TABLE OF THE ELEMENTS (June 2017)

Legend

At. No.	4
Element Symbol	Be
At. Wt.	9.01218
Valence	+2
Density (gms/cu. cm. @ 300 K)	1.85
*Tensile Strength (Mpa)	370
**Thermal Conductivity (W/m K @ 300K)	200
Element Element	Beryllium

* Note: Tensile strengths (@ 300K) may vary widely depending on metallurgical processing (see references above).
* Note: density of gaseous elements refers to the liquid state at its boiling point.
‡ Note: Compressive strength
↑ Note: A red element symbol denotes radioactivity. A red element name denotes a radioactive synthetic element.
X Note: The atomic weights of synthetic radioactive elements are those of the most stable (long lived) isotope.
Note: N/A = Not Available; M.P = Melting Point; ~ = estimate

Periodic table (element symbols by atomic number)

Period																		
1	H (1)																	He (2)
2	Li (3)	Be (4)											B (5)	C (6)	N (7)	O (8)	F (9)	Ne (10)
3	Na (11)	Mg (12)											Al (13)	Si (14)	P (15)	S (16)	Cl (17)	Ar (18)
4	K (19)	Ca (20)	Sc (21)	Ti (22)	V (23)	Cr (24)	Mn (25)	Fe (26)	Co (27)	Ni (28)	Cu (29)	Zn (30)	Ga (31)	Ge (32)	As (33)	Se (34)	Br (35)	Kr (36)
5	Rb (37)	Sr (38)	Y (39)	Zr (40)	Nb (41)	Mo (42)	Tc (43)	Ru (44)	Rh (45)	Pd (46)	Ag (47)	Cd (48)	In (49)	Sn (50)	Sb (51)	Te (52)	I (53)	Xe (54)
6	Cs (55)	Ba (56)	La (57)	Hf (72)	Ta (73)	W (74)	Re (75)	Os (76)	Ir (77)	Pt (78)	Au (79)	Hg (80)	Tl (81)	Pb (82)	Bi (83)	Po (84)	At (85)	Rn (86)
7	Fr (87)	Ra (88)	Ac (89)	Rf (104)	Db (105)	Sg (106)	Bh (107)	Hs (108)	Mt (109)	Ds (110)	Rg (111)	Cn (112)	Nh (113)	Fl (114)	Mc (115)	Lv (116)	Ts (117)	Og (118)
8	Uue (119)	Ubn (120)	Ubu (121)															

Lanthanides: Ce (58), Pr (59), Nd (60), Pm (61), Sm (62), Eu (63), Gd (64), Tb (65), Dy (66), Ho (67), Er (68), Tm (69), Yb (70), Lu (71)

Actinides: Th (90), Pa (91), U (92), Np (93), Pu (94), Am (95), Cm (96), Bk (97), Cf (98), Es (99), Fm (100), Md (101), No (102), Lr (103)

Super-Actinides

The Next Element?

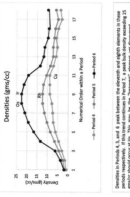

Densities (gms/cc)

Densities in Periods 4, 5, and 6 peak between the eleventh and eighth elements in these periods respectively. If this trend continues in Period 7, a peak bulk density exceeding 25 gms/cc should occur at Hs. This may be the "heaviest" element yet discovered.

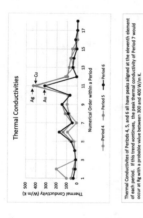

Thermal Conductivities

Thermal Conductivities of Periods 4, 5, and 6 all have peaks aligned at the eleventh element of each period. If this trend continues, the peak thermal conductivity of Period 7 would occur at Rg with a probable value between 300 and 400 W/m K.

Ultimate Tensile Strengths

Ultimate Tensile Strength (UTS) varies regularly across Periods 4, 5, and 6. If this trend continues, it is likely that the UTS in Period 7 will peak at Rg with a bulk value near 2500 MPa, perhaps the "strongest" pure metal yet discovered.

Data in this table comes from a variety of sources:

1) *Periodic Table of the Elements*, Sargent-Welch, 1996.
2) D.E. Gray, *American Institute of Physics Handbook, 3rd ed*, McGraw-Hill, New York, 1972, pp. 2-60 to 2-76.
3) Metallic Elements Properties Data-Matweb.com, www.matweb.com/reference/elements.aspx, accessed 6/24/17.
4) Anon., "Europium", https://en.wikipedia.org/wiki/Europium, accessed 6/24/17.
5) Anon., "Bismuth", http://www.azom.com/properties.aspx?ArticleID=590.

Glossary

Absolute Perfection (Code) [Technical] Code that functions perfectly under all conditions. Absolute perfection is an idealization that can never be realized due to limitations on machine reliability and human understanding, as well as human input errors

Activation Energy [Technical] The minimum amount of energy that must be absorbed by a chemical system to cause it to react

Adjacency Matrix (1) [Technical] A square matrix whose elements specify the number of direct connections between one electrical device, pin, or software instruction and another. Elements are usually "1" (a direct connection exists) or "0" (no direct connection exists), but any integer is possible. (2) [Technical] The matrix of allowable edges

Algebra [Technical] Given a set of objects over a field of scalars for which addition and multiplication is defined, an *Algebra* is formed if "scalar multiplication" and a "distributive law" of multiplication over addition is defined

Arrhenius Relation [Technical] The exponential relationship between reaction rate and Kelvin temperature

Aspect Ratio [Technical] Given a connector with pin length $\mathcal{L}$ and diameter d, the Aspect Ratio (A.R.) of the pin is defined to be $\mathcal{L}$ /d, or a/b, where "a" is the Dimensionless Scaled Pin Length and "b" is the Dimensionless Scaled Pin Diameter

Augmented Matrix [Technical] A way of writing a matrix that represents a set of linear equations

Average Risk per Hazard ($\underline{R}$) [Technical] The quotient of a System's Risk ($\mathcal{R}$) by its total number of hazards N

© Springer Nature Switzerland AG 2020
R. R. Zito, *Mathematical Foundations of System Safety Engineering*,
https://doi.org/10.1007/978-3-030-26241-9

Average Risk per Hazard Contribution (from cell I) ($\underline{R}_I$) [Technical] Contribution to the average risk per hazard from a unit "cell" (area) in the logarithmic severity-probability plane centered on point (S_I, P_I)

Basis Set (of functions) [Colloquial] An *independent* set of functions such that any function of a defined function space can be expressed as a linear combination of members of the set. Verification is by computer trial and error, not a mathematical proof. [Technical] An *independent* set of functions such that any function of a defined function space can be expressed as a linear combination of members of the set. Verification is by mathematical proof

Bivariate Normal (Gaussian) Distribution (BND) [Technical] A Hazard Distribution Function (HDF) that is Gaussian along both the Logarithmic Severity axis and the Logarithmic Probability axis. Also, see Logarithmic Severity, Severity, Logarithmic Probability, Probability, and Hazard Distribution Function

Black Swan [Colloq. & Tech.] A rare (unlikely) mishap for which there is usually no contingency plan for recovery

Bloom [Technical] The oxide of cadmium or zinc on these pure metals

Blue Connector [Technical] An *Ideal Connector*

Built-In-Test (BIT) [Technical] A system function test that is part of a system's circuitry

CAD [Technical] Computer-Aided Design software package

Catastrophic (Hazard) Severity (1) [Colloquial] From MIL-STD-882E. A hazard severity that "could result in one or more of the following: death, permanent total disability, irreversible significant environmental impact, or monetary loss equal to or exceeding \$10M." This definition is colloquial because it is not purely numerical, and the year of dollar value is not specified. (2) [Technical] Modified from MIL-STD-882D. A hazard severity whose total legal, medical, and direct monetary cost expressed in the year 2000 dollars equals or exceeds \$1M

CAVE [Technical] Computer Analysis and Visualization Environment. A stereographic facility

Circuitized Flowchart (CFC) [Technical] A traditional Flowchart (FC) represented as an electronic circuit

Chi-squared (χ^2) [Technical] A statistic used to measure the "Goodness-of-Fit" between two curves

Chimera (1) [Colloquial] A "monster" made of body parts from a variety of animals. (2) [Tech.] In biochemistry, it refers to a macromolecule made up of smaller macromolecules of diverse origin. (3) [Technical] In System Safety, it refers to a system composed of some combination of software, firmware, and hardware

Circuit [Colloquial] Any interconnection of electrical components with conducting tracks (e.g., wires, ribbons, etc.) through which current can flow. Also, see Vague Network

Critical (Hazard) Severity (1) [Colloquial] From MIL-STD-882E. A hazard severity that "could result in one or more of the following: permanent partial disability, injuries or occupational illness that may result in hospitalization of at least three personnel, reversible significant environmental impact, or monetary loss equal to or exceeding \$1M but less than \$10M." This definition is colloquial because it is not purely numerical, and the year of dollar value is not specified. (2) [Technical] Modified from MIL-STD-882D. A hazard severity whose total legal, medical, and direct monetary cost expressed in the year 2000 dollars equals or exceeds \$200 K, but is less than \$1M

Contact Probability [Technical] The probability that a given pin of the male half of a multi-pin connector will contact a neighboring pin if bent

Cost Function [Technical] A function, such as program cost in dollars, that is to be minimized in a Linear Programming problem. The Cost Function is more generally called an Objective Function and may be minimized or maximized, such as when an analyst seeks to minimize the use of dangerous material or maximize heat removal subject to a set of constraints

Dimensionless Scaled Pin Diameter [Technical] Given a connector pin with diameter d and pin separation ℓ, the Dimensionless Scaled Pin Diameter ("b") is defined to be d/ℓ

Dimensionless Scaled Pin Length [Technical] Given a connector pin with length $\mathscr{L}$ and pin separation ℓ, the Dimensionless Scaled Pin Length ("a") is defined to be $\mathscr{L}/\ell$

Doughnut Hole (1) [Colloquial] A hole in any closed surface, especially a torus (or "doughnut"). (2) [Technical] Points of a plane enclosed by a circle, or its homomorphic equivalent in $\mathbf{R}^2$ (Euclidean plane). (3) [Technical] Points of an xy-plane enclosed by the circle $x^2 + y^2 = 1$ on a torus (or "doughnut") defined by $\{(x, y, z) \in \mathbf{R}^3 : \left[(x^2 + y^2)^{1/2} - 2\right]^2 + z^2 = 1\}$, or its homomorphic equivalent in $\mathbf{R}^3$ (Euclidean 3-space), including a tessellated surface defined by the bounding vertices of a toroidal mesh. The set defining the torus is obtained by rotating the circle $(x-2)^2 + z^2 = 1$ in the xz-plane about the z-axis

Edge [Technical] A "line" of a "Graph" representing a bent pin, a wire connection between two electrical devices, or a directional logic flow between two statements in a "Flowchart". Also, see "Graph" and "Flowchart"

Elongation (e) [Technical] The percent of length change from the length of the original tensile test sample before stretching. The maximum elongation is the maximum percent length change before test sample failure

DNREC [Technical] Department of Natural Resources and Environmental Control. A government agency (e.g. the state of Delaware has such an agency)

Entropy [Technical] A measure of a system's disorder and its tendency toward failure or defect expression

Fail Closed/Fail High (FC/FH) [Technical] An analog component is said to fail closed if its impedance becomes temporarily or permanently equivalent operationally to a conducting wire (i.e. impedance $\to 0$ relative to the component's nominal value). A digital component is said to fail high if one or more of its binary outputs incorrectly enters the high state ("1"), either temporarily or permanently

Fail Open / Fail Low (FO/FL) (1)[Technical] A Fail Open/Fail Low (FO/FL) fault removes a conducting component, input/output, or a conducting wire, from one well-defined network, thereby changing it to another well-defined network, or rendering the original well-defined network inoperative. A FO/FL fault is capable of initiating an undesired function or inhibiting a desired function. (2) [Technical]: (Alternate Definition) A FO/FL fault removes an edge, or edges, from a network graph

Fault Tree [Technical] A method of diagraming the steps of a system failure

Feasibility Set [Technical] The set of all points in a Euclidian materials n-space that satisfies all constraints in a materials selection problem. Each n-space axis represents a different material and each n-tuplet a different mix of materials

Flowchart (FC) [Colloq. & Tech.] A graphic means of describing a sequence of operations performed on data

Flowchart Graph (FCG) [Technical] A Circuitized Flowchart (CFC) represented as a connected set of vertices and edges. See Circuitized Flowchart

FOD [Technical] Electrically conducting "Foreign Objects and Debris"

Frequent (Hazard Probability) (1) [Colloquial] From MIL-STD-882E. A Hazard that is "likely to occur often in the life of an item." This definition is colloquial because it is non-quantitative. (2) [Technical] From MIL-STD-882D. A hazard that is "likely to occur often in the life of an item, with a probability of occurrence greater than 10^{-1} in that life."

Fundamental Matrix [Technical] Given a positive integer N, the off-diagonal elements of the Fundamental Matrix $F_N(i, j)$ (where $i \neq j$) are defined by the elements of the Way Matrix, $W_N(i \cdot j)$, and the diagonal elements $F_N(i, i) = \sum_j W(i, j)$ for all j such that the off-diagonal elements of the Adjacency Matrix $A(i,j) \neq 0$ and N-1 > 1, otherwise $F_N(i,i) = 0$. The Fundamental Matrix counts the number of simple loops containing vertex i in any network

Fundamental Theorem of Linear Programming [Technical] A theorem stating that "the Cost Function (Objective Function) of a Linear Programming problem will be minimized, or maximized, at a vertex of the problem's Feasibility Set"

Gaussianally Safe System [Technical] A system with a positive Norm. See Norm

Gaussianally Unsafe System [Technical] A system with a negative Norm. See Norm

Goodness-of-Fit [Technical] A measure of the amount of coincidence between two curves. Commonly, the measure may be the RMS error or the value of Chi-squared

Grain [Technical] A composite energetic material consisting of fuel and oxidizer dispersed uniformly throughout a polymeric binder

Graph [Technical] A connected set of "dots" and "lines" that represents a connector, a "circuit", or a "flowchart". See "Circuit" and "Flowchart"

Green Connector [Technical] A *Strictly Stable Connector*

Hazard Distribution Function (HDF) [Technical] A normalized function F (a 3D surface) defined by the instantaneous fraction of hazards per unit $S \times \mathcal{P}$ area such that $\iint F(S, \mathcal{P}) \, dS \, d\mathcal{P} = 1$ when integrated over the entire positive quadrant of the $S, \mathcal{P}$-plane (logarithmic severity $\times$ probability plane). Also see Logarithmic Severity, Severity, Logarithmic Probability, and Probability

Hard-to-Solve [Technical] A problem is "hard to solve" if its solution can only be obtained by trial and error testing of each element of a universal set of possibilities. In other words, no shortcut or smart algorithm exists for a quick solution. If the universal set of possible solutions is large enough, computer searches could take years, or even exceed the age of the universe. Synonymous terms include NP-Hard-to-Solve (NP stands for "Non- deterministically polynomial") and NP-Complete

Hydrolysis [Technical] Reaction with water

Hypergolic [Technical] Hypergolic fuel spontaneously ignites on contact with its oxidizer, or air. See *Pyrophoric*

Ideal Connector [Technical] A connector whose pin density and pin length are such that no bent pin can contact another pin during a bent pin event

Ideal Pin [Technical] A conducting pin on the male half of a connector that develops no curvature and maintains its length after bending only at its root (base)

Improbable (Hazard Probability) (1) [Colloquial] From MIL-STD-882E. A hazard that is "so unlikely, it can be assumed occurrence may not be experienced in the life of an item." This definition is colloquial because it is non-quantitative. (2) [Technical] From MIL-STD-882D. A hazard that is "so unlikely, it can be assumed occurrence may not be experienced, with a probability of occurrence less than 10^{-6} in that life."

Incidence [Technical] The number of edges that have a given vertex as an endpoint

Independent (set of functions) [Technical] A set of functions such that one cannot be expressed in terms of the other members of the set

Inductive Assertion [Technical] The use of Mathematical Induction to prove code

Interface (1) [Colloquial] A part (or instruction) between two hardware (software) subsystems. (2) [Technical] Undefined since a part or instruction must, by definition, belong to either one subsystem or another, or be considered a subsystem by itself

KSLOC [Technical] Kilo Software Lines of Code

Legacy Subsystem (Part) [Technical] A model of a hardware subsystem, version of a software subsystem, part type, or copy of a code fragment, that has been previously used in another system

Lemon [Colloquial] A system with an abnormally high failure rate, or abnormally high Average Risk (cost) per Hazard ($\underline{R}$). Also, see Average Risk per Hazard

Limited Liability System [Technical] A system with a finite $\underline{R}$. Also see Average Risk per Hazard

Linear Programming [Technical] A method of maximizing, or minimizing, a Cost Function (Objective Function) subject to a set of inequalities called constraints

Logarithmic Probability ($\mathcal{P}$) [Technical] $\mathcal{P} = -\log_{10} P$, where P is the probability of a hazard. Usually, $0 < \mathcal{P} < 6$. See Probability (P)

Logarithmic Severity ($\mathcal{S}$) [Technical] $\mathcal{S} = \text{Log}_{10} S$, where S is the severity of a hazard measured in U.S. dollars of a given year (e.g., the year 2000). Usually, $0 < \mathcal{S} < 6$. See Severity (S)

Marginal (Hazard) Severity (1) [Colloquial] From MIL-STD-882E. A hazard severity that "could result in one or more of the following: injury or occupational illness resulting in one or more lost workday(s), reversible moderate environmental impact, or monetary loss equal to or exceeding \$100K but less than \$1M." This definition is colloquial because it is not purely numerical, and the year of dollar value is not specified. (2) [Technical] Modified from MIL-STD-882D. A hazard severity whose total legal, medical, and direct monetary cost expressed in the year 2000 dollars equals or exceeds \$10K, but is less than \$200K

Marginal Analysis [Technical] A method of evaluating the effect of small changes in the constraint set on the Cost Function (Objective Function) of a Linear Programming problem.

Material Safety Data Sheet (MSDS) [Technical] Contains necessary safety information (e.g., toxicity, flammability and explosion hazard, spill and leak procedure, handling and storage procedure, etc.) for a material, substance, or compound sold by a supplier

Mathematical Lemon [Technical] A system whose Average Risk (cost) per Hazard ($\underline{R}$) is undefined (i.e. ∞). Also, see Average Risk per Hazard

Matrix [Technical] A collection of numbers arranged in rows and columns

Maturity Time [Technical] The time a product (system) was in use after delivery

MCL [Technical] Maximum Contamination Level. The maximum allowable concentration of a toxic substance in drinking water

Mean [Technical] The average value

Meta-Stable Connector A connector in which two conducting pins can be effectively electrically isolated from one another during a bent pin event *only* if they can be separated by a sufficiently large distance. Meta-Stable connectors can form long conducting paths during a bent pin event

Mishap Chain Reaction (MCR) [Colloq. & Tech.] When one failure triggers other failures such that the overall probability of failure is much higher than would be expected on the basis of independent events

Mode [Technical] The peak of a probability density function or un-normalized failure rate curve

Model Infinite System (MIS) [Technical] A system having an infinite number of Rayleigh subsystems whose modes σ are distributed according to a narrow Gaussian with standard deviation s and peak σ_m such that s $\ll$ σ_m

Modified Risk Surface ($\mathfrak{R}$) [Technical] Log_{10} of the Average Risk per Hazard Contribution from cell I ($\underline{R}_I$) of the $\mathcal{S}, \mathcal{P}$ – plane. $\mathfrak{R} \equiv 0$ if $\underline{R}_I \leq \$1$. Also, see Average Risk per Hazard Contribution, Logarithmic Severity, and Logarithmic Probability

Myer's 9 Caveats [Technical] The 9 recognized limitations on the use of Inductive Assertion to prove code

Nearest Neighbor [Technical] Nearest pin(s) to a given pin

Negligible (Hazard) Severity (1) [Colloquial] From MIL-STD-882E. A hazard severity that "could result in one or more of the following: injury or occupational illness not resulting in a lost workday, minimal environmental impact, or monetary loss less than \$100K." This definition is colloquial because it is not purely numerical, and the year of dollar value is not specified. (2) [Technical] Modified from MIL-STD-882D. A hazard severity whose total legal, medical, and direct monetary cost expressed in the year 2000 dollars is less than \$10K

Network [Technical] Any interconnection of conducting pathways, electrical parts, or software instructions, all of which are active at a given time or could be active during a given run

Norm [Technical] The sum, over all cells I of the $S,\mathcal{P}$ − plane, of the difference between the Risk Surfaces for the MIS and a real system. If the Norm is positive, then the real system's Average Risk per Hazard is less than that of the MIS. If the Norm is negative than the real system's Average Risk per Hazard is greater than that of the MIS. Also, see Model Infinite System (MIS) and Risk Surface

Normalized Incidence [Technical] The normalized incidence (or normalized degree) is the incidence of a vertex divided by the total number of vertices in the network graph G, and is denoted by $r_i(v)$ for any vertex v

NP-Complete [Technical] See Hard-to-Solve

NP-Hard-to-Solve [Technical] See Hard-to-Solve

nth Nearest Neighbor [Technical] The nth closest pin(s) to a given pin

Numerical Flowchart (NFC) [Technical] The representation of a FC, CFC, or FCG, as a square matrix of integers

Objective Function [Technical] See Cost Function

Occasional (Hazard Probability) (1) [Colloquial] From MIL-STD-882E. A hazard that is "likely to occur sometime in the life of an item." This definition is colloquial because it is non-quantitative. (2) [Technical] From MIL-STD-882D. A hazard that is "likely to occur sometime in the life of an item, with a Probability of occurrence less than 10^{-2} but greater than 10^{-3} in that life."

Orthogonal Functions (of a set of functions) [Technical] A set of functions such that the definite integral of the square of each function is unity over its range from 0 to ∞, and the definite integral of two different members of the set is zero over the same range

Oxidation Potential (of a chemical environment) [Technical] The ability of a chemical environment to oxidize a molecular species. Typically, a molecule or ion can only exist in a given environment if that environment's oxidation potential (denoted by E or Eh) lies within certain bounds. The oxidation potential is measured in volts on commercially available meters

pH [Technical] The negative logarithm to the base 10 of the hydronium ion concentration (molarity) of a solution. The pH is a measure of acidity and its value can be obtained from commercially available pH meters

Pourbaix Diagram [Technical] A diagram that indicates the oxidation potential and pH necessary for the existence of a given molecule or ion

Practical Lemon [Technical] A system is a Practical Lemon if its Average Risk (cost) per Hazard ($\underline{R}$) exceeds the owner's ability, or willingness, to pay

Practical Perfection (Code) [Technical] Practical Perfection for code is restricted to integer arithmetic on commercially available machines. Human input and requirements understanding errors are allowed

Probability (P) [Technical] P is the probability of a mishap per system lifetime. See System

Probability Density Function (pdf) [Technical] The Probability per unit time (probability density) as a function of time. The pdf has a unit area (probability) under its curve when integrated from 0 to ∞

Probable (Hazard Probability) (1) [Colloquial] A hazard that "will occur several times in the life of an item." This definition is colloquial because it is non-quantitative. (2) [Technical] From MIL-STD-882D. A hazard that "will occur several times in the life of an item, with a probability of occurrence less than 10^{-1} but greater than 10^{-2} in that life."

Pseudo-Weibull Distribution (PWD, TPWD∞) [Technical] A TPWDn when n $\rightarrow \infty$. See Truncated Pseudo-Weibull distribution

Pyrophoric [Technical] Will ignite spontaneously on contact with air. See *Hypergolic*

Receptacle [Technical] An aperture in the female half of a connector. Only a receptacle is connected to power when both halves of a connector are separated

Red Connector An *Unstable Connector*

Redundancy Matrix [Technical] A matrix that counts the number of redundant paths between any two vertices in an electrical or logical network

Redundant Path (Circuit) [Technical] An electrical or logical path that contains loops or multiple edges (switchbacks) (i.e., a vertex can be revisited)

Remote (Hazard Probability) (1) [Colloquial] From MIL-STD-882E. A hazard that is "unlikely, but possible to occur in the life of an item." This definition is colloquial because it is non-quantitative. (2) [Technical] From MIL-STD-882D. A hazard that is "unlikely but possible to occur in the life of an item, with a probability of occurrence less than 10^{-3} but greater than 10^{-6} in that life"

Risk (R) [Technical] The product of the severity of a given hazard with its probability of occurrence. Also, see Severity (S) and Probability (P)

Risk Surface [Technical] The surface produced by plotting $\underline{R}_l$ over the $\mathcal{S}, \mathcal{P} -$ plane. See Logarithmic Severity and Logarithmic Probability

Root [Technical] The location where a pin is mounted on a connector. This is equivalent to a bent pin "vertex"

Row reduction [Technical] A procedure used to simplify an Augmented Matrix or a Tableau so that a solution to a set of linear equations can be easily read

Safety Critical (1) [Colloquial] Defined by the customer. (2) [Technical] Defined by MIL-SPEC-882

Severity (S) [Technical] S is the severity of a hazard measured in U.S. dollars of a given year, usually, year 2000

Simple system (or subsystem) [Technical] A system (or subsystem) having a Rayleigh failure rate law

Simple Path [Technical] An electrical or logical path without loops or switchbacks. Also called a *Way*

Simple Circuit (1) [Technical] A circuit topologically equivalent to a circle. (2) [Technical] A circuit composed of a single loop without any multiple edges (switchbacks)

Simplex Method [Technical] An algebraic method of solving Linear Programming problems

Simplex Tableau [Technical] A tabular way of arranging the equations of a Linear Programming problem

Single Pin Contact Probability [Technical] The probability that any bent pin in a chain of bent pins will make electrical contact with the next pin in the chain

Sneak Circuit [Technical] Any Simple Circuit of a Vague Network that can, under certain conditions, initiate an undesired function or inhibit the desired function. These conditions include, but are not limited to, the state of an electronic device (e.g., integrated circuit, transistor, diode, or switch), the unintentional use of frequencies outside the operational limits of a circuit, the use of discrete components with inappropriate parameters (e.g., capacitance, resistance, etc.), or unexpected current reversal. Sneak Circuits are inadvertently designed into a system because of network complexity. They are not due to hardware failures

Stable Connectors [Technical] Strictly Stable and Meta-Stable connectors

Spanning Set (of functions) [Technical] A set of functions such that all members of a function space can be expressed as linear combinations of elements of the set

Standard Form [Technical] The arrangement of a Simplex Tableau for a Linear Programing *maximization* problem in which all constants in the rightmost column of the tableau are positive

State (of a Component) [Technical] A complete description of which inputs and outputs of a component are conducting, and which are non-conducting

Strictly Stable Connector [Technical] A connector whose conducting path formation during a bent pin event is effectively limited to the shortest and second shortest paths between a given safety-critical pin and a power pin

Subsystem (1) [Colloquial] A subsystem i is a subset S_i of related parts and/or instructions that perform some function i that contributes to the system (**S**) goal such that $\bigcup$(all S_i) $\subseteq$ **S**. Interface parts may exist. (2) [Technical] A subsystem is a subset S_i of parts and/or instructions that have a Rayleigh failure rate law and $\bigcup$(all S_i) $\equiv$ **S**. Interface parts do not exist

System [Colloq. & Tech.] A universal set **S** (total set) of parts and/or instructions that execute a designer's goal(s)

System Risk ($\mathcal{R}$) [Technical] The sum of all risks for a given system

T_{50} [Technical] The operating time T at which a system has accumulated a 50% chance of failure starting from time t = 0

Tableau [Technical] A compact matrix representation of a system of linear equations used to solve a Linear Programming problem

TAL [Technical] Target Analyte List. A list, composed by some government agency, of metals, compounds, ions, or other substances that are routinely monitored for water, or more generally environmental, quality. These metals, compounds, ions, or substances may, or may not, be dangerous to human health

TCL [Technical] Target Compound List. A list, composed by some government, of compounds of environmental concern

Tensile Strength (**S**) [Technical] The maximum tensile stress (force per unit original area) that a material can withstand before failure

Thick Pin [Technical] Any connector pin with diameter d, length $\mathcal{L}$, and pin spacing ℓ such that d is not $\ll \ell$ and $\mathcal{L}$

Thin Pin [Technical] Any connector pin with diameter d, length $\mathcal{L}$, and pin spacing ℓ such that d $\ll \ell$ and $\mathcal{L}$

Tin Whisker [Technical] A crystalline conducting whisker (or needle, or thread) of tin that can grow from a soldered electrical junction containing tin. The whisker can create a short circuit conducting path to neighboring electrical junctions, wires, pins, or parts

TLV [Technical] Threshold Limit Value. The maximum allowable concentration of a toxic substance in air or water.

Trace [Technical]: The sum of the diagonal elements of a matrix

Truncated Pseudo-Weibull Distribution of nth order (TPWDn) [Technical] A system failure rate law that is an nth order expansion in the moments μ_0 through μ_n of a narrow Gaussian distribution of Rayleigh subsystem modes, with standard deviation s and peak at σ_m. The TPWD_n mimics a system Weibull distribution model, but can have a more slowly decaying tail. In the limit as $s/\sigma_m \to 0$, the TPWDn approaches the Weibull distribution for all n

Undefined (1) [Technical] A *term* is undefined because its definition is illogical. (2) [Technical] A *quantity* is undefined because its value is infinite. For example, any number divided by zero is undefined

Unknot [Technical] A sequence of parts or instructions forming a simple loop (without revisited vertices) that is topologically equivalent to a circle. Also, see "vertex"

Unstable Connector [Technical] A connector whose conducting pins are sufficiently long and close-packed such that conducting paths can proliferate without bound during a bent pin even. No two conducting pins are safe from an electrical short circuit in an Unstable Connector during a bent pin event

Vague Network [Technical] Any current or logic carrying interconnection of conducting pathways, electrical parts, or software instructions, some of which may be inactive at any given time, or cannot be active during a given run. One Vague Network may be defined by, or generate, several *Well-Defined Networks*

Vertex [Technical] A "dot" of a "Graph" representing a straight connector pin, an electrical device or junction, or a software instruction. Also, see "Graph"

Way [Technical] A simple electrical or logical path without loops

Way Matrix [Technical] A matrix that counts the number of *ways* (or *simple paths*) between any two different vertices of an electrical or logical network. The Way Matrix has zeros on the main diagonal

Weibull Distribution A failure rate law typical of many hardware systems

Well-Defined Network [Technical] A subset of components and conducting tracks of a vague network through which electrons are actually flowing when its components are in a given conducting state. For software, a well-defined network consists of those instructions that can be activated in a given run

WSESRB [Colloq. & Tech] Weapon System Explosive Safety Review Board

Yellow Connector [Technical] A *Meta-Stable Connector*

Young's Modulus (Elastic Modulus) E [Technical] The ratio of tensile stress to tensile strain for elastic deformation of a material

Zero Sum Game [Technical] A game in which the loss of one player is the gain of another

Index

Printed in the United States
By Bookmasters